Disaster Risk Reduction

Methods, Approaches and Practices

Series Editor

Rajib Shaw, Keio University, Shonan Fujisawa Campus, Fujisawa, Japan

About the Series

Disaster risk reduction is a process that leads to the safety of communities and nations. After the 2005 World Conference on Disaster Reduction, held in Kobe, Japan, the Hyogo Framework for Action (HFA) was adopted as a framework for risk reduction. The academic research and higher education in disaster risk reduction has made, and continues to make, a gradual shift from pure basic research to applied, implementation-oriented research. More emphasis is being given to multi-stakeholder collaboration and multi-disciplinary research. Emerging university networks in Asia, Europe, Africa, and the Americas have urged process-oriented research in the disaster risk reduction field. With this in mind, this new series will promote the output of action research on disaster risk reduction, which will be useful for a wide range of stakeholders including academicians, professionals, practitioners, and students and researchers in related fields. The series will focus on emerging needs in the risk reduction field, starting from climate change adaptation, urban ecosystem, coastal risk reduction, education for sustainable development, community-based practices, risk communication, and human security, among other areas. Through academic review, this series will encourage young researchers and practitioners to analyze field practices and link them to theory and policies with logic, data, and evidence. In this way, the series will emphasize evidence-based risk reduction methods, approaches, and practices.

More information about this series at http://www.springer.com/series/11575

Emily Ying Yang Chan · Rajib Shaw
Editors

Public Health and Disasters

Health Emergency and Disaster Risk Management in Asia

Springer

Editors
Emily Ying Yang Chan
JC School of Public Health
and Primary Care
The Chinese University of Hong Kong
Hong Kong SAR, China

University of Oxford
Oxford, United Kingdom

Rajib Shaw
Graduate School of Media and Governance
Keio University, Shonan Fujisawa Campus
Fujisawa, Kanagawa, Japan

ISSN 2196-4106 ISSN 2196-4114 (electronic)
Disaster Risk Reduction
ISBN 978-981-15-0923-0 ISBN 978-981-15-0924-7 (eBook)
https://doi.org/10.1007/978-981-15-0924-7

This Springer imprint is published by the registered company Springer Nature Singapore Pte Ltd.
The registered company address is: 152 Beach Road, #21-01/04 Gateway East, Singapore 189721, Singapore

Preface

Advancement in science technology in recent decades might have increased our capacity to tackle the adverse human consequences of various kinds of disasters, but the accompanied and interlinking modern phenomena of industrialisation, urbanisation, globalisation, climate change, and widening disparity have created new hazards and exacerbated the old ones, while aggravating people's exposure to these hazards and worsening the vulnerability of the marginalised at the same time, notably in Asia as the most disaster-prone continent. In the face of this dire prospect, the effective, efficient, and synergetic use of available resources for disaster risk reduction could provide an answer. Arguably, human health, or its inadequacy, is an inevitable outcome of disaster, while its enhancement a natural goal of disaster risk reduction (DRR), as enshrined in the landmark United Nations agreements adopted in this decade, including the Sendai Framework for Disaster Risk Reduction 2015–2030, the 2030 Sustainable Development Goals (SDGs), the Paris climate agreement, and the New Urban Agenda (Habitat III). Targeting the health aspect of DRR is thus an obvious route to take. Under the cluster approach instituted by the United Nations in 2006, health is among the 11 key areas where clusters have been established, and the global health cluster is led by the World Health Organization (WHO). Nevertheless, the health sector has traditionally focused on the response to disasters and emergencies.

In recent years, prevention and mitigation have been gradually recognised as a more efficient approach to tackle the health risk associated with emergencies and disasters. Preparedness of the stakeholders and resilience of the health systems based on primary health care at community level to reduce exposure and vulnerability are key concepts and practices under this emerging approach, which means multi-sectoral actors and their coordination are crucial. Consequently, the WHO (2019) has launched the Health Emergency and Disaster Risk Management (Health-EDRM) framework to weave together health, DRR, and other relevant disciplines to fill the theoretical gap and meet the practical needs in the new scene of DRR, with prevention as the pivot in this framework.

Comprising of 9 chapters of theoretical discussion and 14 case studies, this book presents the theoretical framework of Health-EDRM, places it within the global institutional context, and illustrates it within concrete examples from Asia, namely Bangladesh, China, India, Japan, Nepal, Pakistan, Vietnam, and the Philippines. How these case studies can illuminate the Health-EDRM framework and field of study and practice will be summarised before the way forward is explored to wrap up this multi-sectoral dialogue.

This book is intended for advanced undergraduate and postgraduate students, researchers, policy makers, and practitioners in related fields of DRR who are interested in looking at DRR from health in general, as well as its application in Asia. Our special thank goes to Chi Shing Wong of CCOUC for his meticulous support and coordination of this book project, as well as the unfailing editorial support from Asami Komada, Mariko Komaru, Dinesh Natarajan, Yosuke Nishida, Umamagesh Perumal and Taeko Sato at Springer.

Hong Kong SAR, China — Emily Ying Yang Chan
Fujisawa, Japan — Rajib Shaw

Reference

World Health Organization (2019) Health emergency and disaster risk management framework. World Health Organization, Geneva

About This Book

This book presents the research paradigm and research landscape of Health Emergency and Disaster Risk Management (Health-EDRM) with examples from Asia. The intersection of health and disaster risk reduction (DRR) has emerged in recent years as an interdisciplinary field of paramount human consequences. Throughout several landmark United Nations agreements adopted in 2015–2016, including the Sendai Framework for Disaster Risk Reduction 2015–2030, the 2030 Sustainable Development Goals (SDGs), the Paris climate agreement, and the New Urban Agenda (Habitat III), health is recognised as an inevitable outcome and a natural goal of disaster risk reduction, and the cross-over of the two fields is no doubt essential for the successful implementation of the Sendai Framework. Health emergency and disaster risk management, as a joint venture of this cross-over, has emerged as an umbrella field that encompasses emergency and disaster medicine, DRR, humanitarian response, community health resilience, and health system resilience.

Health-EDRM, however, remains to be developed into a coherent enterprise after the launching of the WHO Health-EDRM Framework in 2019. Key challenges with this new field of studies include un-coordinated research, lack of a strategic research agenda, limited development of multi-sectoral and interdisciplinary approaches, deficiency in the science–policy–practice nexus, absence of standardised terminology, and meagre coordination among stakeholders. This book provides a timely and invaluable resource for undergraduate and postgraduate students, researchers, scholars, and frontline practitioners as well as policy makers from across the component domains of Health-EDRM.

Contents

Editors and Contributors

About the Editors

Prof. Emily Ying Yang Chan serves as professor and assistant dean, Faculty of Medicine, and head, Division of Global Health and Humanitarian Medicine, The Chinese University of Hong Kong (CUHK); director, Collaborating Centre for Oxford University and CUHK for Disaster and Medical Humanitarian Response (CCOUC), Centre for Global Health (CGH) and Centre of Excellence (ICoE-CCOUC), Integrated Research on Disaster Risk (IRDR); co-chairperson, WHO Thematic Platform for Health Emergency and Disaster Risk Management Research Network; member, Asia Pacific Science Technology and Academia Advisory Group (APSTAAG); visiting professor, Oxford University Nuffield Department of Medicine; and fellow, FXB Center, Harvard University. Her research interests include disaster and humanitarian medicine, climate change and health, global and planetary health, human health security and Health Emergency and Disaster Risk Management (Health-EDRM), remote rural health, implementation and translational science, ethnic minority health, injury and violence epidemiology, and primary care. Awarded the 2007 Nobuo Maeda International Research Award of the American Public Health Association, Prof. Chan has published more than 300 international peer-reviewed academic, technical, and conference articles and 10 academic books. She also had extensive experience as a frontline emergency relief practitioner in the mid-1990s spanning 20 countries.

Rajib Shaw is a Japanese national of Indian origin, he is the professor in Graduate School of Media and Governance in Keio University's Shonan Fujisawa Campus (SFC). Earlier, he was the Executive Director of the Integrated Research on Disaster Risk (IRDR), a decade-long research programme co-sponsored by the International Council for Science (ICSU), the International Social Science Council (ISSC), and the United Nations International Strategy for Disaster Reduction (UNISDR). He is also the senior fellow of Institute of Global Environmental Strategies (IGES) Japan, and the chairperson of SEEDS Asia, a Japanese NGO. Previously, he served as a professor in the Graduate School of Global Environmental Studies of Kyoto University. His expertise includes community-based disaster risk management, climate change adaptation, urban risk management, and disaster and environmental education. He is a member of the United Nations global Science Technology Advisory Group (STAG) and is the co-chair of the Asia Pacific Science Technology Academic Advisory Group (APSTAAG). He is the editor of a book series on disaster risk reduction, published by Springer. Prof. Shaw has published more than 45 books and over 300 academic papers and book chapters.

Contributors

Md. Anwarul Abedin Department of Soil Science, Bangladesh Agricultural University, Mymensingh, Bangladesh

Muhammad Abid Centre for Climate Research and Development (CCRD), COMSATS University, Islamabad, Pakistan

Emily Ying Yang Chan Collaborating Centre for Oxford University and CUHK for Disaster and Medical Humanitarian Response, JC School of Public Health and Primary Care, The Chinese University of Hong Kong, Shatin, Hong Kong

Gloria Kwong Wai Chan JC School of Public Health and Primary Care, The Chinese University of Hong Kong, Shatin, Hong Kong

Kennedy E. Espina Ateneo de Manila University, Quezon City, Philippines

Maria Regina E. Estuar Ateneo de Manila University, Quezon City, Philippines

Maria Regina Justina E. Estuar Ateneo de Manila University, Quezon City, Philippines

Satoshi Furuya Makioka Hospital, Yamanashi, Japan

Tran Thi Tuyet Hanh Hanoi University of Public Health, Hanoi City, Vietnam

Janice Y. Ho JC School of Public Health and Primary Care, The Chinese University of Hong Kong, Shatin, Hong Kong

Zhe Huang JC School of Public Health and Primary Care, The Chinese University of Hong Kong, Shatin, Hong Kong

Heidi Hung JC School of Public Health and Primary Care, The Chinese University of Hong Kong, Shatin, Hong Kong

Jingzhong Ye College of Humanities and Development Studies (COHD), China Agricultural University, Beijing, People's Republic of China

Md. Khalequzzaman Department of Public Health and Informatics, Bangabandhu Sheikh Mujib Medical University, Dhaka, Bangladesh

Sakiko Kanbara Faculty of Nursing, University of Kochi, Kochi, Japan

Fahmida Afroz Khan Department of Public Health and Informatics, Bangabandhu Sheikh Mujib Medical University, Dhaka, Bangladesh

Mohammad Golam Kibria Department of Soil Science, Bangladesh Agricultural University, Mymensingh, Bangladesh;
School of Agriculture and Environment, University of Western Australia, Perth, Australia

Yuichi Koido National Disaster Medical Center, Tokyo, Japan

Takeshi Komino CWS Japan, Japan CSO Coalition for DRR, Tokyo, Japan

Supriya Krishnan National Institute of Public Finance and Policy, New Delhi, India

Tatsuhiko Kubo Hiroshima University, Hiroshima, Japan

Holly Ching Yu Lam JC School of Public Health and Primary Care, The Chinese University of Hong Kong, Shatin, Hong Kong

Hyeon Ju Lee Faculty of Nursing, University of Kochi, Kochi, Japan

Tran Nu Quy Linh Ha Tinh, Center for Disease Control and Prevention, Ha Tinh, Hatinh Province, Vietnam

Sida Liu Collaborating Centre for Oxford University and CUHK for Disaster and Medical Humanitarian Response, JC School of Public Health and Primary Care, The Chinese University of Hong Kong, Shatin, Hong Kong

Asta Yi Tao Man JC School of Public Health and Primary Care, The Chinese University of Hong Kong, Shatin, Hong Kong

Akiko Matsuyama Tsuda University, Tokyo, Japan

Shoko Miyagawa Keio University, Fujisawa, Japan

Hiroyuki Miyazaki University of Tokyo, Tokyo, Japan

Virginia Murray WHO Thematic Platform Health and Emergency Disaster Risk Management Research Network, Geneva, Switzerland;
Public Health England, London, UK;
Integrated Research on Disaster Risk (IRDR) Scientific Committee, Geneva, Switzerland;
WHO Collaborating Centre on Mass Gatherings and Global Health Security, London, UK;
UNU-International Institute of Global Health, Kuala Lumpur, Malaysia

Apsara Pandey Nursing Association of Nepal, Kathmandu, Nepal

Ila Patnaik National Institute of Public Finance and Policy, New Delhi, India

Christian E. Pulmano Ateneo Block Chain Laboratory, Quezon City, Philippines

Bhaswati Ray Sivanath Sastri College, University of Calcutta, Kolkata, India

Roy Miguel G. Romarate Factset Philippines, Inc, Taguig City, Philippines

Flavio Salio World Health Organization, Geneva, Switzerland

Miho Satoh Yokohama City University, Yokohama, Japan

Ashfaq Ahmad Shah School of Management Science and Engineering, Nanjing University of Information Science and Technology (NUIST), Jiangsu, People's Republic of China;
College of Humanities and Development Studies (COHD), China Agricultural University, Beijing, People's Republic of China

Rajib Shaw Graduate School of Media and Governance, Keio University, Fujisawa, Japan

Charles Ventura Factset Philippines, Inc, Taguig City, Philippines

Chi Shing Wong Collaborating Centre for Oxford University and CUHK for Disaster and Medical Humanitarian Response, JC School of Public Health and Primary Care, The Chinese University of Hong Kong, Shatin, Hong Kong

Yoko Yamada CANNUS, Ishinomaki, Japan

Chapter 1
Overview of Health-EDRM and Health Issues in DRR: Practices and Challenges

Emily Ying Yang Chan and Rajib Shaw

Abstract This chapter outlines the Health-EDRM framework as an emerging multidisciplinary subject of enquiry. This chapter explains key concepts that bridge health with disaster risk reduction and highlights the pivotal idea of disaster prevention in Health-EDRM. It discusses the challenges of this discipline, including the diverse research landscape, unified terminologies and data collection issues.

Keywords Disaster prevention · Disaster risk reduction · Health Emergency and Disaster Risk Management (Health-EDRM) · Health risk · Vulnerability

1.1 Introduction

Natural hazards, in the absence of risk reduction strategy, preparedness and management, might result in disaster events and lead to catastrophic human and economic impact. In 2015, an estimated 22,500 deaths and 100 million affected were reported to be attributable to disaster events globally (UNISDR 2015). Human impacts of disasters and emergencies are complex. Emergencies and disasters may cause direct and indirect impacts to health (examples of the latter include the disruption of life-line infrastructure, health systems and facilities). However, regardless of impact dimensions, adverse outcomes from disasters may cause major barriers to human development and progress.

Published literature and reports have already indicated that hazards, vulnerabilities, capacities and risks in emergency responses distribute unequally globally. However, impact of emergencies and disasters frequently affect disproportionally on vulnerable populations such as poor, extreme of ages, disabilities and ethnic

Emily Ying YangChan (✉)
JC School of Public Health and Primary Care, The Chinese University of Hong Kong, Shatin, Hong Kong
e-mail: emily.chan@cuhk.edu.hk

R. Shaw
Graduate School of Media and Governance, Keio University, Fujisawa, Japan

E. Ying Yang Chan and R. Shaw (eds.), *Public Health and Disasters*, Disaster Risk Reduction, https://doi.org/10.1007/978-981-15-0924-7_1

minority group (Chan 2017). As developing countries tend to have less resilient systems and intrinsic capacity, when emergencies strike (e.g. climate-related ones), the human impact will likely to be more profound than in developed context. Even within a developed nation, the intra-nation variation in socioeconomic characteristics will result in different levels of vulnerabilities and results. People living in poverty are often considered as under high risks and "poverty reduction" has been proposed as essential components of vulnerability reduction for emergency. Health of extremes of age, i.e. children and older people, are particularly vulnerable in emergency and extreme disaster events. For instance, children represented 30–50% of mortality in disasters. In addition to socio-demographic vulnerabilities, context may also contribute to emergency risks. Highly crowded density, restricted point of access, limited crowd control, sudden change of weather, and lack of preparedness and medical support may all present as potential health risks in mass gathering events (WHO 2017a, b, c, d, e, f, g, h). Environmental and site preparedness, crowd safety, mass casualty preparedness for mass gathering requires good preparation for risk reductions and management. Meanwhile, it is almost impossible to report true impact of emergency on communities due to the lack of appropriate indicators to capture and the existing statistics reports might only reflect human suffering superficially. To protection population from the adverse health impact of crisis, proactive risk management has always been the objectives of people working in public health and related fields (Chan and Shi 2017).

Despite the challenges, with proper preparedness, coordination and resources, health risks encountered in emergencies and disasters might be reduced and minimized at all levels. Reports indicated that health sectors in only around 100 WHO member states have set aside separate budgets and resources for emergency preparedness and response (WHO 2008). Factors affecting response capacity might include suboptimal disaster risk management systems, lack of resources and knowledge and ongoing insecurity arising from conflict (WHO 2008). At the individual level, health risks might be conceptualized and addressed by looking into the vulnerability, exposure to a hazard or hazardous context, and how people might be exposed to, respond and manage such risks.

1.2 Health Emergency and Disaster Risk Management: Mainstreaming Health in Disaster Risk Reduction

The intersection of health and disaster risk reduction (DRR) has emerged as an interdisciplinary field of paramount human consequences in recent years. In a number of landmark United Nations agreements adopted in 2015–2016, including the Sendai Framework for Disaster Risk Reduction 2015–2030, the 2030 Sustainable Development Goals (SDGs), the Paris climate agreement, and the New Urban Agenda (Habitat III), health is recognized as an inevitable outcome and a natural goal of disaster risk reduction, and the crossover of the two fields is no doubt crucial for the successful implementation of the Sendai Framework. Health Emergency and Disaster Risk Management (Health-EDRM), as a joint venture of this crossover to engage all relevant practitioners and researchers, has emerged as an overarching field encompassing emergency and disaster medicine, disaster risk reduction (DRR), humanitarian response, community health resilience, and health system resilience (CCOUC n.d.; Chan and Murray 2017; Lo et al. 2017; WHO 2019).

Health-EDRM is an academic paradigm that is actively being developed and evolved since 2009 (WHO 2008). The discipline aims to examine health and disaster risks and applies public health tools to engage in the management of health and disaster risk. In contrast to the traditional medical emergency and disaster approaches that are often response-based, the Health-EDRM paradigm targets systematic analysis and management of health risks. It emphasizes emergency preparedness and disaster risk reduction (Aitsi-Selmi and Murray 2015; Chan and Murray 2017) by adopting the preventive public health approach that addresses risks to reduce potential adverse impact and harm for all-hazard throughout the emergency cycle (WHO 2017g). A focus on prevention can also provide opportunities for developing research infrastructure in normal times. Research and alignment of actions in Health-EDRM may identify relevant health risks associated with a context, emergency and disaster situation (Chan et al. 2019). Its actions implement scientific evidence-based solutions and adopt policies that support preparedness, response and rehabilitation capacity building to enhance the resilience of a health system and its associated supporting systems.

Health-EDRM aims to consider all areas of risk management and determinants that might affect health (WHO 2017a, b, c, d, e, f, g, h). It argues that comprehensive, multidisciplinary approach should be used to analysis all-hazard risks within the emergency management cycle. This discipline invites the engagement of both health and non-health actors to understand the theories and frameworks that describe how human well-being, health risks and outcomes might be affected by hazards and disaster events. Health-EDRM also embraces multidisciplinary actors and adopts an inter-sectoral action approach to proactively prevent disaster- and emergency-related health risks. It also advocates for both top-down and bottom-up approaches as of equal importance to maximize impact management. As a scientific discipline, it aims to build relevant inter-sectoral and multidisciplinary frameworks and gather evidence to reduce health risks and impacts in crises and emergencies. The WHO launched the strategic framework in 2019 (WHO 2019) and in years to come, this book would serve as a useful resource and reference in this new field of study.

1.3 Key Challenges for Health-EDRM

The focuses as well as challenges of Health-EDRM research include the agreement within the academic and research community to adopt:

1. an all-hazards approach that incorporates the full spectrum of hazards that may cause disasters and crises;
2. a holistic all-needs approach, including physical, mental, and psycho-social health and well-being to support planning and examine outcomes;
3. disaster risk identification for populations with specific health needs such as children, people with disabilities, and the elderly;
4. identify evidence-based interventions facilitated during all phases of a disaster; and
5. research on and the building of health resilience in all communities. (CCOUC, n.d.).

As a new discipline, the fragmented nascent field of Health-EDRM, however, remains to be developed into a coherent enterprise (WHO 2019). Key challenges include non-alignment of research tools, lack of a strategic overarching research agenda, suboptimal development of multisectoral and interdisciplinary approaches, absence of the science–policy–practice nexus, deficiency in standardized terminology, and meagre coordination among stakeholders (Lo et al. 2017).

Moreover, there are general uncertainties about the agreed health indicators for Health-EDRM, as well as the absence of an agreed all-hazard and disasters classification for the purpose of Health-EDRM data collection. There is also a lack of consensus to account for thresholds relating to temporality (slow-onset versus protracted events), attribution (direct versus indirect causes of morbidity and mortality), and baseline data, working epidemiological definitions are urgently needed (Chan and Murray 2017).

When revisiting the research landscape in terms of the application of the preventive concept in research planning, implementation, and research ethics, Health-EDRM research needs to evolve beyond guidelines, codes, and approval processes. In view of the relatively neglected emergency health-related studies in the current research systems, the proactive adoption of the Health-EDRM approach may help create the conditions for conducting research in emergency and disaster situations in productive as well as ethical ways (Chan et al. 2019).

1.4 Application of Health-EDRM Framework in Asian Case Studies

This book attempts to provide the scientific, academic and practitioner communities an overview of the underlying concepts of Health-EDRM. It describes the key public health principles which will facilitate the conceptualization of health risks in times of emergencies and crisis. Specifically, these discussions aim to facilitate and familiarize the understanding of non-health readers towards key principles of public health and how health risks might be considered. It also describes actual case examples of how interventions and policies might be conceptualized under this paradigm. It highlights the current gaps in infrastructure and knowledge of the discipline (Chan and Murray 2017; Chan et al. 2019).

This book has four main sections and 23 Chapters. *Section One* provides an overview of Health-EDRM in practices and challenges. *Section Two* includes six chapters that describe the important public health principles and theories that underlie the paradigm. Chapter 2 explains public health prevention hierarchy for the conceptualization of disaster context within a preventive perspective. Chapter 3 highlights how demographic and epidemiological transitions may present a complex picture to understanding health risks and community vulnerability in the twenty-first century. Chapter 4 analyses the current evidence gaps in bottom-up efforts for building a health resilient community. Chapters 5–7 delineate how Health-EDRM might address the four important global agendas of Sendai Framework for Disaster Risk Reduction, Paris climate agreement, United Nations Sustainable Development Goals and New Urban Agenda (Habitat III). *Section Three* consists of 14 Chapters (Chaps. 8– 21) that describe policies and programme initiatives that aim to address Health-EDRM in each of the contexts. Contributing authors share various actual policy and programme examples of how Health-EDRM might be applied in programmes and policy planning in disaster risk reduction of health. Chapter 22 of *Section Four* summarizes how these cases might contribute to the Health-EDRM discipline. The last chapter is the Epilogue of this book, including a discussion of the current gaps remained unaddressed in the paradigm.

References

Aitsi-Selmi A, Murray V (2015) The Sendai framework: disaster risk reduction through a health lens. Bull World Health Organ 93:362. https://doi.org/10.2471/BLT.15.157362

Collaborating Centre for Oxford University and CUHK for Disaster and Medical Humanitarian Response (CCOUC) (n.d.) Health Emergency and Disaster Risk Management (H-EDRM). Retrieved from http://ccouc.org/h-edrm

Chan EYY (2017) Public health response in humanitarian crisis. Routledge, London

Chan EYY, Murray V (2017) What are the health research needs for the Sendai framework? Lancet 390(10106):e35–e36. https://doi.org/10.1016/S0140-6736(17)31670-7

Chan EYY, Shi P (2017) Health and risks: integrating health into disaster risk reduction, risk communication, and building resilient communities. Int J Disaster Risk Sci 8(2):107–108

Chan EYY, Wright K, Parker M (2019) Health-emergency disaster risk management and research ethics. Lancet 393:112–113

Lo STT, Chan EYY, Chan GKW, Murray V, Abrahams J, Ardalan A, Kayano R, Yau JCW (2017) Health emergency and disaster risk management (health-EDRM): developing the research field within the Sendai framework paradigm. Int J Disaster Risk Sci 8:145–149. https://doi.org/10.1007/s13753-017-0122-0

UNISDR (2016) 2015 disasters in numbers. Available at: http://www.unisdr.org/files/47804_2015disastertrendsinfographic.pdf

World Health Organization (2008) Global assessment of national health sector emergency preparedness and response. Available at https://www.who.int/hac/about/Global_survey_inside.pdf

World Health Organization (2017a) Public Health England, & partners. Health emergency and disaster risk management: child health
World Health Organization (2017b) Public Health England, & partners. Health emergency and disaster risk management: communicable diseases
World Health Organization (2017c) Public Health England, & partners. Health emergency and disaster risk management: injury prevention and trauma care
World Health Organization, Public Health England, & partners (2017d) Health emergency and disaster risk management: mass gatherings
World Health Organization, Public Health England, & partners (2017e) Health emergency and disaster risk management: non-communicable diseases
World Health Organization, Public Health England, & partners (2017f) Health emergency and disaster risk management: nutrition
World Health Organization (2017g) Public Health England, & partners. Health emergency and disaster risk management: overview
World Health Organization, Public Health England, & partners (2017h) Health emergency and disaster risk management: water, sanitation and hygiene
World Health Organization (2019) Health emergency and disaster risk management framework. World Health Organization, Geneva

Chapter 2
Public Health Prevention Hierarchy in Disaster Context

Emily Ying Yang Chan and Chi Shing Wong

Abstract The public health prevention hierarchy—namely primary, secondary and tertiary prevention, is one of the most important public health principles which guide policy and programme development. Primary prevention attempts to prevent the onset of disease or reduce health risks. Strategies may include health protection and health promotion. Health protection can be carried out through the establishment of policies, regulations and programmes (e.g. vaccinations), while health promotion mainly involves health education and information sharing. Secondary prevention refers to stopping disease progression. Related activities aim to detect disease early and thus increase the opportunity for early intervention to prevent progression and symptom development. Screening is a classic example of secondary prevention. Tertiary prevention focuses on the rehabilitation of patients with an established disease to minimise residual disabilities and complications. It aims to restore bodily functions that have been impaired by the disease and impact. Treatment, rehabilitation and palliative care are examples of the tertiary prevention services. This chapter applies public health prevention hierarchy in disaster prevention and response. The application of these prevention concepts to support Health-EDRM in establishing disaster mitigation strategies, response programmes and post-disaster recovery policies may enhance individual survival and protect communities from adverse health outcomes in natural disasters in a cost-effective way.

Keywords Disaster risk management · Disaster risk reduction · Health Emergency and Disaster Risk Management (Health-EDRM) · Prevention hierarchy

---

E. Y. Y. Chan (✉) · C. S. Wong
JC School of Public Health and Primary Care, The Chinese University of Hong Kong, Shatin, Hong Kong
e-mail: emily.chan@cuhk.edu.hk

E. Ying Yang Chan and R. Shaw (eds.), *Public Health and Disasters*, Disaster Risk Reduction, https://doi.org/10.1007/978-981-15-0924-7_2

2.1 Introduction

Health-EDRM aims to put people's health at the centre of emergency and disaster risk management. Internationally, it is advocated by World Health Organization (WHO) and enshrined in such disaster-related international policy frameworks as the Sendai Framework for Disaster Risk Reduction 2015–2030, Sustainable Development Goals and Paris Agreement on climate change. The framework demands a stronger role for science and for all stakeholders and groups (including women, children, people with disabilities and older people) in disaster risk management. In addition to fitting into this multi-stakeholder bottom-up approach to disaster risk management, the public health prevention concept arguably also provides one of the key hinges to unify this emerging field of Health-EDRM as a crossover between health and disaster risk reduction encompassing the disciplines of emergency and disaster medicine, DRR, humanitarian response, community health resilience, and health system resilience. World Health Organization suggests the goal for Health-EDRM as minimising the health impact of emergencies and disasters, while the prevention concept captures the crux of cost-effectiveness behind various means to this end WHO has suggested, e.g. safe hospitals to mitigate negative public health consequences post-disaster, safe water supply to reduce exposure to hazards, vaccinations to minimise vulnerabilities, mass casualty response plans to strengthen local capacities for response and recovery, and community health care to build local health resilience (WHO 2015).

Under the Health-EDRM framework, emergency and disaster risk management policies, activities and programmes involving multidisciplinary sectors can help avoid or reduce the health impacts of disaster, such as deaths, injuries, diseases, disabilities and psychosocial problems. WHO highlights Health-EDRM as referring to the systematic analysis and management of health risks posed by emergencies and disasters. Through reduction in hazard, exposure and vulnerability, better preparedness, response, and recovery could be expected. Although the traditional focus of the health sector in emergencies and disasters has been on the clinical on-site response to and survival in emergencies and disasters, Health-EDRM will re-direct this focus to enhancing the upstream aspects of preparedness and hazard, exposure and vulnerability reduction. It emphasises prevention and capacity development of community and country to provide timely, planned and resource-effective response and recovery, as well as building resilient health systems based on community-level primary health care. These attempts will reduce community vulnerability, protect health facilities and services, and scale-up health response to meet the surging health needs post-disaster (Chan and Murray 2017; WHO et al. 2017).

The prevention-focused Health-EDRM also echoes with the Sendai Framework's expected outcome ("The substantial reduction of disaster risk and losses in lives, livelihoods and health"), goal ("Prevent new and reduce existing disaster risk through the implementation of integrated and inclusive economic, structural, legal, social, health, cultural, educational, environmental, technological, political and institutional measures that prevent and reduce hazard exposure and vulnerability to disaster, increase preparedness for response and recovery, and thus strengthen resilience.") and three of its seven global targets ("Reduce disaster mortality, reduce the number of affected people, and reduce disaster damage to critical infrastructure and disruption of basic services, including health facilities") (United Nations Office for Disaster Risk Reduction 2015).

This chapter applies public health prevention hierarchy in disaster context to provide a conceptual skeleton for Health-EDRM. The application of these prevention concepts in establishing disaster mitigation strategies, response programmes and post-disaster recovery policies may facilitate better individual survival and protect communities from adverse health outcomes in natural disasters.

2.2 Hierarchy of Prevention

In public health, prevention is often divided into *primary prevention* (the prevention of disease from the source in the first place, for example, through providing clean water or immunisation against infectious diseases), *secondary prevention* (the early detection of an infectious disease at an early stage when it can be treated or contained, through periodic health surveillance) and *tertiary prevention* (prevention of disability, for example, by providing rehabilitation services after an injury).

The classical framework of Leavell and Clark (1958) suggests three levels of health prevention, namely primary, secondary and tertiary prevention. **Primary prevention** concerns measures that prevent the onset of diseases at the source. Strategies may include health protection and health promotion, the former can be carried out through the establishment of health policies and regulations as well as the provision of clean water or vaccinations against infectious diseases, while the latter mainly involves health education. **Secondary prevention** refers to blocking the progression of a disease after its onset. It aims at an early detection of a disease by such means as screening to increase the opportunity for successful interventions to prevent the progression of the disease and the emergence of symptoms. Another example is the detection of an infectious disease at an early stage when it can be treated or contained, through periodic health surveillance. **Tertiary prevention** focuses on rehabilitating patients suffering from an established disease to minimise complications and disabilities. It aims to restore bodily functions that have been impaired by the disease. Interventions in this category include treatment, rehabilitation and palliative care. The translation of these prevention concepts in Health-EDRM may enhance individuals' survival and protect communities from adverse health outcomes brought about by natural disasters (Leavell and Clark 1958; Department of Health of the Government of the Hong Kong Special Administrative Region 2008).

Health protection and health promotion are two major strategies to achieve the goal of disease prevention. Health protection strategies focus on controlling or removing health hazards, slowing down disease progression and reducing the impacts of established diseases with a wide range of primary, secondary and tertiary prevention strategies; while health promotion strategies focus on encouraging healthy behaviours and lifestyle to improve well-being and to reduce the risk of developing diseases, which are mostly in the realm of primary prevention.

2.3 Disasters

There are various ways of classifying disasters. One of the common ways to classify disasters is based on the nature of the triggering cause or hazard, namely, natural disasters, human-caused disasters and complex emergencies. Meanwhile, classification based on cause has the merit of targeting prevention efforts under the Health-EDRM framework.

Natural disasters are disastrous events triggered by hazards of natural origin. The Centre for Research on the Epidemiology of Disasters (CRED) of the Université catholique de Louvain in Belgium further classifies natural disasters into six major subcategories (CRED n.d.; Below et al. 2009; Chan 2017):

- **Geophysical disasters**: Events originating from the earth's solid surface and its interior;
- **Meteorological disasters**: Events caused by short-term, microscopic atmospheric processes;
- **Hydrological disasters**: Events caused by deviations in the normal water cycle or overflow of bodies of water caused by strong wind;
- **Climatological disasters**: Events caused by long-term, meso- to macro-scale processes;
- **Biological disasters**: Events caused by the exposure of humans or livestock to germs or toxic substances;
- **Extraterrestrial disasters**: Events caused by asteroids, meteoroids and comets which pass near the earth or strike the earth, or any changes in interplanetary conditions that affect the earth's magnetosphere.

Specifically, climatological, hydrological and meteorological disasters can be grouped together as hydro-meteorological disasters. Together with biological disasters, they are natural disasters with direct associations with the climate system and thus affected by global climate change.

Human-caused disasters are disastrous events arising from human-related hazards, either unintentionally (e.g. traffic accidents, industrial accidents, nuclear accidents and hazardous material spills) or intentionally (e.g. wars and terrorist attacks) (United States Federal Emergency Management Agency [FEMA] 2008, D-32). Human-caused disasters may range from technological disasters to bioterrorism. Technological disasters can result from human errors or breakdown of technological systems, which can be further categorised into: industrial accidents (collapses, explosions, fires, gas leaks, poisoning, radiation and others), transport accidents (rail, road, water and air) and miscellaneous accidents (Chan 2017).

A complex emergency is "[a] multifaceted humanitarian crisis in a country, region or society where there is a total or considerable breakdown of authority resulting from internal or external conflict and which requires a multi-sectoral, international response that goes beyond the mandate or capacity of any single agency and/or the ongoing UN country programme" (United Nations Office for the Coordination of Humanitarian Affairs [OCHA] 2003, p. 9). Complex emergencies often occur in settings having experienced protracted disruptions to livelihoods (by warfare, civil disturbance and large-scale population movements). It might induce violence and illegal actions. In these settings, many people are forced to leave their homes to seek refuge elsewhere or escape from the destruction of their homes, hunger, diseases and persecution. These people become internally displaced persons (IDPs) when they moved to other places inside their own country, or become refugees when they cross national borders.

While UNISDR suggests that disaster prevention refers to "[t]he outright avoidance of adverse impacts of hazards and related disasters… through actions taken in advance" (United Nations International Strategy for Disaster Reduction [UNISDR] 2009, p. 22), the public health approach to prevention could enrich the content of and deepen the understanding towards disaster prevention by delineating three levels of prevention.

Figure 2.1 displays the hierarchy of prevention in Health-EDRM. In disaster settings, primary prevention may be represented as the lowest level of the pyramid. These activities target the wider community and cover the largest proportion of preventable health impact. Secondary prevention focuses on a smaller population directly affected by a disaster. Tertiary prevention concerns only those who have already experienced the health impact of a disaster, which constitute a small portion of the affected population. The per capita cost of the preventive measures tends to increase as intervention choices migrate up the hierarchy. Typically, prevention at the primary level tends to be the most cost-effective.

In the context of disaster preparedness, **primary prevention** proactively addresses the potential health risks associated with hazards and disaster events beforehand, so as to prevent the onset of negative health impact. Its activities aim to prepare for and enhance resilience before a disaster. For example, in flood-prone areas, heavy rainfall might bring about stagnant water that can become breeding sites for mosquitoes and hence increase the risk of vector-borne diseases, such as malaria, dengue fever, Japanese encephalitis and West Nile fever (World Health Organization Regional Office for the Eastern Mediterranean [WHO-EMRO] 2005). Primary prevention activities include building structures that avoid the formation of water traps and the accumulation of stagnant water, as well as the promotion of activities and raising community awareness of the disease risks associated with stagnant water. In earthquake-prone areas, design and building codes of seismic-resistant hospitals is another example of primary prevention to minimise the direct health impact of disasters and the health risk associated with disruption of health services by disasters (World Health Organization (WHO), United Kingdom Health

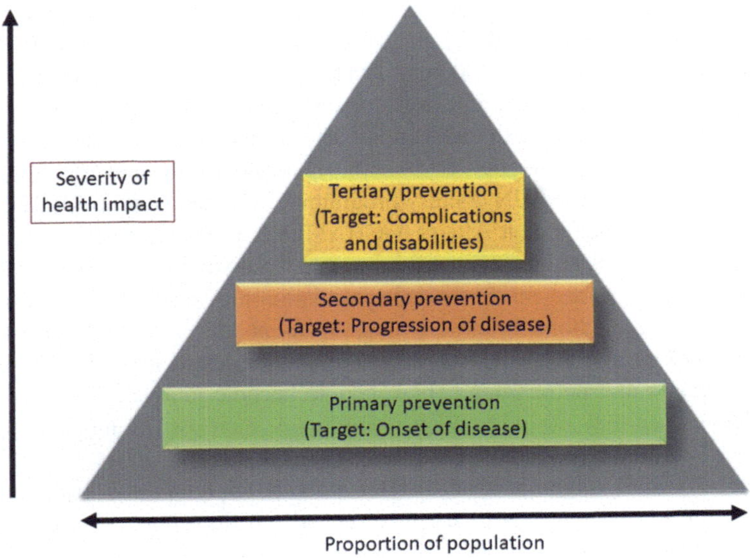

Fig. 2.1 Hierarchy of prevention pyramid in Health-EDRM

Protection Agency (HPA), & partners 2011c). The World Health Organization recognises the crucial role of hospitals in a disaster and made specific guidelines to promote safe hospitals (World Health Organization Regional Office for the Western Mediterranean [WHO-WPRO] 2010).

For technological disasters, preventive measures that aim at the primary level should attempt to reduce the negative health outcomes (e.g. mortality and morbidity in terms of injuries, disabilities or chronic diseases). Specific guidelines and emergency-related training should be implemented in industries. For instance, to prepare for a chemical release in chemical incidents or radiation emergencies, guidelines and training should cover relevant scenario analyses, impact assessment, training as well as equipping responders to deal with loss of containment (World Health Organization (WHO), United Kingdom Health Protection Agency (HPA), & partners 2011a). For radiation emergencies, continuous monitoring of the environment and high-risk groups should be applied (World Health Organization (WHO), United Kingdom Health Protection Agency (HPA), & partners, 2011b).

Secondary prevention in disaster settings refers to actions addressing the health risk and its response during the immediate aftermath of a disaster, including blocking the spread of diseases (or other negative health conditions) and their adverse health impact. For instance, to avoid an increased burden of clinical consultations after the onset of a disaster, the health needs of people with chronic disease conditions (e.g. drugs and specific diet requirements) should be proactively managed (e.g. via maintenance of drug supply) to avoid medical complications due to lack of management. Even with limited resources and capacity in post-disaster settings, there are always ways to support populations with chronic conditions after disasters. For example, giving health advice incurs relatively little operation cost but has potential long-term implications for disease prevention. In another example of nuclear reactor accidents where radioactive materials may be released into the environment, the World Health Organization has developed guidelines for iodine prophylaxis during nuclear accidents and recommended that iodine tablets should be distributed among the affected population to minimise the potential harm arising from radiation, due to the concerns over radioactive contamination of food and water and the associated risk of thyroid cancer (WHO 1999). To implement meaningful preventive-based relief programmes, however, it is pertinent to emphasise the need to collect relevant demographic profiles, health information, knowledge, attitudes and behaviour information during health needs assessment so as to design and implement relevant programmes according to the project needs.

In a disaster context, **tertiary prevention** aims to minimise the impact on and damage to human health post-disaster. Tertiary prevention measures target people who have already suffered from the negative health impact of the disaster, but aim to prevent poor outcomes (such as death). For example, after an earthquake, patients might suffer from orthopaedic trauma and require operations. While rapid clinical operations could save lives, it is also important to offer early post-operational physiotherapies to maximise functional recovery potential of the patients (e.g. amputees).

Figure 2.2 illustrates the application of preventive hierarchy in a Health-EDRM framework in an example of managing the health risk of cholera outbreak post-disaster.

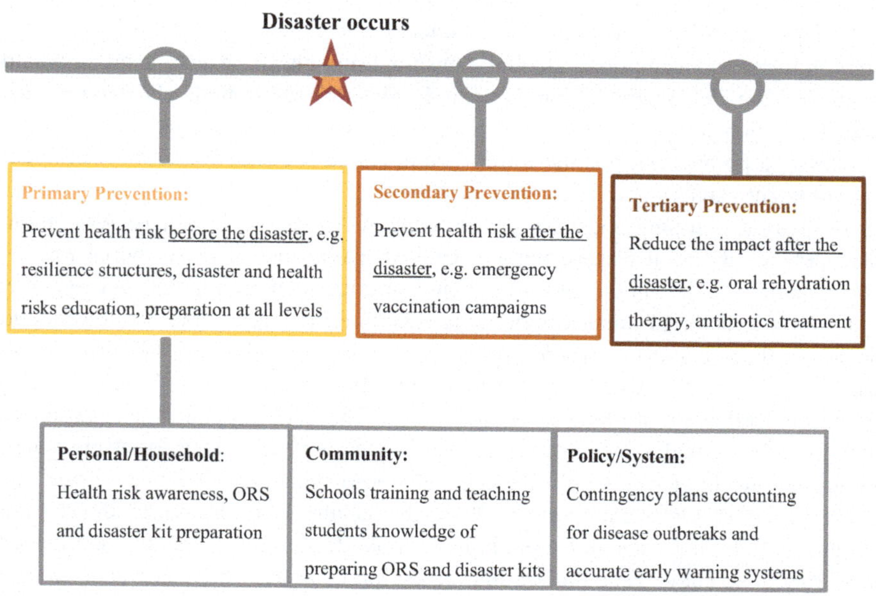

Fig. 2.2 Components of the prevention concept in the Health-EDRM framework against potential cholera outbreak post-disaster

2.4 Health Risk Transition

The hierarchy of prevention model is further supplemented by the notion of "health risk transition" to help prioritise Health-EDRM policies and actions in countries at different development stages.

Health risk transition can be considered a logical extension of the double transitions, i.e. demographic and epidemiological transitions. The classical demographic transition is the transition from high birth and mortality rates to lower birth and mortality rates as a country undergoes industrialisation (Caldwell et al. 2006). During the demographic transition as a result of industrialisation and modernisation, the mortality rate of a society starts to drop as a result of improved food access and nutrition on the one hand, and advances in public health and medicine on the other. The decline in birth rate picks up later and results in lower birth and mortality rates, as well as longer life expectancy. As a result of better hygiene, nutrition and healthcare, there sees the epidemiological transition where communicable diseases start to recede while chronic non-communicable diseases constitute an increasing portion of the disease burden. The mechanism of this epidemiological transition is further explicated by a third parallel transition, the health risk transition.

World Health Organization proposed the concept of risk transition in its *World Health Report 2002*. Risk is defined as "a probability of an adverse outcome, or a factor that raises this probability" (WHO 2002, p. 7) and the report described the amount of disease, disability and mortality in the world that could be attributed to some common modern risks to human health, i.e. the top ten risks in terms of the burden of disease they cause. According to the report (WHO 2002), the ten leading risk factors globally were found to be: underweight; unsafe sex; high blood pressure; tobacco consumption; alcohol consumption; unsafe water, sanitation and hygiene; iron deficiency; indoor smoke from solid fuels; high cholesterol; and obesity, which accounted for more than one-third of total deaths in the world. However, it pointed out that these risk factors varied between countries of different development stages.

In many developing countries in sub-Saharan Africa and Southeast Asia, more than 30% of the total disease burden resulted from three of these global top ten risks, namely underweight, unsafe sex and unsafe water, sanitation and hygiene. The top risk factor of underweight alone accounted for more than 3 million child deaths in these countries, where poverty was a strong underlying health determinant. Besides, it was also estimated that more than 99% of the HIV infections in Africa in 2001 were attributable to unsafe sex. The WHO (2002) report estimates that about an annual 1.7 million deaths worldwide were attributed to unsafe water, sanitation and hygiene, mainly through infectious diarrhoea. Around 90% of the deaths were children and almost all of the deaths were found in developing countries. In both Africa and Asia, indoor smoke from solid fuels, zinc deficiency and iron deficiency were the leading risks for disease, which are also closely associated with poverty.

In more developed middle-income countries like China and those in Latin America, five risk factors (alcohol consumption, high blood pressure, tobacco consumption, underweight and overweight) have been shown to cause more than one-sixth of all disease burden. In the most developed countries in North America, Europe and the Asia-Pacific, at least one-third of all disease burden was attributable to tobacco consumption, high blood pressure, alcohol consumption, high cholesterol and obesity, which are related to "overconsumption" at the other end of the risk factor scale as compared with poverty and are mostly lifestyle related. However, these risk factors and the diseases associated with them are becoming more prevalent in developing countries, where they have created a double disease burden in addition to the remaining communicable diseases that have been afflicted poorer countries. The more developed middle-income countries suffered a double burden of risks, with their top six risk factors including all the top five overconsumption-related risks and the top poverty-related risk factor. The WHO (2002) report estimated that the number of deaths attributable to tobacco consumption in the year 2000 was 4.9 million, over 1 million more than it was in 1990. Although most of the smoking-related disease burden remained to be found in industrialised countries, the increase was most marked in developing countries, which demonstrates a shift of health risk. Similarly, while alcohol contributed to a higher proportion of death and disease burdens in the Americas and Europe, alcohol consumption has been increasing in developing countries.

This transition of health risks suggests that these health risks have shifted as a result of the development and modernisation of a society. WHO (2002) reported that the five overconsumption-related risk factors of high blood pressure, tobacco consumption, alcohol consumption, high cholesterol and obesity were part of a "risk transition" associated with marked changes in lifestyles in many parts of the world. In many developing countries, rapid increases in body weight have been observed, particularly among children, adolescents and young adults. From 1980 to 2000, obesity rates have risen threefold or more in some parts of the developed North America, Australasia and Eastern Europe, as well as the developing China, the Middle East and the Pacific Islands. While eating fruit and vegetables can help prevent cardiovascular diseases and some cancers, the low intake of them was responsible for almost 3 million deaths a year from those diseases during the turn of the millennium. The emergence of these health risks is attributable to changes in food processing and production and in agricultural and trade policies, which have changed people's daily diet and are in increasing trends among those in developing countries.

Meanwhile, changes in living and working patterns as a result of industrialisation and modernisation have led to less physical activity and less physical labour. The WHO report (2002) finds that physical inactivity was associated with about 15% of some cancers, diabetes and heart disease. The less-regulated marketing of tobacco and alcohol in developing countries also exposed more people to such products, which pose serious long-term health risks. For example, studies found that the death rates among smokers of all ages were two or three times higher than those of non-smokers.

In summary, health risk transition has to be taken into account when formulating preventive Health-EDRM policies and actions for countries at different development stages.

2.5 Conclusion

This chapter has outlined how prevention hierarchy framework may support Health-EDRM-based actions and policies. As a general framework, it can apply to various types of disasters. Various levels of preventive measures for disaster health risk reduction under the Health-EDRM framework can be tailored and prioritised based on the different stages of development in different countries, particularly in many developing Asian countries facing the double burden of poverty- and overconsumption-related health risks.

References

Below R, Wirtz A, Guha-Sapir D (2009) Disaster category classification and peril terminology for operational purposes (Working Paper No. 264). Centre for Research on the Epidemiology of Disasters (CRED), Brussels, Belgium & Munich Reinsurance (Munich RE), Germany. http://cred.be/sites/default/files/DisCatClass_264.pdf

Caldwell JC, Caldwell BK, Caldwell P, McDonald PF, Schindlmayr T (2006) Demographic transition theory. Springer, Dordrecht, The Netherlands

Centre for Research on the Epidemiology of Disasters (CRED), Université catholique de Louvain (n.d.) EM-DAT: the international disaster database: general classification. https://www.emdat.be/classification

Chan EYY (2017) Public health humanitarian responses to natural disasters. Routledge, London

Chan YYE, Murray V (2017) What are the health research needs for the Sendai framework? Lancet 390(10106):e35–e36

Department of Health of the Government of the Hong Kong Special Administrative Region (2008) Promoting health in Hong Kong: a strategic framework for prevention and control of non-communicable diseases. http://www.change4health.gov.hk/filemanager/common/image/strategic_framework/promoting_health/promoting_health_e.pdf

Leavell HR, Clark EG (1958) Preventive medicine for the doctor in his community: an epidemiologic approach, 2nd edn. McGraw-Hill, New York

United Nations International Strategy for Disaster Reduction (2009) 2009 UNISDR terminology on disaster risk reduction. http://www.unisdr.org/files/7817_UNISDRTerminologyEnglish.pdf

United Nations Office for Disaster Risk Reduction (2015) Sendai framework for disaster risk reduction. https://www.unisdr.org/we/coordinate/sendai-framework

United Nations Office for the Coordination of Humanitarian Affairs (2003) Glossary of humanitarian terms in relation to protection of civilians in armed conflict. United Nations, New York, NY

United States Federal Emergency Management Agency (FEMA) (2008) Producing emergency plans: a guide for all-hazard emergency operations planning for state, territorial, local, and tribal governments (Interim Version 1.0). http://www.training.fema.gov/hiedu/docs/cgo/week%203%20-%20producing%20emergency%20plans.pdf

World Health Organization (1999) Guidelines for iodine prophylaxis following nuclear accidents: Update 1999. http://www.who.int/ionizing_radiation/pub_meet/Iodine_Prophylaxis_guide.pdf
World Health Organization (2002) The world health report 2002: reducing risks, promoting healthy life. http://www.who.int/whr/2002/en/
World Health Organization (2015) WHO's six-year strategic plan to minimize the health impact of emergencies and disasters. http://www.who.int/hac/ercm_strategic_plan_web.pdf
World Health Organization Regional Office for the Eastern Mediterranean (2005) Vector-borne diseases: addressing a re-emerging public health problem (technical report). http://apps.who.int/iris/bitstream/10665/122334/1/EM_RC52_3_en.pdf?ua=1
World Health Organization Regional Office for the Western Mediterranean (2010) Safe hospitals in emergencies and disasters: structural, non-structural and functional indicators. http://www.wpro.who.int/emergencies_disasters/documents/SafeHospitalsinEmergenciesandDisastersweb optimized.pdf
World Health Organization (WHO), United Kingdom Health Protection Agency (HPA), & partners (2011a) Chemical safety (disaster risk management for health fact sheet) [Internet]. Retrieved from http://www.who.int/hac/events/drm_fact_sheet_chemical_safety.pdf
World Health Organization (WHO), United Kingdom Health Protection Agency (HPA), & partners (2011b) Radiation emergencies (disaster risk management for health fact sheets) [Internet]. Retrieved from http://www.who.int/hac/events/drm_fact_sheet_radiation_emergencies.pdf
World Health Organization, United Kingdom Health Protection Agency (HPA), partners (2011c) Safe hospitals: prepared for emergencies and disasters (disaster risk management for health fact sheet). http://www.who.int/hac/events/drm_fact_sheet_safe_hospitals.pdf
World Health Organization, Public Health England, partners (2017) Health emergency and disaster risk management: overview. http://www.who.int/hac/techguidance/preparedness/who-factsheet-overview-december2017.pdf

Chapter 3
Key Public Health Challenges for Health-EDRM in the Twenty-First Century: Demographic and Epidemiological Transitions

Emily Ying Yang Chan and Heidi Hung

Abstract The phenomenon of demographic transition suggests that societies experiencing modernization transform from pre-modern demography of high fertility and high mortality to a post-modern one in which both are low. An increase in life expectancy and the aging of populations also lead to an epidemiological transition that shifts disease profile from communicable diseases to non-communicable diseases (NCDs) in many countries. With aging populations and the increasing behavioral health risk factors associated with rapid urbanization and lifestyle modernization in many developing countries, the mortality and disease burden in developing countries will continue to shift from communicable disease-based to predominantly NCD-related in the remaining decades of this century. Globally, over 60% of all deaths in 2005 were caused by chronic NCDs such as hypertension and diabetes mellitus. While these chronic NCDs constituted significant disease burden in developed countries, more than 80% of the NCD-related deaths occurred in low-income and middle-income countries. For many low-income countries experiencing these double transitions in the twenty-first century, a double burden of disease poses a major and severe public health challenge during disasters. In a disaster context, due to limited resources, many aid agencies overlook the importance of addressing both acute and underlying disease patterns (such as NCDs) of the affected community. As a result, disaster victims may have issues in accessing therapies and medicines that may potentially reduce morbidity and mortality.

Keywords Chronic disease · Demographic transition · Epidemiological transition · Health Emergency and Disaster Risk Management (Health-EDRM) · Older people

---

E. Y. Y. Chan (✉) · H. Hung
JC School of Public Health and Primary Care,
The Chinese University of Hong Kong, Shatin, Hong Kong
e-mail: emily.chan@cuhk.edu.hk

E. Ying Yang Chan and R. Shaw (eds.), *Public Health and Disasters*,
Disaster Risk Reduction, https://doi.org/10.1007/978-981-15-0924-7_3

3.1 Introduction

Public health is "the science and art of preventing disease, prolonging life and promoting health through organized efforts of society" (Acheson 1988), and public health in practice generally encompasses health protection, health improvement and health services (Griffiths et al. 2005). To fully appreciate the scope of public health activities, one must bear in mind the definition of health, which is "a state of complete physical, mental and social well-being and not merely the absence of disease or infirmity" (International Health Conference 1946). The quantity, nature, and type of health needs have been changing dramatically in the twenty-first century, following the demographic transition and epidemiological transition experienced by human population across the world. The implications of the two transitions have dominated most health-related dialogues and policies, such as health care financing, universal health coverage, and the discipline of health emergency and disaster risk management in recent years (Health-EDRM). Specifically, Health-EDRM refers to "the systematic analysis and management of health risks, posed by hazardous events, including emergencies and disasters, through a combination of hazard, exposure, and vulnerability reduction to prevent and mitigate risks, preparedness, response, and recovery" (World Health Organization and Public Health England 2017).

The discipline of Health-EDRM highlighted the unique role of health in disaster risk management. As witnessed by the strong reference to health resilience in a number of key international agreements, in particular the Sendai Framework for Disaster Risk Reduction 2015–2030 (Sendai Framework), the 2030 Agenda for Sustainable Development (SDG), Health-EDRM is a core strategy for sustainable development in today's world, with disaster risk reduction being part of the development process involving individuals, communities, governments at all levels. As Health-EDRM is concerned with managing health risks in disaster and emergency settings, it should consider both in theory and in practice to the demographic and epidemiological transitions. This chapter will analyze the key public health challenges for Health-EDRM arising from the demographic transition and epidemiological transitions, and we will start by looking at the relationship between public health and disaster management.

3.2 Public Health and Disaster Management

There are different ways to define a disaster, and the focus varies among different disciplines. One of the most widely adopted definitions of disaster in the field of epidemiology describes a disaster as "a situation or event which overwhelms local capacity, necessitating a request to a national or international level for external assistance; an unforeseen and often sudden event that causes great damage, destruction and human suffering. Though often caused by nature, disasters can have human origins" (Centre for Research on the Epidemiology of Disasters [CRED] 2018a). Therefore, from the perspective of public health and epidemiology, a disaster must have a human impact, must have been unpredictable, and must have required external assistance. Disasters can be broadly classified into three categories based on their causes, which are given as follows:

(a) Natural disasters are those with natural origin (e.g., earthquake),
(b) Human-caused disasters are those occurred due to human-related causes (e.g., nuclear explosion), and
(c) Complex emergencies which are a multifaceted humanitarian crisis (e.g., war combined with persistent famine and drought).

For natural disasters, CRED classified them into six groups according to the nature of the hazard involved:

(a) Geophysical disaster originating from earth (e.g., volcanic eruption),
(b) Meteorological disaster due to atmospheric processes (e.g., storm),
(c) Hydrological disaster caused by abnormal water cycle or water flow (e.g., flood),
(d) Climatological disaster caused by various climate-related processes (e.g., drought), (e) Biological disaster caused by germs and toxic substances (e.g., epidemic), and
(e) Extra-terrestrial disaster caused by interplanetary conditions (e.g., space weather) (Centre for Research on the Epidemiology of Disasters [CRED] 2018b).

The classification of disasters has major implications as specific health patterns/health outcomes are often associated with different types of disasters (e.g., disease outbreak is a major concern after flooding but not earthquake), and such analysis allows preparedness planning and disaster response, hence reducing negative health impacts of disasters. Disasters very often bring about serious public health consequences. As public health is concerned with managing the health and well-being of populations, as opposed to the health of individuals; the public health impacts of disasters are centered at the community level:

(a) Overwhelmed local response capacity: unexpected mortality and morbidity is the main reason for the local emergency and relief response system not being able to cope with the disaster. Number and nature of mortality and morbidity resulting from a disaster are affected by the type, duration, and severity of the disaster involved, the demographic and epidemiological characteristics of the affected population;
(b) Destroyed health infrastructure and disrupted provision of health services: after disaster strike, the demand for health services at all levels shoot up, from informal care, primary care, secondary to tertiary care; the resilience of the health infrastructure and health care system, therefore, is an important determinant of the public health impact;
(c) Adverse impact on population following environmental damage: many environmental impacts of disasters (e.g., damage of forests, release of toxic

materials) have major health implications on population health, which include acute risk from release of hazardous materials (e.g., nuclear plant explosion following tsunami), damage of natural resources (e.g., fish supplies), environmental cost of relief operations (e.g., improper management of expired medicine);

(d) Psychological and social problems: psychosocial health of disaster victims are affected by traumatizing experience, stress related to loss of loved ones, properties, livelihood, destruction of social networks, local economy, law and order, violence, and mental health issues such as post-traumatic stress disorder, depression and anxiety have been observed; and

(e) Undesirable long-term consequences: Types of long-term impacts vary with the nature of the disasters and characteristics of the affected community. Long-term consequences include all aspects of the physical, mental and social wellbeing of the population, e.g., excess cases of chronic diseases, birth defects, psychological illnesses, left-behind orphans and widows. Economic cost of a disaster also has strong impact on public health, e.g., reduced access to care following loss of livelihood of individuals, reduced health care spending by governments. (Chan 2017).

In order to reduce the negative public health impacts of disasters described above, public health practices and measures have a critical role to play in all phases of the disaster management cycle. Disaster management can be divided into three phases: (a) pre-impact phase, (b) impact phase, and (c) post-impact phase. The pre-impact stage is concerned mostly with disaster preparedness; disaster response is the key during the impact phase; while the post-impact phase focuses on disaster recovery and rehabilitation. The pre-impact phase, in particular, has the greatest potential for improving the health outcome of disasters. Health-EDRM emphasized, in particular, the role of health not only during the impact phase but also the reduction of disaster risk and its impact on health in the pre-impact phase. In other words, disaster risk management is no longer "response-driven" but "risk-driven" (Aitsi-Selmi and Murray 2016). The paradigm shift represented by the Health-EDRM discourse means that health is not only a responsive component in disaster response, but plays a proactive role in disaster risk reduction.

As explained earlier, the public health impact of a disaster is determined not only by the nature of the disaster but also the demographic and epidemiological characteristics of the affected community. Therefore, changes in the demography and epidemiology patterns must be taken into account in disaster preparedness and response to minimize public health impacts and to meet the health needs. In the following section, the demographic transition and the epidemiological transition that have characterized the human population profile of the twenty-first century will be discussed in detail.

3.3 Demographic Transition: Aging Population

Demography is the scientific study of human population structures and dynamics, which are affected mainly by the three demographic processes of birth, death/aging, and migration. The demographic characteristics of a population generally include the pattern in age, sex, race, fertility rate, immigration, and emigration. The classic demographic transition was formulated by Notestein in 1945, who hypothesized that the fertility transition and morality transition observed in high-income countries in Europe and North America would occur also in the middle- and low-income countries. The classical formulation characterizes the transition into three phases represented by an inverted U-shape population change, driven largely by industrialization, urbanization, advancement of technology (in particular, medical advancement and public hygiene), and rising cost of living (Defo 2014):

(a) Pre-transition stage: high mortality and high fertility which result in either modest increase or decrease in population size;
(b) Transition stage: declining mortality followed by declining fertility causing population growth, with young people making up a large proportion of the population; and
(c) Post-transition stage: low mortality and low fertility result in population decrease, with longer life expectancy and aging population gradually sets in.

Since demographic transition began much earlier in the high-income countries than in the middle- and low-income countries, countries across the world, therefore, are now with demographic profile hence facing different implications from the transition, e.g., for the middle-and-low-income countries, many are at the stage of seeing a growing middle-age population which requires abundant employment opportunities; while high-income countries are having unprecedented pressure on its healthcare and pension systems with an increasing elderly population and declining working population. It is expected that many low-income countries will experience a much faster transition as compared with the high-income countries in the 1800s and 1900s, due to better technology, nutrition and public health structures (Chan 2017). While it took almost 150 years in France for its proportion of population aged over 60 to change from 10 to 20%, it will take countries like Brazil, China, and India only over 20 years to experience a similar change (World Health Organization 2018a). Table 3.1 illustrates the trend in demographic transition globally and in Asia, expressed by changes in old-age dependency ratio.

In 2015, there are 900 million people in the world aged 60 and above, and by 2050, the number is expected to jump to 2 billion (i.e., from 12 to 22% of world's population), with 80% of them living in low- and middle-income countries (World Health Organization 2018a). The size and age structure of a population affects the development and even survival of any society in every way, e.g., economic development, welfare structure, housing needs, health needs, etc. The last stage of the demographic transition, i.e., aging population, is the biggest challenge faced by many high-income countries and is what the middle- and low-income countries

Table 3.1 Changes and projected changes in old-age dependency ratio (per 100) from 1950 to 2100

	1950	1960	1970	1980	1990	1995	2000	2005	2010	2015	2020	2025	2030	2040	2050	2060	2070	2080	2090	2100
World	8	9	9	10	10	11	11	11	12	13	14	16	18	22	25	29	30	33	35	38
Asia	7	6	7	8	8	9	9	10	10	11	13	15	18	23	28	34	37	40	44	46

Credit United Nations, Department of Economic and Social Affairs, Population Division (2017). World Population Prospects: The 2017 Revision, Volume I: Comprehensive Tables (ST/ESA/SER.A/399) have to prepare for. From the public health angle, aging population means ever-increasing demand on health care. For example, a recent study projected that care needs will grow by 25% between 2015 and 2025 in England and Wales following aging population and related physical and mental morbidities, and associated burden of disability will increase: life expectancy at 65 years is projected to increase by 1.7 years, with 0.7 of these years to be with disability (Guzman-Castillo et al. 2017). In addition, quality of care should not be compromised, with dignity and respect upheld for both the elderly and their carers (The Lancet Public Health 2017). While regional differences in the magnitude of change in health service demands are projected at this stage, it is not a challenge that any country could avoid. To support such expansion in health care demand, an inseparable discussion concerns health care financing. The World Bank projected that for many middle- and low-income countries, there would be an overall increase of 37–62% in health spending between 2005 and 2025, just to accommodate demographic changes (Gottret and Schieber 2006).

3.4 Epidemiological Transition: Chronic Disease Burden

The theory of epidemiological transition was put forward by Omran in 1971 focusing "on the complex change in patterns of health and disease and on the interactions between these patterns and their demographic, economic and sociologic determinants and consequences", and the epidemiological transition described and predicted was one in which "degenerative and man-made diseases displace pandemics of infection as the primary causes of morbidity and mortality (Omran 1971). The transition was set out in five propositions (Omran 1971):

(a) Proposition One: "*The theory of epidemiologic transition begins with the major premise that mortality is a fundamental factor in population dynamics*": it was argued that mortality (as opposed to fertility) was the key determinant of population growth;
(b) Proposition Two: "*During the transition, a long-term shift occurs in mortality and disease patterns whereby pandemics of infection are gradually displaced by degenerative and man-made diseases as the chief form of morbidity and primary cause of death*": it was argued that the transition from infectious to degenerative disease predominance was observed in Europe and Japan, with the change more distinct since 1945, while the transition began later in developing countries.
(c) Proposition Three: "*During the epidemiologic transition the most profound changes in health and disease patterns obtain among children and young women*": it was argued that since children and females in the adolescent and reproductive age periods were more susceptible to infectious diseases, they benefit the most in terms of health from the transition;

(d) Proposition Four: "*The shifts in health and disease patterns that characterize the epidemiologic transition are closely associated with the demographic and socioeconomic transitions that constitute the modernization complex*": it was argued sequential demographic changes occur during epidemiological transition, including age and sex structure and the dependency ratios, with high old dependency ratio at the end of the transition; and
(e) Proposition Five: "*Peculiar variations in the pattern, the pace, the determinants and the consequences of population change differentiate three basic models of the epidemiologic transition: the classical or western model, the accelerated model and the contemporary or delayed model*": three models of epidemiological transition were developed in the first formulation of the theory, applicable to the western and non-western world in light of the differences in their level of developments. Omran later has increased the number of models to six in attempt to take into account the huge variations among countries (Omran 1998).

Omran's theories of epidemiological transition developed at different time periods have been subject to various criticisms (Caldwell 2001), but it has been no doubt one of the most powerful and influential theories shaping the development of public health discourse, practices, and policies in recent years. Proposition Two, which described the shift from infectious disease to chronic disease predominance with modernization and development, reflects exactly the situation in developed countries at present. For WHO European Region, Region of the Americas and West Pacific Region, chronic diseases are the main causes of death (World Health Organization 2016). Chronic diseases now account for 41 million deaths, or 71% of all deaths globally, with cardiovascular diseases being the top killing (17.9 million people annually), followed by cancers (9.0 million), respiratory diseases (3.9 million), and diabetes (1.6 million) (World Health Organization 2018b). The importance of infectious diseases, however, should not be underplayed, including the large burden of HIV/AIDS, the re-emerging infectious diseases like multi-drug resistant tuberculosis. Indeed, many developing countries, notably in Africa and South-east Asia, are facing a "double burden of disease", under which countries are facing the emerging epidemic of chronic diseases while some major infectious diseases remain high on the agenda (World Health Organization 1999). As infectious diseases and chronic diseases have adverse interaction on individuals' health, e.g., diabetic patients predisposed to infections due to compromised immune systems there have been calls that these countries should try to adopt interventions that address the two jointly (Bygbjerg 2012). With globalization of behavioral risk factors, in particular tobacco use, physical inactivity, unhealthy diet and the harmful use of alcohol, chronic diseases are expected to increase in developing countries. In fact, chronic diseases are already disproportionately affecting people in middle-and-low-income countries, which took up more than three-quarters of deaths arising from chronic diseases worldwide (World Health Organization 2018b), and are considered the "key barriers to poverty alleviation and sustainable development" (World Health Organization 2014). Figure 3.1 compares the

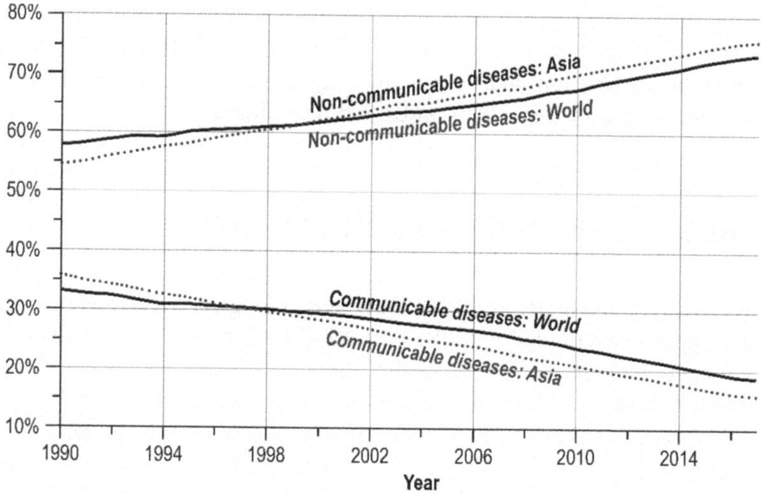

Fig. 3.1 Changes in the causes of death by percentage between 1990 and 2017—Global and Asia. *Credit* Global Burden of Disease Studies (http://ghdx.healthdata.org/gbd-2017)

percentage of deaths caused by communicable and non-communicable diseases globally and in Asia since 1990.

A shift in disease pattern obviously requires a corresponding change in the set up of the health system. In other words, the health systems should be well-resourced and geared towards the needs of chronic disease patients, while maintaining the necessary capacity for managing infectious diseases. In many developing countries, however, the transformation in the health system is behind, leaving needs of chronic disease patients largely unmet (Kushitor and Boatemaa 2018). The long-term medical needs and crucial role of self-care of chronic disease patients also highlight the significance of a strong primary care system, and health promotion and education. It is clear that the epidemiological transition has brought about a different set of health needs which the health system must respond to.

3.5 Double Transition

The demographic transition and the epidemiological transition are not two separate phenomena but are occurring in tandem in the world, though at different stages and pace for countries with different levels of development. Demographic transition means the world is moving towards an aging community, and it coupled with the epidemiological transition, which points to the predominance of chronic diseases in terms of disease burden. With globalization, urbanization, and the spread of lifestyle/behavioral health risks, the two transitions are taking place at faster pace in the developing world. Aging population and increasing burden of chronic diseases are closely linked, in particular in the case of the health needs of the older adults. All countries, developed or developing, are faced with the challenge of building and financing a health system that meets the needs of an older population with increasing prevalence of chronic diseases and disabilities.

3.6 Public Health Challenges in Health-EDRM

From the explanations on demographic and epidemiological transitions, we can see that aging population coupled with the increasing burden of chronic diseases is the most imminent challenge for public health globally. Disaster risk is determined by the interaction between a hazard and peoples' exposure to the hazard, and their vulnerability discounted by their capacity to cope. We can see from the analysis of the demographic and epidemiological transitions that older adults and chronic disease patients are two major vulnerable groups in today's world. Health-EDRM must ensure that the health needs of these subgroups are met, and to build up their capacity to cope. In this section, we will explain the vulnerabilities and health needs of older adults and chronic disease patients in disaster settings. While the analysis will be divided into two sections, it must be pointed out that there are overlaps in the needs of these two vulnerable subgroups, as illustrated in cases of older adults with chronic conditions. For disaster-pronged developed countries like Japan, which has over 30% of its population aged over 65 (Ministry of Internal Affairs and Communications of Japan 2018); and cancers, cardiovascular diseases and diabetes constituting the highest burden of disease (World Health Organization 2015a), disaster risk reduction planning and operations must prioritize jointly the needs of older adults and chronic disease patients.

3.6.1 Vulnerabilities and Health Needs of Older Adults

Older people are disproportionately affected by disasters: more than half of the deaths consistently occurred among people over 60 years of age in five major natural disasters in the past (World Health Organization 2015b). This happens even in countries with an established disaster response and health systems, e.g., of the 14,800 deaths during the 2003 heatwave in France, 70% were people over 75 years old (Hutton 2008); over 70% who died in the wake of Hurricane Katrina in 2005 was 60 years old and above (Inter-Agency Standing Committee 2008), over 60% of those killed in 2011 Japan earthquake aged over 60 (United Nations Population Fund and HelpAge International 2012). There are arguments that it is not the advancement of age alone that causes the vulnerabilities, but the physical disabilities, diminished mental capacity, pre-existing health conditions, and social and economic isolation associated with this group that has hampered their disaster preparedness and coping capacity (Fernandez et al. 2002). Therefore, recognizing the potential contributions of older people in disaster management is equally important. Older people have the knowledge, wisdom, and experience of previous disasters, and their "sense of history" is valuable to the community (Chan 2019; HelpAge International 2002). A cross-culture study found a strong sense of independence expressed among older people having experienced disasters, and they would like to be consulted on disaster response planning (Duggan et al. 2010). After the 2010 Haiti earthquake, displaced older people living in camps often acted as "focal points" for the affected older population: they assisted in identifying the most vulnerable ones, collecting health data, and delivering aid to the less mobile members (United Nations Population Fund and HelpAge International 2012).

Disasters can affect older people's health and chance of survival in a number of ways, e.g., physical injuries, mental and psychological issues, worsening of pre-existing conditions. The lack of support for older people in emergencies was one of the concerns raised in a large-scale consultation with older people conducted by the WHO around the world (Chan 2019; World Health Organization 2007a). An understanding of the major vulnerabilities of older people in disaster context is necessary to design public health measures that can meet their health needs.

(a) Poorer health: Older adults are often in a more vulnerable state of health as a result of physiological changes, e.g., reduced mobility and energy, sensory deterioration, mental problems, disabilities, pre-existing conditions, and compromised immune system. With such health conditions, they are more susceptible to diseases commonly associated with disasters and minor conditions can easily become significant handicaps, e.g., hypothermia caused by lack of blanket, immobility due to loss of eyeglasses and walking cranes. Those with chronic conditions may have their conditions worsened if access to required medication is interrupted following a disaster;

(b) Economic and social marginalization: Poverty is one of the main threats to the wellbeing of older people, especially for women (United Nations Population Fund and HelpAge International 2012). In 15 out of 35 members of the Organisation for Economic Co-operation and Development (OECD), poverty rates are higher for older people than for the population as a whole (OECD 2017). In addition to material deprivation, intergenerational dependencies are increasingly eroded following urbanization and migration (Hutton 2008), and older persons are often left behind without traditional family support (United Nations Population Fund and HelpAge International 2012). The lack of economic and social resources leaves older people at higher health risks when a disaster strikes.

(c) Limited access to information: Early warning and public health information are crucial for empowering older people to protect themselves from the negative health impacts of disasters. Their ability to access, understand and act on the information, however, requires much attention. Whether such information is accessible and effective for the older population depends on its delivery, its mode of communication, message content, etc. A recent study in Hong Kong on weather information acquisition behavior during cold waves revealed that while smartphone was the most preferred channel for the general population, old age was significantly associated with less smartphone apps use and preference (Chan et al. 2017).

Weather Information Acquisition and Health Significance During Extreme Cold Weather in a Subtropical City: A Cross-sectional Survey in Hong Kong (Chan et al. 2017)

Potential of Smartphone Among Older People

- A population-based telephone survey was conducted in Hong Kong with 1017 respondents at the beginning of 2016, immediately after the day recorded with the lowest minimum temperature since 1957.
- Study objectives are (1) to examine how the public in the subtropical metropolis of Hong Kong acquires its weather information, (2) to identify the sociodemographic patterns of information acquisition behavior, and (3) to establish the public health significance of using smartphone apps during a cold surge.
- Television (50.1%) and smartphone app (32%) was the most popular channels for seeking weather information in the extreme temperature event.
- However, in terms of the preferred information-seeking channel, 16.5% of all respondents had indicated the intention to switch to the use of weather smartphone apps, meaning that smartphone app would be the most preferred channel (45.6%), followed by television (36.3%).
- Older age and lower education levels were significantly associated with less smartphone apps use and preference.

3.6.2 *Vulnerabilities and Health Needs of Chronic Disease Patients*

In a disaster context, a rising burden of chronic disease means that the conventional focus on infectious disease outbreaks and injury management is no longer sufficient. A significant number of post-disaster deaths are now caused by insufficient care of pre-existing conditions (World Health Organization 2011) and chronic disease management has become an obvious health need following a disaster (Aitsi-Selmi and Murray 2016). A study conducted after the 2008 Sichuan Earthquake in China found that 77% of the patients attended an emergency triage clinic had at least one underlying medical condition, dominated by hypertension and diabetes mellitus; and 54% of them had clinical complications due to either disease exacerbation or lack of medication following the disaster (Chan and Kim

2011). Unfortunately, a study on the Kashmir earthquake in 2005 found that the greatest gap in health services at all sites was non-communicable disease management (Chan and Griffiths 2009). Chronic disease management in disaster and emergency settings has yet to be integrated into the disaster risk reduction plans in most countries, meaning that many disaster victims do not have access to medication or medical services required, leading to deterioration or complications of their chronic conditions. Several reasons for chronic disease management to remain a forgotten issue in disaster include: there is often a lack of appropriate human resources like general medical teams specializing in chronic diseases at the scene, hospitals and clinics are not geared towards the needs of the chronic disease patients, there is a lack of operational mandate among relief agencies to handle chronic diseases, aid organizations may not prioritize resources for chronic diseases over acute diseases or injuries (Chan and Southgate 2014). On a brighter note, one encouraging development in recent years is the clear and specific reference to chronic disease management before, during and after a disaster in the Sendai Framework:

> People with life-threatening and chronic disease, due to their particular needs, should be included in the design of policies and plans to manage their risks before, during and after disasters, including having access to life-saving services. (paragraph 30(k))

Chronic conditions could be exacerbated, and new conditions may arise due to physical injuries, change in living conditions or interruption of care during a disaster (UN Interagency Taskforce on NCDs and World Health Organization 2016). The most important health need of people with chronic disease in a disaster is uninterrupted access to the medication and care required, e.g., supply of insulin for some diabetic patients, dialysis service for patients with severe kidney diseases. A disaster may interrupt access to medication and worsen the conditions of a chronic disease patient in the following way:

(a) Interruption of care: A disaster may destroy health infrastructure, e.g., hospitals and clinics, surveillance system, communication network (Ryan et al. 2016), limit access to medication (e.g., oxygen supplies for people with respiratory diseases, insulin for people with diabetes), disrupt medical supplies, reduce number of healthcare providers;
(b) Degradation of living conditions: Inadequate nutrition or unavailability of special diet (e.g., low sugar low sodium diet for people with diabetes), limited clean water, unhygienic shelter, exposure to infectious diseases (UN Interagency Taskforce on NCDs and World Health Organization 2016);
(c) Physical health: With weaker immune systems, chronic disease patients are more susceptible to infections and injuries, and are more vulnerable to the stresses brought about by disasters (Aldrich and Benson 2008), which present higher risks for surgical interventions; and
(d) Mental health implications: An area of chronic disease management in disaster settings that merits more attention is the related mental health issues of the patients. Mental health impact of disasters is well-recognized, with mental

problems including stress, depression, anxiety, post-traumatic stress disorders, found to be associated with population affected by certain disasters, e.g., hurricane (Neria and Shultz 2012), flooding (Stanke et al. 2012); and WHO issued recommendations on mental health care in emergencies (World Health Organization 2013a). There is also increasing evidence showing association and co-morbidities between major chronic diseases and mental disorders, e.g., cancer (Caruso et al. 2017), cardiovascular diseases (Fornaro et al. 2017), diabetes (Petrak et al. 2015). The cross-over between the two, i.e., management of health risk of patients with both chronic and mental health conditions in disaster context, however, has hardly been researched on.

Having analyzed in detail the disaster-related health risks and vulnerabilities of the two subgroups resulted from the demographic transition and the epidemiological transition, i.e., older adults and chronic disease patients, the next section will be devoted to examine how health risk of the two groups are to be managed in the disaster context under the concept of Health-ERDM.

3.7 Health-EDRM: Disaster Risk Management for Older People and Chronic Disease Patients

Health-EDRM is concerned with the analysis and management of health risks through reduction in hazard, exposure, and vulnerability in every phase of the disaster management cycle. Resilience-building is a key concept for minimizing the health risks of older people and chronic disease patients. Resilience is defined as "the ability of a system, community or society exposed to hazards to resist, absorb, accommodate to and recover from the effects of a hazard in a timely and efficient manner, including through the preservation and restoration of its essential basic structures and functions" (United Nations Office for Disaster Risk Reduction [UNISDR] 2009). It is recommended that resilience should be built at both individual, community, and system levels in order to handle the public health challenges posed by the increasing population of older people and chronic disease patients.

3.7.1 Resilience at Individual and Community Levels

Resilience of older people and chronic disease patients in disaster settings could be built through empowerment initiatives to improve their health outcomes. **Self-care** by the population concerned should also be promoted and facilitated. One practical way is to encourage or make accessible personal emergency kits taking into account the special health needs of older people and chronic disease patients (Pickering et al. 2018). Items that should be included are medications for chronic diseases, medical records, and prescriptions, walking sticks, eyeglasses, portable medical equipment (e.g., blood sugar measuring device), self-monitoring checklist, important contacts (e.g., family members, doctors), etc. For chronic disease patients, they should also have knowledge on how to use their medication (e.g., type of insulin used, insulin self-injection kit with instructions). For people with multiple drug prescriptions, it will be important for them to understand which ones are critical, and the key contraindications of their regular medications. In the event that health facilities and medical supplies are interrupted, it will be life-sustaining for chronic disease patients to have stockpile of preferably 10–14 days' supply of medications (Arrieta et al. 2008). Extensive effort is required to promote emergency preparedness among chronic disease patients, as a systematic review published in 2014 found that a considerable number of chronic disease patients lost their medication and medical aids during evacuation, many did not bring prescriptions with them when evacuated, making it difficult to fill in prescriptions, and that medication loss and prescription loss pose significant burden on the medical relief teams (Ochi et al. 2014). For improved community resilience and **bottom-up response capacity**, older people's associations or patients' groups should be set up to facilitate these subgroups to participate and support disaster management, from vulnerability mapping to participation in drills (Fernandez et al. 2002). This will ensure that their special health needs are mainstreamed in disaster reduction and management plans. Humanitarians and community workers must be aware of the health impacts of disasters on older people and chronic disease patients. Community partnership is crucial and those involved should be sensitive in choosing the most appropriate health interventions, e.g., adverse drug interactions, unsuitable diets for people with diabetes (Chan and Sondorp 2007).

3.7.2 Health System Resilience

Health system resilience, which is one of the focuses in the Sendai Framework, is an indispensable Health-EDRM strategy for managing the health risks of older adults and chronic disease patients. WHO describes health systems in terms of six "building blocks": (a) service delivery, (b) health workforce, (c) health information systems, (d) access to essential medicines, (e) financing, and (f) leadership/governance (World Health Organization 2007b), and a set of indicators have also been published to track progress and evaluate performance (World Health Organization 2010). The six building blocks concept has been applied to building climate-resilient health system by the WHO (World Health Organization 2015c), and an extension of it to the disaster context has been advocated (Olu 2017). A basic requirement of a resilient health system is that health professionals must understand the needs of the older people and chronic disease patients during and after a disaster, and disruption to the health services and medication they require following a disaster must be kept to the minimum. Based on the meaning of health system as sketched out by the six building blocks, possible strategies for building up the health system resilience for better Health-EDRM are discussed below.

Disaster management as part of the overall policy: Health-EDRM should be integrated into the overall policy frameworks related to older adults and chronic disease patients at global, national and local levels; and policy-makers should be sensitized on the health needs and vulnerabilities of these vulnerable subgroups. For example, disaster management is a theme in WHO's guide on age-friendly cities (World Health Organization 2007a) but not mentioned in its global action plan on chronic diseases (World Health Organization 2013b).
Safe health facilities: Safety of health facilities, including hospitals and clinics, is the foundation for uninterrupted medical services during a disaster. Hospitals and care facilities (e.g., clinics) should have response plans and basic features to withstand relevant disasters or crises. The Comprehensive Safe Hospital Framework and the Safe Hospital Index issued by the WHO provide elaborate guidance in this area and countries should try to adopt them as far as possible (World Health Organization 2015d).
Establish surveillance system with register: A public health surveillance system should be established to record essential information on the health needs of vulnerable groups to facilitate disaster planning and effective response, e.g., age, medical history, disabilities, chronic conditions and other health conditions, life-maintaining medications, drug and food allergy, nutritional needs, address, and contact information. The register should be able to generate age and sex-disaggregated data to facilitate need assessment, stockpiling of drugs and supplies, reaching out to the older adults and chronic disease patients, and as baseline for evaluating short-term and long-term health outcomes of the disaster (World Health Organization 2011) (Khan et al. 2014).
Service planning: The health needs of older adults and chronic disease patients in disaster context should be mainstreamed into health service planning. The planning must include surge capacity to cope with the sudden rise in service demand from these groups, to ensure continuous service provision and accessibility, including mobile clinics, outreach services (Watson et al. 2013).
Strengthen primary care: Primary care plays an exceptionally critical role in managing the health risks of older people and chronic disease patients in a disaster context, and it could reduce the burden on secondary and tertiary care. Community-based healthcare services at district level should be equipped and resourced to handle basic medical needs and facilitate self-care; chronic disease management should be integrated into primary care with reference to the Package of Essential Non-communicable Disease Interventions (PEN) (Slama et al. 2017); caretakers and humanitarians should be educated on the health impacts of disasters and the special needs of older people and chronic disease patients (Demaio et al. 2013); and follow-up and rehabilitation services should be built-in.
Healthcare workforce preparedness: Healthcare personnel should be aware of the health needs of older people and chronic disease patients in a disaster context. In addition, staff planning is critical to provide interrupted service and to facilitate

self-care by elderly and chronic disease patients (Ryan et al. 2016). For health care workers who have underlying chronic conditions, they must be prepared for their own medical needs.

Information dissemination: Health authorities should be involved in the dissemination of disaster information, to ensure that the health needs of the older people and chronic disease patients are specifically addressed, e.g., reminder on medication needs, self-care equipment.

3.8 Conclusion

The changes in the demography and disease burdens among human population in the last two centuries, termed as demographic transition and epidemiological transition, have transformed the health and health care needs and response requirements across the world, and called for adaptations in all aspects of the wider health system, including health risk management in disaster setting. Put it simply, the implication of the demographic transition is an aging population and an increasing number of older people; while the epidemiological transition described and rightly predicted the predominance of chronic diseases as the main disease burden in an increasing number of countries, and eventually globally. Our health system and disaster risk reduction system must respond to the health needs and vulnerabilities of these two groups of population for us to see the best health outcomes possible.

Demographic transition and epidemiological transition are not two separate phenomena but with similar drivers and related consequences, and there is a significant overlap in the health needs and vulnerabilities of the two resultant subgroups, namely the older adults and chronic disease patients. Older people and chronic disease patients are sometimes marginalized economically and socially, due to their inability to work, reduced mobility and healthcare burden. While some older people are able to stay healthy physically and mentally in spite of advancing age, many are living with poverty, disability and poor health. The underlying vulnerabilities of some older people and chronic disease patients mean they are at a higher risk of being affected by disasters; and indeed, research shows that disaster affects disadvantaged groups disproportionately. These disparities mean that the needs of these vulnerable populations must be highlighted and addressed.

Health-EDRM emphasizes the reduction and management of health risks in all phases of the disaster management cycle. Health-EDRM strategies must be able to meet the challenges posed by the demographic transition and epidemiological transition, through addressing the health needs of the older people and chronic disease patients. The main approaches adopted under Health-EDRM are to promote self-care and bottom-up response capacity and to build up the resilience of the health system. Resilient individuals and communities would be able to draw on available resources and demonstrate their capacity to cope in the face of disasters, and a resilient health system would be able to anticipate and respond to the physical and mental health needs of the older people and chronic disease patients, in addition to infectious disease outbreaks and injuries, conventionally associated with disasters. It must be borne in mind that future Health-EDRM research and practices must take into account the needs of the older population and chronic disease patients for them to be relevant and effective.

References

Acheson D (1988) Public health in England: the report of the committee of inquiry into the future development to the public health function. London, UK

Aitsi-Selmi A, Murray V (2016) Protecting the health and well-being of populations from disasters: health and health care in the Sendai framework for disaster risk reduction 2015–2030. Prehosp Disaster Med 31:74–78. https://doi.org/10.1017/S1049023X15005531

Aldrich N, Benson WF (2008) Disaster preparedness and the chronic disease needs of vulnerable older adults. Prev Chronic Dis 5:A27

Arrieta MI, Foreman RD, Crook ED, Icenogle ML (2008) Insuring continuity of care for chronic disease patients after a disaster: key preparedness elements. Am J Med Sci 336:128–133. https://doi.org/10.1097/MAJ.0b013e318180f209

Bygbjerg IC (2012) Double burden of noncommunicable and infectious diseases in developing countries. Science (80-) 337:1499–1501. https://doi.org/10.1126/science.1223466

Caldwell JC (2001) Population health in transition. Bull World Health Organ 79:159–160

Caruso R, Nanni MG, Riba MB, Sabato S, Grassi L (2017) The burden of psychosocial morbidity related to cancer: patient and family issues. Int Rev Psychiatry 29:389–402. https://doi.org/10.1080/09540261.2017.1288090

Centre for Research on the Epidemiology of Disasters (CRED) (2018a) EM-DAT glossary. https://www.emdat.be/Glossary#letter_d. Accessed 2 Jul 2018

Centre for Research on the Epidemiology of Disasters (CRED) (2018b) EM-DAT general classification. https://www.emdat.be/classification. Accessed 2 Jul 2018

Chan EYY (2017) Public health humanitarian responses to natural disasters, first. Routledge, London

Chan EYY (2019) Disaster public health and older people. Routledge, London

Chan EYY, Griffiths S (2009) Comparision of health needs of older people between affected rural and urban areas after the 2005 Kashmir, Pakistan earthquake. Prehosp Disaster Med 24:365–371

Chan EYY, Kim J (2011) Chronic health needs immediately after natural disasters in middle-income countries: the case of the 2008 Sichuan, China earthquake. Eur J Emerg Med 18:111–114. https://doi.org/10.1097/MEJ.0b013e32833dba19

Chan E, Sondorp E (2007) Medical interventions following natural disasters: missing out on chronic medical needs. Asia Pacific J Public Heal 19:45–51. https://doi.org/10.1177/101053950701901S08

Chan E, Southgate R (2014) Responding to chronic disease needs following disasters: a rethink using the Human Security approach. In: Hobson C, Bacon P, Cameron R (eds) Human security and natural disasters, 1st edn. Routledge, London

Chan EYY, Huang Z, Mark CKM, Guo C (2017) Weather information acquisition and health significance during extreme cold weather in a subtropical city: a cross-sectional survey in Hong Kong. Int J Disaster Risk Sci 8:134–144. https://doi.org/10.1007/s13753-017-0127-8

Defo BK (2014) Demographic, epidemiological, and health transitions: are they relevant to population health patterns in Africa? Glob Health Action 7:22443. https://doi.org/10.3402/gha.v7.22443

Demaio A, Jamieson J, Horn R, de Courten M, Tellier S (2013) Non-communicable diseases in emergencies: a call to action. PLoS Curr. https://doi.org/10.1371/currents.dis.53e08b951d59ff913ab8b9bb51c4d0de

Duggan S, Deeny P, Spelman R, Vitale CT (2010) Perceptions of older people on disaster response and preparedness. Int J Older People Nurs 5:71–76. https://doi.org/10.1111/j.1748-3743.2009.00203.x

Fernandez LS, Byard D, Lin C-C, Benson S, Barbera JA (2002) Frail elderly as disaster victims: emergency management strategies. Prehosp Disaster Med 17:67–74. https://doi.org/10.1017/S1049023X00000200

Fornaro M, Solmi M, Veronese N, De Berardis D, Buonaguro EF, Tomasetti C, Perna G, Preti A, Carta MG (2017) The burden of mood-disorder/cerebrovascular disease comorbidity: essential neurobiology, psychopharmacology, and physical activity interventions. Int Rev Psychiatry 29:425–435. https://doi.org/10.1080/09540261.2017.1299695

Gottret P, Schieber G (2006) Health financing revisited a practioner's guide. Washington, D.C.

Griffiths S, Jewell T, Donnelly P (2005) Public health in practice: the three domains of public health. Public Health 119:907–913. https://doi.org/10.1016/j.puhe.2005.01.010

Guzman-Castillo M, Ahmadi-Abhari S, Bandosz P, Capewell S, Steptoe A, Singh-Manoux A, Kivimaki M, Shipley MJ, Brunner EJ, O'Flaherty M (2017) Forecasted trends in disability and life expectancy in England and Wales up to 2025: a modelling study. Lancet Public Heal 2: e307–e313. https://doi.org/10.1016/S2468-2667(17)30091-9

HelpAge International (2002) Older people in disasters and humanitarian crises: Guidelines for best practice. London

Hutton D (2008) Older people in emergencies: Considerations for action and policy development

Inter-Agency Standing Committee (2008) Humanitarian action and older persons: an essential brief for humanitarian actors

International Health Conference (1946) Constitution of the World Health Organization

Khan Y, Schwartz B, Johnson I (2014) Surveillance and epidemiology in natural disasters: a novel framework and assessment of reliability. PLoS Curr. https://doi.org/10.1371/currents.dis.6773eb9d5e64b733ab490f78de346003

Kushitor MK, Boatemaa S (2018) The double burden of disease and the challenge of health access: evidence from access, bottlenecks, cost and equity facility survey in Ghana. PLoS ONE 13: e0194677. https://doi.org/10.1371/journal.pone.0194677

Ministry of Internal Affairs and Communications of Japan (2018) Population estimates by age (five-year group) and Sex. In: Stat. Japan. http://www.stat.go.jp/english/data/jinsui/tsuki/index.html

Neria Y, Shultz JM (2012) Mental health effects of Hurricane Sandy: characteristics, potential aftermath, and response. JAMA 308:2571–2572. https://doi.org/10.1001/jama.2012.110700

Ochi S, Hodgson S, Landeg O, Mayner L, Murray V (2014) Disaster-driven evacuation and medication loss: a systematic literature review. PLoS Curr 6. https://doi.org/10.1371/currents.dis.fa417630b566a0c7dfdbf945910edd96

OECD (2017) Old-age income poverty. In: Pensions at a glance 2017: OECD and G20 indicators. OECD Publishing, Paris

Olu O (2017) Resilient health system as conceptual framework for strengthening public health disaster risk management: an african viewpoint. Front Public Heal 5. https://doi.org/10.3389/fpubh.2017.00263

Omran A (1971) The epidemiologic transition: a theory of the epidemiology of population change. Milbank Mem Fund Q 29:509–538

Omran A (1998) The epidemiologic transition theory revisited thirty years later. World Heal Stat Q 51:99–119

Petrak F, Baumeister H, Skinner TC, Brown A, Holt RIG (2015) Depression and diabetes: treatment and health-care delivery. Lancet Diabetes Endocrinol 3:472–485. https://doi.org/10.1016/S2213-8587(15)00045-5

Pickering C, O'Sullivan T, Morris A, Mark C, McQuirk D, Chan E, Guy E, Chan G, Reddin K, Throp R, Tsuzuki S, Yeung T, Murray V (2018) The promotion of 'grab bags' as a disaster risk reduction strategy. PLOS Curr Disasters

Ryan BJ, Franklin RC, Burkle FM, Aitken P, Smith E, Watt K, Leggat P (2016) Reducing disaster exacerbated non-communicable diseases through public health infrastructure resilience: perspectives of australian disaster service providers. PLoS Curr 8. https://doi.org/10.1371/currents.dis.d142f36b6f5eeca806d95266b20fed1f

Slama S, Kim H-J, Roglic G, Boulle P, Hering H, Varghese C, Rasheed S, Tonelli M (2017) Care of non-communicable diseases in emergencies. Lancet 389:326–330. https://doi.org/10.1016/S0140-6736(16)31404-0

Stanke C, Murray V, Amlôt R, Nurse J, Williams R (2012) The effects of flooding on mental health: outcomes and recommendations from a review of the literature. PLoS Curr. https://doi.org/10.1371/4f9f1fa9c3cae

The Lancet Public Health (2017) Ageing: a 21st century public health challenge? Lancet Public Heal 2:e297. https://doi.org/10.1016/S2468-2667(17)30125-1

UN Interagency Taskforce on NCDs and World Health Organization (2016) Noncommunicable diseases in emergencies

United Nations Office for Disaster Risk Reduction (UNISDR) (2009) 2009 UNISDR terminology on disaster risk reduction. Geneva

United Nations Population Fund and HelpAge International (2012) Ageing in the twenty-first century: A celebration and a challenge

Watson S, Rudge J, Coker R (2013) Health systems' "surge capacity": state of the art and priorities for future research. Milbank Q 91:78–122. https://doi.org/10.1111/milq.12003

World Health Organization (1999) The double burden: Emerging epidemics and persistent problems. The world health report 1999. Geneva

World Health Organization (2007a) Global age-friendly cities: a guide. Geneva

World Health Organization (2007b) Everybody business : strengthening health systems to improve health outcomes : WHO's framework for action

World Health Organization (2010) Monitoring the building blocks of health systems: a handbook of indicators and their measurement strategies

World Health Organization (2011) Disaster risk management for health: non-communicable diseases

World Health Organization (2013a) Building back better: sustainable mental health care after emergencies

World Health Organization (2013b) Global action plan for the prevention and control of NCDs 2013–2020). Geneva

World Health Organization (2014) Global status report on noncommunicable diseases 2014. Geneva

World Health Organization (2015a) Japan: WHO statistical profile. http://www.who.int/gho/countries/jpn.pdf?ua=1. Accessed 20 Jul 2018

World Health Organization (2015b) World report on ageing and health. Geneva

World Health Organization (2015c) Operational framework for building climate resilient health systems

World Health Organization (2015d) Comprehensive safe hospital framework

World Health Organization (2016) Global health observatory data—Top 10 causes of death. http://www.who.int/gho/mortality_burden_disease/causes_death/top_10/en/. Accessed 13 Jul 2018

World Health Organization (2018a) Ageing and health. http://www.who.int/en/news-room/fact-sheets/detail/ageing-and-health. Accessed 22 Jul 2018

World Health Organization (2018b) Noncommunicable diseases. http://www.who.int/news-room/fact-sheets/detail/noncommunicable-diseases. Accessed 12 Jul 2018

World Health Organization and Public Health England (2017) Health emergency and disaster risk management: overview. http://www.who.int/hac/techguidance/preparedness/who-factsheet-overview-december2017.pdf?ua=1. Accessed 1 Jun 2018

Chapter 4
Evidence Gaps in Community Resilience Building of Health-EDRM in Asia

Emily Ying Yang Chan and Gloria Kwong Wai Chan

Abstract As a crucial element of Health-EDRM, community resilience should be built at individual, household, as well as community levels. This chapter will first explore the concept of resilience and its presence in two landmark disaster-related international frameworks, followed by a discussion of the importance of community participation in resilience building under the Health-EDRM framework. Evidence gaps in community disaster resilience building will then be identified, supported by several relevant case studies in Asian region, which will help direct efforts in the development of theories and practices in community disaster resilience building as well as Health-EDRM framework in general.

Keywords Bottom-up resilience · Community participation · Community resilience · Evidence gap · Risk communication · Health Emergency and Disaster Risk Management (Health-EDRM)

4.1 Introduction

Health-EDRM has a strong community focus and community participation is central to its effective implementation. In an emergency or a disaster, community members are the primary respondents as well as victims. The resilience of communities can be enhanced by assisting them to locate relevant hazards and vulnerabilities and build their capacities to mitigate, prepare for, respond to and recover from emergencies and disasters. Participation of communities in risk assessments to identify local hazards and vulnerabilities can also reduce risks prior to a health emergency and a disaster occurring (WHO 2017a, b). Community-level surveillance, household preparedness, first-aid training are what the community can contribute and engage actively (Chan et al. 2016).

E. Y. Y. Chan (✉) · G. K. W. Chan
JC School of Public Health and Primary Care, The Chinese University of Hong Kong, Shatin, Hong Kong
e-mail: emily.chan@cuhk.edu.hk

E. Ying Yang Chan and R. Shaw (eds.), *Public Health and Disasters*, Disaster Risk Reduction, https://doi.org/10.1007/978-981-15-0924-7_4

Urban living is the predominant living pattern of the twenty-first century. Global urban-based population will continue to soar. While 55% of the population in the world lives in urban areas in 2018, such proportion is anticipated to rise to 68% by 2050. Asia is home to 54% of the world's urban population lives in 2018. All the way up to 2050, India and China together with Nigeria in Africa, will account for 35% of the projected growth of global urban population (UN DESA 2018). The wave of urbanization is unfolding the vulnerability of people living in hazard-exposed countries. The World Bank (2009) estimated that urban population exposed to cyclones would increase from 310 million to 680 million by 2050, while exposure to major earthquakes from 370 to 870 million. The impact of disasters with a greater concentration of people can be devastating. In 2011 in Japan, a 9.0 M earthquake hit Tohoku and triggered a tsunami with waves up to 29 km (USGS 2011), leaving over 16,000 dead and missing (National Police Agency of Japan 2018). In the Philippines in 2013, Typhoon Haiyan landed as a Category five super typhoon with wind speeds up to 235 kph, affected more than 16 million people and killed 6300 (National Disaster Risk Reduction and Management Council 2013). Nepal was hit by a 7.8 M earthquake in 2015 affecting 864,000 and killing 8702 (OCHA 2015). In 2018, Central Sulawesi province of Indonesia was hit with a 7.4 M earthquake that resulted in a large tsunami, killing over 2000 people (IFRC 2018). Urban community resilience is thus an important focus for disaster risk reduction efforts.

Preparing cities to reduce disaster risks and strengthening resilience are crucial to ensuring sustainable development and poverty reduction. The poor are particularly vulnerable in disasters, especially when a significant proportion of them reside in more hazardous settlements and lack the safety nets to recover from environmental or economic devastations (The World Bank 2018a). It is estimated that 10% of the world's population, that is 735.9 million people, lived on less than US$1.90 a day in 2015 despite noted progress in poverty reduction from 2013. The poverty rate was 12.4% in South Asia, 2.3% in East Asia and 1.5% in Central Asia, compared to the highest 41.1% in Sub-Saharan Africa (The World Bank 2018b).

The concept of disaster resilience has been gaining a wider interest and popularity, particularly after the adoption of the international *Hyogo Framework for Action 2005–2015: Building the Resilience of Nations and Communities to Disasters* (hereafter 'Hyogo Framework'), wherein the main goal of hazard planning, and disaster prevention, preparedness and mitigation shift from reducing vulnerability to building community resilience. The *Sendai Framework for Disaster Risk Reduction 2015–2030* (hereafter 'Sendai Framework') succeeding the Hyogo Framework significantly shifted further to have a stronger emphasis on disaster risk management with reduction of disaster risk as an expected outcome, and health resilience is recurrently promoted throughout the framework.

4.2 Definition of Resilience

'Resilience' is a buzz word used in many different settings. Dictionary definitions of the word refer to the act of rebounding off a person or an object, after a disturbance. It has a Latin root from 'resilire' which means 'to recoil or jump back'. It has been applied in a number of fields, from ecology, physics, engineering and psychology, to nowadays gaining the popularity in the multi-disciplinary field emergency and disaster risk management.

The United Nations International Strategy for Disaster Reduction (UNISDR, later United Nations Office for Disaster Risk Reduction, UNDRR) defines resilience as '*the ability of a system, community or society exposed to hazards to resist, absorb, accommodate, adapt to, transform and recover from the effects of a hazard in a timely and efficient manner, including through the preservation and restoration of its essential basic structures and functions through risk management.*' (UNISDR 2009). A similar definition is embodied in the Hyogo Framework and renewed in the Sendai Framework, to consolidate the effort in building disaster resilience across the globe. Also a UN body, the United Nations Development Programme (UNDP) takes building resilience as '*an inherent as well as an acquired condition achieved by managing risks over time at individual, household, community and societal levels in ways that minimize costs, build capacity to manage and sustain development momentum and maximize transformative potential.*' (UNDP 2013). Compared with UNISDR, UNDP takes resilience building to carry a transformative perspective and can be described in other words—'build back better'.

Resilience is pivotal in enhancing disaster preparedness for effective response and building back better in the later phase of recovery, rehabilitation and reconstruction. It can be both a process and an outcome. It is noted that there is a gradual change in the concept of disaster resilience to more process-focused in recent years. This can be reflected from an excellent overview of the definition of resilience provided by Manyena (2006) capturing across institutions and research fields.

In practice, community resilience and disaster resilience are largely synonymous terms. Together with the European Union that defines resilience as '*The ability of an individual, a household, a community, a country or a region to withstand, to adapt and to quickly recover from stresses and shocks*' (European Commission 2012), it is worth to note that intergovernmental organizations including UNISDR, UNDP, WHO and EU all capture the community element in the resilience concept (Table 4.1). The recognition of community resilience emphasizes that many protective factors enhancing the psychological resilience of individuals in adversity operate at community level. Tolerance of the disruption of lifeline infrastructure due to disaster, including power supply, clean water, communication networks, buildings and bridges, roads, railways, etc., all contribute to community resilience (Castleden et al. 2011).

Table 4.1 Resilience defined by some intergovernmental organisations

Organisation	Definition
UNISDR (2004)[a]	The capacity of a system, community or society potentially exposed to hazards to adapt, by resisting or changing in order to reach and maintain an acceptable level of functioning and structure This is determined by the degree to which the social system is capable of organizing itself to increase this capacity for learning from past disasters for better future protection and to improve risk reduction measures
UNISDR (2009)[b]/WHO[c]	The ability of a system, community or society exposed to hazards to resist, absorb, accommodate, adapt to, transform and recover from the effects of a hazard in a timely and efficient manner, including through the preservation and restoration of its essential basic structures and functions through risk management
European Commission (2012)	The ability of an individual, a household, a community, a country or a region to withstand, to adapt and to quickly recover from stresses and shocks
IPCC (2012)/The World Bank (2013)[d]	The ability of a system and its component parts to anticipate, absorb, accommodate or recover from the effects of a hazardous event in a timely and efficient manner, including through ensuring the preservation, restoration or improvement of its essential basic structures and functions
UNDP (2013)	An inherent as well as an acquired condition achieved by managing risks over time at individual, household, community and societal levels in ways that minimize costs, build capacity to manage and sustain development momentum and maximize transformative potential

[a]UNISDR adopts this definition of resilience in Hyogo Framework for Action 2005–2015
[b]UNISDR adopts this definition of resilience in Sendai Framework for Disaster Risk Reduction 2015–2030
[c]World Health Organization (WHO) adopts the UNISDR definition of resilience in strengthening health systems and primary health care (WHO) (WHO and UNICEF 2018)
[d]The World Bank adopts the IPCC definition of resilience

Promoting Resilience in Conflict

'Resilience' is increasingly being used in emergency and disaster prevention and preparedness, and joined the humanitarian lexicon as a highlight by donors and UN agencies. In the backdrop of its broad and vague definition, it refers to anything that 'mitigates negative coping strategies' in many instances and some humanitarians rebrand existing work as resilience-building programmes at the field level (IFRC 2016). However, questions arise in applying the concept of resilience in humanitarian conflict response.

The integration of strategies for 'building back better' into comprehensive disaster risk management frameworks is considered essential to help communities seize opportunities for building resilience. Most of such strategies apply to the strengthening of countries' and communities' resilience to natural shocks through stronger, faster and more inclusive post-disaster reconstructions (The World Bank 2018c). For armed conflict that the pre-conflict system was most of the time corrupt, weak and violent, it is not desirable to seek resilience to bounce back to such an incapable condition. Some argue that resilience should carry a transformative element.

The European Union considers the added value of promoting resilience in armed conflict or fragility, where poor government performance, a history of social exclusion and the legacy of conflict tend to linger, resides in addressing not only the symptoms of the crisis but also its root causes. This is especially important when history tends to repeat itself with one in two countries return in conflict situation within five years after a peace agreement is made (European Commission 2015). Some peace advocates even assert that resilience should be viewed as a way to catalyze communities to manage change and transform negative peace to positive peace. Transformation, instead of incremental change, should be discussed in an integral manner in stabilization strategies (Menkhaus 2013).

Nevertheless, going down to the root cause of conflict, that is, the underlying causes of chronic vulnerability for long-lasting conflicts, may create incompatibility with the humanitarian imperative or principles of impartiality and neutrality when the suffering mostly relates to political violence. For example, IFRC argues that contradictions exist between the common understanding of resilience-building and humanitarian principles in armed conflicts, which need to be acknowledged. To have resilience contributed meaningfully to humanitarian operations in violent context, humanitarian practitioners must not be expected to address the root causes of conflict, which are largely politically fueled. Instead, robust analysis going beyond needs assessments can serve as evidence of resilience to give humanitarians a deeper understanding of how a community has adapted to conflict and the extent these coping strategies are compatible with war or peace (IFRC 2016).

4.3 From Hyogo to Sendai, and Further

To help communities build the capacity to bounce back from inevitable shocks they face, resilience needs to be built on top of a reduction of disaster risk. The Hyogo Framework was the first framework for explaining, describing and detailing the work that is required from all different sectors and actors to reduce disaster losses. It outlined five priorities for action in order to substantially reduce disaster losses by 2015 by building the resilience of nations and communities to disasters (UNISDR 2004). In that period, measures were taken by some states to reduce the risk to disasters through awareness-raising campaigns, risk assessments, early warning systems, emergency response capacities, legislation, policies, programmes and projects at local, national and international levels (Briceño 2015). The UNISDR created an online self-assessment tool to assist governments in assessing their progress in building resilience to disaster, helping governments understand their qualitative capabilities (UNISDR 2018).

New commitments were made in 2015 in the Sendai Framework with seven targets and four priorities for action. Like the Hyogo Framework, the Sendai Framework keeps the people-centered focus, engagement and partnership with all of society. It recognizes the strong connection between health and disasters, aiming to reduce disaster risk and losses including lives and livelihoods (UNISDR 2015a, b). It has a bigger emphasis on science and technology, noting the importance of evidence-based policy and scientific activity, drawing attention to critical topics such as health, climate change and sustainable development (Aitsi-Selmi et al. 2015). The Sendai Framework emphasizes an all-hazard preventive approach combining economic development and climate change, capturing risk, vulnerability and exposure (Aitsi-Selmi et al. 2016).

Priority 3 of Sendai Framework specifically focuses on investing in disaster risk reduction for resilience. It states 'public and private investment in disaster risk prevention and reduction through structural and non-structural measures are essential to enhance the economic, social, health and cultural resilience of persons, communities, countries and their assets, as well as the environment' (UNISDR 2015a, b). Priority 3 hopes to invest in long-term health solutions and helps create the ability to bounce back better from inevitable disasters with even the most vulnerable in mind. The Sendai Framework shifts from a response driven to risk-driven approach, emphasizing resilience for critical infrastructures with valued cooperation between stakeholders. It recognizes that weak institutional arrangements are drivers of risk, and therefore disaster risk reduction is central to all new investments (Calkins 2015).

There are 26 key focused actions within and across sectors by states at local, national, regional and global levels in priority 3, including 'public and private investment in disaster risk prevention and reduction through structural and non-structural measures'. These are cost-effective drivers of innovation, growth and job creation that enhance the economic, social, health and cultural resilience of individuals, communities, countries and the environment (UNISDR 2015a, b). Focused actions include investing in safe hospitals and health facilities, health system resilience and disaster risk management for health, basic health care services, life-threatening and chronic diseases, ecosystem and environmental health, animal health and implementing the Sendai Framework.

The Sendai Framework also came at a time when other world agreements were being formed, marking 2015 as a landmark UN agreement year for coherence across different and overlapping policy areas. Shortly after the Sendai Framework was signed in March 2015 (UNISDR 2015a, b) and adopted by 187 UN member states (UNISDR 2015a, b), the Sustainable Development Goals (SDGs) were signed by 193 countries in September 2015 succeeding the end the Millennium Development Goals (MDGs) (United Nations 2015), and Paris Agreement on

Climate Change was signed by 195 countries in December 2015 (United Nations Climate Change 2015), holistically strengthening a larger resilience agenda for the global community.

Among the 17 SDGs, six goals with eight targets also support building resilience to eradicate poverty, ending hunger and reducing the vulnerability of disaster and climate change as crucial components of disaster risk management (Table 4.2) (UN 2015).

Table 4.2 Sustainable development goals capturing the building of resilience (UN 2015)

SDGs	Description
Goal 1. End poverty in all its forms everywhere	
1.5	By 2030, build the resilience of the poor and those in vulnerable situations and reduce their exposure and vulnerability to climate-related extreme events and other economic, social and environmental shocks and disasters
Goal 2. End hunger, achieve food security and improved nutrition and promote sustainable agriculture	
2.4	By 2030, ensure sustainable food production systems and implement resilient agricultural practices that increase productivity and production, that help maintain ecosystems, that strengthen capacity for adaptation to climate change, extreme weather, drought, flooding and other disasters and that progressively improve land and soil quality
Goal 9. Build resilient infrastructure, promote inclusive and sustainable industrialization and foster innovation	
9.1	Develop quality, reliable, sustainable and resilient infrastructure, including regional and transborder infrastructure, to support economic development and human well-being, with a focus on affordable and equitable access for all
9.a	Facilitate sustainable and resilient infrastructure development in developing countries through enhanced financial, technological and technical support to African countries, least developed countries, landlocked developing countries and small island developing States
Goal 11. Make cities and human settlements inclusive, safe, resilient and sustainable	
11.b	By 2020, substantially increase the number of cities and human settlements adopting and implementing integrated policies and plans towards inclusion, resource efficiency, mitigation and adaptation to climate change, resilience to disasters and develop and implement, in line with the Sendai Framework for Disaster Risk Reduction 2015–2030, holistic disaster risk management at all levels
11.c	Support least developed countries, including through financial and technical assistance, in building sustainable and resilient buildings utilizing local materials
Goal 13. Take urgent action to combat climate change and its impacts	
13.1	Strengthen resilience and adaptive capacity to climate-related hazards and natural disasters in all countries
Goal 14. Conserve and sustainably use the oceans, seas and marine resources for sustainable development	
14.2	By 2020, sustainably manage and protect marine and coastal ecosystems to avoid significant adverse impacts, including by strengthening their resilience and take action for their restoration in order to achieve healthy and productive oceans

4.4 Community Participation and Resilience in Health-EDRM

As discussed in the previous section, the promotion of health resilience is at the heart of the Sendai Framework. The importance of health as a core dimension in disaster risk reduction was further illustrated in details through the adoption of 'Bangkok Principles for the implementation of the health aspects of the Sendai Framework for Disaster Risk Reduction 2015–2030', which was agreed upon in March 2016 in Bangkok, Thailand. To enhance cooperation between health authorities and other stakeholders in strengthening national capacity for Health-EDRM, implementing the International Health Regulations (2005) and building resilient health systems, the promotion of health resilience was suggested as one of the key actions that can strengthen health emergency and disaster risk management capacities at all levels, including policies and legislation, planning and coordination, human and financial resources, monitoring and evaluation, information management, health infrastructure and logistics, health-related services, risk communication and community capacity development. On a similar but different token, the Bangkok Principles suggested that public and private investment in emergency and disaster risk reduction (including in health facilities and infrastructure) should be stimulated and the safety functionality and resilience of critical health infrastructure and facilities be enhanced by safety assessments, strengthened implementation of the Safe Hospital Initiative, and adopting 'building back better' principle in recovery and reconstruction in coordination with communities (WHO 2016).

To engage all relevant parties, health emergency and disaster risk management (Health-EDRM) has emerged as a nascent field of inquiry derived from the disciplines of risk management, emergency management, outbreaks preparedness and response and health systems strengthening (Chan and Murray 2017). Health-EDRM promotes a focus on the intersection of health and disaster risk reduction and supports implementing the health aspects of the Sendai Framework in an integrated approach (Lo et al. 2017).

Improvement of overall health status in the community is a crucial contributing factor to community resilience. Besides a high baseline coverage for essential health services, this also includes improvements in health behaviour, a range of health outcomes including reduced rates of chronic disease and maternal and child health, enhancement in accessing to health services and health literacy, which are the contributions of a resilient primary health care system (WHO and UNICEF 2018). This, in turn, builds community resilience and provides the foundation for responding to wide-ranging health needs in emergencies and disasters (WHO 2017a, b).

Bangladesh: Building Overall Health Status Crucial in Emergency and Disaster Response

Bangladesh is vulnerable to many recurring natural hazards such as extreme cyclones and flooding, due to its flat topography, low-lying and climatic features, population density and socioeconomic environment (Islam et al. 2017). In 1970, Cyclone Bhola struck, killing 250,000–500,000 people (Hossain 2018). Bangladesh made significant efforts in disaster resilience to reduce its vulnerability, especially among its coastal areas. Disaster risk reduction has been applied to some communities, in which residents state that the impact of disasters is much less than in the past because they are better informed on how to protect themselves, including having warning forecasts and evacuating ahead of time (IFRC 2017). Studies show that disaster risk management is working: cyclone-related mortality in Bangladesh has decreased by more than 100-fold in the last 40 years, from 500,000 deaths in 1970–4234 deaths in 2007, due to improved defensive measures, increased awareness and communications (Haque et al. 2012).

Bangladesh faced another emergency situation in 2017 when there was a sudden influx of Rohingya people from Myanmar. Over 600,000 people fled to Bangladesh from the violence in the Rakhine State of Myanmar in a 3-month period, living in camp settlements. There were reported measles and diphtheria cases in the camp area, bringing mortality and healthcare concerns (Chan et al. 2018a). Measles can be prevented through the building of herd immunity by vaccination in the community, while diphtheria was once eliminated in Bangladesh. The outbreak of the diseases in the camp settlements showed an inadequate coverage of immunization of Rohingya people in Myanmar (WHO Bangladesh 2017). They are also threatened by low coverage of routine immunization, poor access to clean water and sanitation facilities and malnutrition (WHO 2017a, b).

The Philippines: Disaster Resilience and Preparedness Against Typhoons

The Philippines is one of the most at-risk nations in the world for climate-related natural hazards. Typhoon Haiyan touched down in the Philippines in 2013, killing 6329, becoming one of the strongest tropical cyclones in record (National Disaster Risk Reduction and Management Council 2013). Since then, the Philippines has been engaged in numerous activities in community-based resilience to prevent another tragedy, such as assessments of hazards, vulnerability and risk; risk transfer mechanisms; disaster preparedness capacity building; early warning systems and evacuations; response and relief operations and efforts in rehabilitation, recovery and reconstruction (Alcayna et al. 2016). As a result, many communities perceive themselves to be resilient to recurring natural hazards (Usamah et al. 2014). People living in areas frequented by typhoons have the highest levels of preparedness, and those who have prior experience dealing with a disaster cited that as a reason for being prepared for future disasters. It was found that many Filipinos feel self-reliant in disasters, preparing food, drinking water and clothes (Bollettino et al. 2018).

In 2018, Typhoon Mangkhut with similar strength as Typhoon Haiyan led to over 70 fatalities (OCHA 2018). The government had applied lessons of Typhoon Haiyan and got better prepared, including evacuating more than 105,000 people (Beech and Paddock 2018) to temporary shelters, placing stronger shelters on high ground and delivering food and clean water ahead of the storm as lifesaving goods (Erickson 2018).

4.5 Evidence Gaps in Community Disaster Resilience Building

The concept of community disaster resilience offers a people-centred approach to serve as one of the key constructs of disaster management. However, with the broad definition of community disaster resilience and the multi-disciplinary nature of Health-EDRM, collecting evidence to support decision making can be challenging. There is a lack of standardized measurements in disaster resilience research (Witt and Lill 2018). Baseline disaster data and health data may not be available, which means a lack of meaningful reference to assess the data of disaster damage and loss (Maini et al. 2017) and the same difficulty applies to justify resilience building. Measurements and indicators for evaluating the progress of resilience building and the data required for such measurements are crucial for communities to formalize the resilience concept and integrate into the disaster management system. Establishing the link between disaster resilience measurements and disaster risk is also important for the building of robust evidence (UNISDR 2016a).

The area of disaster risk management research becomes more complex when the interdependencies among hazards grow (UNISDR 2016b). Earthquake may trigger a tsunami, and the risks of disease outbreak escalate in urban disasters where the health system and infrastructures are poor. Under the all-hazards approach, hazardscapes get increasingly broad with growing complexity when the impact of climate change soars (IPCC 2012). Resulting from climate change-induced sea-level rise, impacts of typhoons or cyclones can be anticipated to become more severe (Mimura 2013). How climate adaptations in the community synergize with resilience building is yet to be explored.

The building of community disaster resilience involves a co-production of knowledge between at-risk communities and disaster responders and/or researchers. The development of community competence through encouraging community action and collective self-efficacy to promote positive health outcomes in affected communities is a gap to be addressed. There are major uncertainties of Health-EDRM evidence-based interventions that may be applied at the field level (Chan et al. 2019). At the moment, the lack of generalizability, Health-EDRM solution and evidence to scale up the implementation of strategies (van Kessel et al. 2014) affect potential efficacy of empowering community members to facilitate the disaster governance process, which in turn supports community decision making (van Kessel et al. 2014) (see the three cases below).

Communicating effectively, including risk communication, is a critical function of Health-EDRM. It is so important that people at-risk can have real-time access and exchange of information, advice and opinions for informed decisions to be made to prevent, mitigate and respond to hazards (WHO 2017a, b). This synchronizes with the building of community disaster resilience through the provision of information about risks. The influence on resilience of effective access to information during major infrastructure collapse and communication technology breakdown, media strategies (or public information activities) in the pre-disaster and disaster phases, and the role of information during recovery phase should be explored and investigated more in details (van Kessel et al. 2014).

Hong Kong: Household-Based Preparedness for Public Health Emergencies

Densely-populated Asian cities are among the highest at-risk localities for natural hazards and health-related emergencies. The community can reduce their risks by practicing household-based preparedness tasks include first-aid training, stockpiling equipment and supplies, preparing emergency plans such as evacuation and meeting points, and practicing drills. In Hong Kong, a study in 2012 showed that household preparedness practice and risk literacy were low. 82.3% of respondents did not perceive Hong Kong as a disaster-susceptible city, 54.6% thought that the local population had lower disaster awareness than counterparts in other global cities, and only 26.1% of respondents participated in basic first-aid training. Commonly perceived hazards were infectious disease outbreaks (74%), typhoons (12.9%), and fires (7.3%). 75.1% of the respondents had three out of five preparedness measures at home and 28.3% had all five items: first-aid kit, basic aid supplies, emergency drinking water and food, basic medication and a fire extinguisher (Chan et al. 2016). Enhancing general public's disaster health risk literacy is important and should be prioritized in order to reduce disaster risks.

In a study in 2014, in both general and specific infectious-disease emergencies, 59.2% of respondents were found to have good household preparedness levels, with at least five items stocked up such as first-aid kits, food and drinking water, basic medication, essential long-term medication, masks, alcohol hand rub and antivirals (Tam et al. 2018). Compared to other countries and cities, Hong Kong citizens were found to have a relatively good household preparedness level due to risks easily communicated through weather warnings and warning systems that are accompanied by precautionary measures. This may indicate general disaster preparedness improvement over the years. Consistent with previous studies, good household preparedness is associated with higher education and socioeconomic status. The number of those prepared can still increase, as many respondents wish more to be done in terms of risk communication. Risk communication and public health surveillance were mostly conducted via television, the internet and telephone.

A high level of disaster-related community mitigation knowledge may facilitate better preparedness enhancement programs. Educational programmes through talks and group discussions from health promoters raise awareness. For those lacking preparedness, such as those in low-income households, a tailored preparedness health campaign program is necessary. Official indices for infectious-disease outbreaks should be developed in order to simplify risk communication messages in a direct and timely manner (Tam et al. 2018).

Nepal: Mapping Project Enhances Community Disaster Resilience

Kathmandu, Nepal is the world's most seismically at-risk urban area [1905 Kangra Ms 7.8; 1934 Bihar Mw 8.1; 1950 Assam Mw 8.6; 2011 Sikkim Mw 6.9 (Dixit et al. 2013)]. The threat of destructive earthquakes spurred the government to implement the Open Cities Project in 2012 mapping over 130,000 buildings (The World Bank 2018d). When two earthquakes struck within three weeks of each other in 2015, with magnitudes 7.8 and 7.3, killing 8702 (OCHA 2015), the information gathered during mapping helped in recovery efforts. The Nepal military, Red Cross and other organizations used the Open Cities data and integrated it into long-term disaster preparedness planning exercises. In addition, Kathmandu was ready for a large earthquake, with structured collaboration and coordination among national and international response agencies, robust healthcare systems and disaster management preparation including structured earthquake drills. Compared with the Richter scale magnitude 7 earthquake that struck Port-au-Prince, Haiti in January 2010 causing the death of 160,000–200,000 people and the injury of more than 300,000, the scale of the double earthquake that hit Nepal in 2015 was comparable, if not greater. While every life counts, disaster preparedness in Nepal with a robust health care system and communication system, as well as community resilience in protection and response in disaster, were recognized as effective (Auerbach 2015).

China: CCOUC Ethnic Minority Health Project—A Bottom-Up Resilience-Building Programme in Rural China

While constituting half of the country's population, China's rural communities face the highest natural disaster risk in the world. People living in these disaster-prone rural communities are often ethnic-minorities, living in extreme poverty (less than US$1.25/person/day) and in lack of resources or technical capacity for disaster preparedness. After the onset of disasters, these rural communities are often left to manage their own health risks and post-disaster development challenges alone. The development of disaster risk resilience from the bottom is thus very important.

The Ethnic Minority Health Project

The Collaborating Centre for Oxford University and CUHK for Disaster and Medical Humanitarian Response (CCOUC), a joint academic unit between the University of Oxford in the United Kingdom and The Chinese University of Hong Kong (CUHK), developed the Ethnic Minority Health Project (EMHP) in 2009 to examine health outcomes and to construct bottom-up disaster resilience programmes in remote, extremely poor and disaster-prone ethnic minority villages in rural China (Chan 2018; Collaborating Centre for Oxford University and CUHK for Disaster and Medical Humanitarian Response, n.d.). Across these years, the project found that even in communities with a relatively young population who had received formal education, health and disaster risk awareness were often very limited. CCOUC's survey results showed in general, 40–50% of the communities perceived their health as poor.

Project objectives

The primary goal for EMHP is to identify evidence-based solutions to support bottom-up health and disaster resilience-building efforts in remote and rural communities. Five major project objectives include: (1) empowering 'vulnerable communities in rural and remote settings to prepare and mitigate the adverse impact of natural disasters'; (2) bringing science 'to the people by adapting technical "know-how" developed in academic settings to concrete practice in the field'; (3) developing 'human resources to work in rural and remote communities by offering practice and field-based trainings'; (4) raising 'global awareness of issues related to disaster impact and preparedness among remote communities in developing countries' and (5) documenting 'empirical findings to support future development of related intervention for other rural, community-based health projects' (Collaborating Centre for Oxford University and CUHK for Disaster and Medical Humanitarian Response, n.d.).

Site selection and needs assessment

Site selection for the EMHP project has four criteria: geographical remoteness, ethnicity, economic vulnerability and disaster proneness. Hence, the EMHP targets ethnic minority communities in remote parts of rural China who live in extreme poverty (an income of less than US$1.25 per person per day). By December 2018, the project team has established more than 16 project village sites across eight provinces in China. More than 600 public health, medical and related professionals have participated in the project, which has also outreached to more than 40,000 villagers in remote communities of southwest, northwest and northeast Chan, as well as the Tibetan Plateau.

Project structure

The CCOUC project team collaborates with multi-disciplinary partners and the community leaders to locate specific health risks and disaster preparedness solutions that might be relevant and appropriate for their local conditions, so as to build disaster resilience from bottom-up. In-depth health needs assessments are conducted in each selected site through household interviews, focus groups and observation and analysis of the living environment. Information related to general socioeconomic, demographic and health status, as well as community- and household-based disaster preparedness (e.g. availability of household disaster kits, knowledge of ORS preparation, water and food security, infectious-disease proneness and healthcare service access). A team will commit 18–24 months to each of the project sites in a project cycle, which comprises four main phases, namely site selection, needs assessment, health intervention and impact evaluation. At least one visit to the village is required in each phase of the cycle.

Main areas of interventions

The CCOUC project team has gained many valuable insights for bottom-up health and disaster resilience building in remote areas of China since 2009 (Chan 2018). Various rural interventions implemented at the field level in the EMHP are within five main areas of health and disaster preparedness (water, indoor environment, waste management, health behaviour and disaster preparedness and disaster risk reduction). The following are some key findings that might be relevant for programme developers or policymakers in these areas.

Inadequate protection of the water source

In many rural villages in China, although water shortage may not be the primary issue, water quality and safety pose significant risk of waterborne diseases. Drinking water in rural villages may be polluted by various human factors and practices. For example, farmers may allow their livestock to drink from human drinking sources, leading to cross-contamination and disease risk dissemination.

Lack of knowledge on drinking water management

Although most knew that drinking boiled water can protect health, only 74.7% of households in CCOUC project sites practised this habit. Especially in villages with adequate water supply, villagers perceive mountain water may be clean and frequently less than half of the villagers drank boiled water.

Lack of basic hygiene infrastructure

CCOUC study found toilet availability in remote areas was low. On average, 43% of households interviewed did not have a designated place for urination and defaecation. Unprocessed and directly exposed human excreta provides a breeding ground for vectors (e.g. flies, mosquitoes and rodents), thereby increasing the likelihood of the outbreak of vector-borne communicable diseases.

Lack of proper personal hygiene habits

While good personal hygiene is crucial for disease prevention and health protection, reports from project sites found 30% of households did not have handwashing habits before food consumption or a meal. Therefore, even with economic development and infrastructure advancement, basic hygiene practice promotion remains core objective of public health programme.

Lack of waste management knowledge

A majority of the villages visited by CCOUC did not have the economic or logistic capacity to manage the ever-increasing amount of waste. CCOUC's household survey found that less than 10% of the village communities are aware of the concept 'Reduce, Reuse and Recycle' and 82.3% of the households had the habit of indiscriminately burning waste (including batteries and plastics) at home with no awareness of the potential health impacts associated with such practices.

Lack of knowledge and management on chronic diseases

CCOUC's survey indicates that only half of the 43.4% of studied households with members who required long-term medication for chronic illness control could afford to acquire medication on a regular basis. When affected individuals finally decided to seek medical help, complications might have already been developed and the effectiveness of treatment was often seriously compromised.

Lack of disaster preparedness and risk reduction awareness at household and individual levels

Only 20–30% of households in CCOUC project sites believed that they had the capability of keeping their family members and themselves safe and less than 80% of household who lived in disaster-prone communities had ever thought about preparing for disasters.

Sources

Chan (2018), Chan et al. (2017), Chan et al. (2018b), Chan et al. (2014), Chan et al. (2018c) and Collaborating Centre for Oxford University and CUHK for Disaster and Medical Humanitarian Response (n.d.).

4.6 Conclusion

Building community and bottom-up resilience is a core component of Health-EDRM. As highlighted in this chapter, there are major evidence-based gaps that need to be urgently filled to facilitate the implementation and advancement of disaster risk reduction in health.

References

Aitsi-Selmi A, Egawa S, Sasaki H, Wannous C, Murray V (2015) The Sendai framework for disaster risk reduction: renewing the global commitment to people's resilience, health, and well-being. *International Journal of Disaster Risk Science, 6*(2), 164–176. https://link.springer.com/article/10.1007%2Fs13753-015-0050-9

Aitsi-Selmi A, Blanchard K, Murray V (2016) Ensuring science is useful, usable and used in global disaster risk reduction and sustainable development: a view through the Sendai framework lens. Palgrave Communications. https://doi.org/10.1057/palcomms.2016.16

Alcayna T, Bollettino V, Dy P, Vinck P (2016) Resilience and disaster trends in the Philippines: Opportunities for national and local capacity building. PLoS Currents 8:ii

Auerbach P (2015) Preparedness explains some differences between Haiti and Nepal's response to earthquake. BMJ 350:h3059. https://doi.org/10.1136/bmj.h3059

Beech H, Paddock R (2018) Typhoon Mangkhut wreaks havoc in Philippines, leaving at least 25 dead. The New York Times. Retrieved from https://www.nytimes.com/2018/09/15/world/asia/typhoon-mangkhut-philippines-death-toll.html

Bollettino V, Alcayna T, Enriquez K, Vinck P (2018) Perceptions of disaster resilience and preparedness in the Philippines. Program on Resilient Communities, Harvard Humanitarian Initiative. Retrieved from https://hhi.harvard.edu/sites/default/files/publications/prc-phillippine-report-final_0.pdf

Briceño S (2015) Looking back and beyond Sendai: 25 years of international policy experience on disaster risk reduction. Int J Disaster Risk Sci 6(1):1–7. https://doi.org/10.1007/s13753-015-0040-y

Calkins J (2015) Moving forward after Sendai: how countries want to use science, evidence and technology for disaster risk reduction. PLOS Curr Disasters. https://doi.org/10.1371/currents.dis.22247d6293d4109d09794890bcda1878

Castleden M, McKee M, Murray V, Leonardi G (2011) Resilience thinking in health protection. J Public Health 33(3):369–377. https://doi.org/10.1093/pubmed/fdr027

Chan EYY (2018) Building bottom-up health and disaster risk reduction programmes. Oxford University Press, Oxford

Chan EYY, Kim JH, Lin C, Cheung EYL, Lee PPY (2014) Is previous disaster experience a good predictor for disaster preparedness in extreme poverty households in remote Muslim minority based community in China? J Immigr Minor Health 16(3):466–72. Available from http://www.ncbi.nlm.nih.gov/pubmed/23264186

Chan E, Murray V (2017) What are the health research needs for the Sendai Framework? Lancet 390(10106):E35–E36. https://doi.org/10.1016/S0140-6736(17)31670-7

Chan E, Yue J, Wang S (2016) Socio-demographic predictors for urban community disaster health risk perception and household based preparedness in a Chinese Urban City. PLOS Curr Disasters

Chan EYY, Guo C, Lee P, Liu S, Mark CKM (2017) Health Emergency and Disaster Risk Management (Health-EDRM) in remote ethnic minority areas of rural China: the case of a flood-prone village in Sichuan. Int J Disaster Risk Sci. https://doi.org/10.1007/s13753-017-0121-1

Chan E, Chiu C, Chan G (2018a) Medical and health risks associated with communicable diseases of Rohingya refugees in Bangladesh 2017. Int J Infect Dis 68:39–43

Chan EY, Ho JY, Huang Z, Kim JH, Lam HC, Chung PP, Wong CK, Liu S, Chow S (2018b) Long-term and immediate impacts of Health Emergency and Disaster Risk Management (Health-EDRM) education interventions in a rural chinese earthquake-prone transitional village. Int J Disaster Risk Sci 9(3):319–330

Chan EY, Lam HC, Chung PP, Huang Z, Yung TK, Ling KW, Chan GK, Chiu CP (2018c) Risk perception and knowledge in fire risk reduction in a dong minority rural village in China: a Health-EDRM Education Intervention Study. Int J Disaster Risk Sci 9(3):306–318

Chan EYY, Man AYT, Lam HCY (2019) Scientific evidence on natural disasters and health emergency and disaster risk management in Asian rural-based area. British Med Bull ldz002:1–15. https://doi.org/10.1093/bmb/ldz002

Collaborating Centre for Oxford University and CUHK for Disaster and Medical Humanitarian Response (n.d.). Ethnic minority health project [Internet]. Available at: http://ccouc.org/home-5

Dixit AM, Tatabe R, Dahal RK, Bhandary NP (2013) Public school earthquake safety program in Nepal. Geomat Nat Hazards Risk 293–319

Erickson A (2018) How the Philippines learned from disaster. The Washington Post. Retrieved 3 Dec 2018, from https://www.washingtonpost.com/world/2018/09/17/how-phillipines-learned-disaster/?noredirect=on&utm_term=.24fa5434c49f

European Commission (2012). The EU approach to resilience: learning from food security crises. European Commission

European Commission (2015) Note No. 4. Promoting resilience in situations of conflict and fragility. Reference document no 17 "Operating in situations of conflict and fragility: An EU staff handbook". Publications Office of the European Union, Luxembourg

Haque U, Hashizume M, Kolivras K, Overgaard H, Das B, Yamamoto T (2012) Reduced death rates from cyclones in Bangladesh: what more needs to be done? Bull World Health Organ 90 (2):150–156. https://doi.org/10.2471/BLT.11.088302

Hossain N (2018) The 1970 Bhola cyclone, nationalist politics, and the subsistence crisis contract in Bangladesh. Disasters 187–203

IFRC (2016) In: Sanderson D, Sharma A (eds) World disasters report 2016. Resilience: saving lives today, investing for tomorrow. International Federation of Red Cross and Red Crescent Societies, Geneva

IFRC (2017) Disaster preparedness makes a difference in a flood-prone Bangladeshi District. Retrieved 1 Nov 2018, from International Federation of Red Cross and Red Crescent Societies: https://media.ifrc.org/ifrc/2017/08/11/disaster-preparedness-makes-difference-flood-prone-bangladeshi-district/

IFRC (2018) Indonesia: Sulawesi earthquake and tsunami—emergency plan of action operation MDRID013 Update no 6. Retrieved from International Federation of Red Cross and Red Crescent Societies: https://reliefweb.int/report/indonesia/indonesia-earthquakes-and-tsunami-emergency-plan-action-operation-mdrid013-update-2

IPCC (2012) Managing the risks of extreme events and disasters to advance climate change adaptation. Special report of the intergovernmental panel on climate change. Intergovernmental panel on climate change. Cambridge University Press, New York

Islam R, Walkerden G, Amati M (2017) Households' experience of local government during recovery from cyclones in coastal Bangladesh: resilience, equity, and corruption. Nat Hazards 85(1):361–378. https://doi.org/10.1007/s11069-016-2568-6

Lo S, Chan E, Chan G, Murray V, Abrahams J, Ardalan A, Yau J (2017) Health emergency and disaster risk management (Health-EDRM): developing the research field within the sendai framework paradigm. Int J Disaster Risk Sci 8(2):145–149. https://doi.org/10.1007/s13753-017-0122-0

Maini R, Clarke L, Blanchard K, Murray V (2017) The Sendai framework for disaster risk reduction and its indicators—where does health fit in? Int J Disaster Risk Sci 8(2):150–155. https://doi.org/10.1007/s13753-017-0120-2

Manyena SB (2006) The concept of resilience revisited. Disasters 30(4):433–450

Menkhaus K (2013) Making sense of resilience in peacebuilding contexts: approaches, applications, implications. Retrieved from Geneva peacebuilding platform: https://www.gpplatform.ch/sites/default/files/PP%2006%20-%20Resilience%20to%20Transformation%20-%20Jan.%202013_3.pdf

Mimura N (2013) Sea-level rise caused by climate change and its implications for society. In: Horikawa K (ed) Proceedings of the Japan Academy, Series B physical and biological sciences, Vol. 89, No. 7, pp. 281–301. https://doi.org/10.2183/pjab.89.281

National Disaster Risk Reduction and Management Council (2013) Final report re effects of Typhoon "Yolanda" (Haiyan). Retrieved from http://ndrrmc.gov.ph/attachments/article/1329/FINAL_REPORT_re_Effects_of_Typhoon_YOLANDA_%28HAIYAN%29_06-09NOV2013.pdf

National Police Agency of Japan (2018) Police countermeasures and damage situation associated with 2011 Tohoku district—off the Pacific Ocean Earthquake. Retrieved from https://www.npa.go.jp/news/other/earthquake2011/pdf/higaijokyo_e.pdf

OCHA (2015) Nepal: Earthquake 2015. Situation report no. 20. Retrieved from https://reliefweb.int/sites/reliefweb.int/files/resources/OCHANepalEarthquakeSituationReportNo.20%283June2015%29_Final.pdf

OCHA (2018, October). Humanitarian bulletin Philippines: Issue 9. Retrieved from https://reliefweb.int/sites/reliefweb.int/files/resources/OCHAPhilippines_Humanitarian_Bulletin_No9_October_2018_FINAL.pdf

Tam G, Huang Z, Chan EY (2018) Household preparedness and preferred communication channels in public health emergencies: a cross-sectional survey of residents in an asian developed urban city. Int J Environ Res Public Health

The World Bank (2009) Density and disasters. Economics of Urban Hazard Risk. Policy Research Working Paper 5161, The World Bank, Finance, Economics, and Urban Development Department & Environment and Energy, Development Research Group. Retrieved from http://documents.worldbank.org/curated/en/456301468148146720/pdf/WPS5161.pdf

The World Bank (2013) Building resilience. Integrating climate and disaster risk into development. The World Bank, Washington DC

The World Bank (2018a) City resilience program. Retrieved from understanding poverty. Disaster Risk Management. https://www.worldbank.org/en/topic/disasterriskmanagement/brief/city-resilience-program

The World Bank (2018b) Poverty. Retrieved 1 Nov 2018, from The World Bank: https://data.worldbank.org/topic/poverty

The World Bank (2018c) Building back better. The World Bank. Retrieved from https://www.gfdrr.org/sites/default/files/publication/Building%20Back%20Better.pdf

The World Bank (2018d) Mapping to build resilience in Kathmandu. Mapping local communities to reduce disaster risk. The World Bank. Retrieved from https://www.gfdrr.org/sites/default/files/publication/REVISED%20FINAL%20-%20Results%20in%20Resilience%20-%20Open%20Cities%20Kathmandu%20Project%20-%204.2.18.pdf

UN (2015) Transforming our world: the 2030 agenda for sustainable development. United Nations

UNDP (2013) UNDP strategic plan 2014–2017. United Nations Development Programme (UNDP), New York

UN DESA (2018) 2018 revision of world urbanization prospects. Retrieved 3 Dec 2018, from Population Division. United Nations Department of Economic and Social Affairs. https://population.un.org/wup/

UNISDR (2004) Hyogo framework for action 2005–2015. United Nations International Strategy for Disaster Reduction

UNISDR (2009) 2009 UNISDR terminology on disaster risk reduction. United Nations International Strategy for Disaster Reduction, Geneva

UNISDR (2015a) The Sendai framework for disaster risk reduction 2015–2030. United Nations International Strategy for Disaster Reduction

UNISDR (2015b, March 18) World conference adopts new international framework for disaster risk reduction after marathon negotiations. Sendai, Japan

UNISDR (2016a) Sharing innovations to improve implementation and reporting of the Sendai Framework for disaster risk reduction 2015–2030. Launching UNISDR science and technology partnership and the science and technology road map to 2030. Short concept note: Work Stream 3, Working Group 3. United Nations International Strategy for Disaster Reduction, Geneva. Retrieved from https://www.preventionweb.net/files/45270_unisdrcnws3wg3sharinginnovations.pdf

UNISDR (2016b) Research gaps. Launching UNISDR science and technology partnership and the science and technology road map to 2030. Short concept note: work stream 4, working group 3. Geneva. Retrieved from https://www.preventionweb.net/files/45270_unisdrcnws4wg3researchgaps.pdf

UNISDR (2018). HFA monitoring & review. Retrieved from PreventionWeb: https://www.preventionweb.net/english/hyogo/hfa-monitoring/

United Nations (2015, September 25–27) Sustainable development goals offically adopted by 193 countries. New York, United States

United Nations Climate Change (2015, December 12) Historic paris agreement on climate change: 195 nations set path to keep temperature rise well below 2 degrees celsius. Paris, France

Usamah M, Handmer J, Mitchell D, Ahmed I (2014) Can the vulnerable be resilient? Co-existence of vulnerability and disaster resilience: informal settlements in the Philippines. Int J Disaster Risk Reduct 10:178–189. https://doi.org/10.1016/j.ijdrr.2014.08.007

USGS (2011) Magnitude 9.1: near the east coast of Honshu, Japan. Retrieved from https://earthquake.usgs.gov/earthquakes/eventpage/official20110311054624120_30/executive

van Kessel G, MacDougall C, Gibbs L (2014) Resilience-rhetoric to reality: a systematic review of intervention studies after disasters. Disaster Med Public Health Preparedness 8(5):452–460. https://doi.org/10.1017/dmp.2014.104

WHO (2016) Bangkok principles for the implementation of the health aspects of the Sendai framework for disaster risk reduction 2015–2030. World Health Organization, Bangkok

WHO (2017a) Diphtheria outbreak response update. Cox's Bazar, Bangladesh. Update #2. Retrieved from http://www.searo.who.int/bangladesh/bancxbdiphtheriaresponseupdate2271217.pdf?ua=1

WHO (2017b) Health emergency and disaster risk management overview. World Health Organization

WHO, UNICEF (2018) A vision for primary health care in the 21st century: towards universal health coverage and the sustainable development goals. World Health Organization (WHO) and the United Nations Children's Fund (UNICEF)

WHO (n.d.) Health systems strengthening glossary. Retrieved 1 Nov 2018, from World Health Organization (WHO). http://www.who.int/healthsystems/hss_glossary/en/

WHO Bangladesh (2017, November 9) Mortality and morbidity weekly bulletin (MMWB). Cox's Bazar, Bangladesh. 6. Retrieved from http://www.searo.who.int/bangladesh/mmwbvol6.pdf

Witt E, Lill I (2018) Methodologies of contemporary disaster resilience research. Proc Eng 212:970–977. https://doi.org/10.1016/j.proeng.2018.01.125

Chapter 5
Health-EDRM in International Policy Agenda I: Sendai Framework for Disaster Risk Reduction 2015–2030

Emily Ying Yang Chan, Heidi Hung, Virginia Murray and Rajib Shaw

Abstract The Sendai Framework for Disaster Risk Reduction 2015–2030 has been directing global efforts in disaster risk reduction since its adoption in 2015, and its unprecedented emphasis on health has affirmed unequivocally the critical role of health throughout the disaster cycle. The Sendai Framework calls for the strengthening of resilience through the prevention and reduction of hazard exposure and vulnerability, and increased preparedness for response and recovery. While health resilience is one of the core themes in the Sendai Framework, with it being recognized as both a determinant and outcome of disaster risk reduction at the Global Platform for Disaster Risk Reduction in Cancun, Mexico in 2017, its impact at implementation level requires sustained efforts on all fronts. The aspirations and recommendations set out in the Bangkok Principles on the implementation of health aspects of the Sendai Framework have not yet been fully addressed. To synergize the concerted efforts all relevant practitioners, Health Emergency and Disaster Risk Management (Health-EDRM) has emerged as a discipline that cuts across emergency and disaster medicine, disaster risk reduction, humanitarian response,

E. Y. Y. Chan (✉) · H. Hung
JC School of Public Health and Primary Care, The Chinese University of Hong Kong, Shatin, Hong Kong
e-mail: emily.chan@cuhk.edu.hk

E. Y. Y. Chan · V. Murray
WHO Thematic Platform Health and Emergency Disaster Risk Management Research Network, Geneva, Switzerland

V. Murray
Public Health England, London, UK

Integrated Research on Disaster Risk (IRDR) Scientific Committee, Geneva, Switzerland

WHO Collaborating Centre on Mass Gatherings and Global Health Security, London, UK

UNU-International Institute of Global Health, Kuala Lumpur, Malaysia

R. Shaw
Keio University, Fujisawa, Japan

E. Ying Yang Chan and R. Shaw (eds.), *Public Health and Disasters*,
Disaster Risk Reduction, https://doi.org/10.1007/978-981-15-0924-7_5

community health resilience, and health systems resilience. Health-EDRM highlights the intersection of health and disaster risk reduction and has been playing a key role in supporting the implementation of the health aspects of the Sendai Framework. Major challenges however remain for Health-EDRM to develop into a well-established field of inquiry, with the quality of data and research methodology requiring urgent attention.

Keywords Build back better · Disaster risk reduction (DRR) · Health Emergency and Disaster Risk Management (Health-EDRM) · Safe hospital · Sendai Framework

5.1 Introduction

The Sendai Framework for Disaster Risk Reduction 2015–2030 (Sendai Framework) was adopted by 187 Member States on 18 March 2015 at the third World Conference for Disaster Risk Reduction in Sendai, Japan (United Nations Office for Disaster Risk Reduction 2015) and adopted by the UN General Assembly in June 2015 by all UN member states. The significance of the Sendai Framework must be understood not only in its own right as the driving force for global disaster reduction efforts, but it being part of the global policy agenda on various different but connected areas. The symbolic and practical significance of the synchronous adoption of the Sendai Framework, the Sustainable Development Goals (SDGs), the Paris Agreement within the United Nations Framework Convention on Climate Change, the Agenda for Humanity and The New Urban Agenda together form a far-reaching resilience agenda across the areas of disaster risk reduction, health, development, climate change and humanitarian efforts, involving global, national and local actors.

The Sendai Framework is the successor to the Hyogo Framework for Action (2005–2015) (Hyogo Framework) and charts the global direction in disaster risk reduction until 2030. The expected outcome of the Sendai Framework is '*the substantial reduction of disaster risk and losses in lives, livelihoods and health and in the economic, physical, social, cultural and environmental assets of persons, businesses, communities and countries.*' (UNISDR 2015, Paragraph 16) To achieve this, the Sendai Framework calls for the strengthening of resilience, to be brought about by prevention and reduction of hazard exposure and vulnerability, and increased preparedness for response and recovery. It calls for a stronger role of national governments and all other stakeholders in reducing and managing disaster risks and identifies the following priority areas for action:

(a) Understanding disaster risk.
(b) Strengthening disaster risk governance to manage disaster risk.
(c) Investing in disaster risk reduction for resilience.

(d) Enhancing disaster preparedness for effective response, and to "Build Back Better" in recovery, rehabilitation and reconstruction.

The Sendai Framework sets out one expected outcome, one goal, seven global targets (with 38 indicators developed subsequently), and four priority areas (see Box 1), with the actions under the four priority areas forming the bulk of the document. The list of priorities for action was introduced to ensure focused action by relevant actors, based on experience from the implementation of the Hyogo Framework.

Box 1 Purpose and Scope, Expected Outcome, Goal and Global Targets Under the Sendai Framework

Purpose and Scope—The Sendai Framework applies to the risk of small-scale and large-scale, frequent and infrequent, sudden and slow-onset disasters, caused by natural or manmade hazards as well as related environmental, technological and biological hazards and risks. It aims to guide the multi-hazard management of disaster risk in development at all levels as well as within and across all sectors.
Expected Outcome—The substantial reduction of disaster risk and losses in lives, livelihoods and health and in the economic, physical, social, cultural and environmental assets of persons, businesses, communities and countries.
Goal—Prevent new and reduce existing disaster risk through the implementation of integrated and inclusive economic, structural, legal, social, health, cultural, educational, environmental, technological, political and institutional measures that prevent and reduce hazard exposure and vulnerability to disaster, increase preparedness for response and recovery, and thus strengthen resilience.

Global Targets

(a) Substantially reduce global disaster mortality by 2030, aiming to lower average per 100,000 global mortality rate in the decade 2020–2030 compared to the period 2005–2015.
(b) Substantially reduce the number of affected people globally by 2030, aiming to lower average global figure per 100,000 in the decade 2020–2030 compared to the period 2005–2015.
(c) Reduce direct disaster economic loss in relation to global gross domestic product (GDP) by 2030.
(d) Substantially reduce disaster damage to critical infrastructure and disruption of basic services, among them health and educational facilities, including through developing their resilience by 2030.
(e) Substantially increase the number of countries with national and local disaster risk reduction strategies by 2020.

(f) Substantially enhance international cooperation to developing countries through adequate and sustainable support to complement their national actions for implementation of this Framework by 2030.
(g) Substantially increase the availability of and access to multi-hazard early warning systems and disaster risk information and assessments to the people by 2030.

Credit: The Sendai Framework for Disaster Risk Reduction 2015–2030, adopted at the Third UN World Conference on Disaster Risk Reduction in Sendai, Japan, on March 18, 2015.

In the Foreword to the Sendai Framework Margareta Wahlstrom UN Special Representative to the UN Secretary General notes that '*health resilience is strongly promoted throughout*' (UNISDR 2015, Foreword). In 2016, The International Conference on the Implementation of the Health Aspects of the Sendai Framework for Disaster Risk Reduction 2015–2030 was held in Bangkok. The Conference called for a central place of health resilience in disaster management. The 'Bangkok Principles' were adopted to guide the implementation of the health aspects of the Sendai Framework (Box 2) (International Conference on the Implementation of the Health Aspect of the Sendai Framework for Disaster Risk Reduction 2015–2030 2016). The Bangkok Principles now provide the most important directions guiding international and national efforts in improving the health outcomes of disasters.

Box 2 Seven Recommendations on the Implementation of the Health Aspects of the Sendai Framework for Disaster Risk Reduction 2015–2030

1. Promote systematic integration of health into national and sub-national disaster risk reduction policies and plans and the inclusion of emergency and disaster risk management programmes in national and sub-national health strategies.
2. Enhance cooperation between health authorities and other relevant stakeholders to strengthen country capacity for disaster risk management for health, the implementation of the International Health Regulations (2005) and building of resilient health systems.
3. Stimulate people-centered public and private investment in emergency and disaster risk reduction, including in health facilities and infrastructure.
4. Integrate disaster risk reduction into health education and training and strengthen capacity building of health workers in disaster risk reduction.
5. Incorporate disaster-related mortality, morbidity and disability data into multi-hazards early warning system, health core indicators and national risk assessments.

6. Advocate for, and support cross-sectoral, transboundary collaboration including information sharing, and science and technology for all hazards, including biological hazards.
7. Promote coherence and further development of local and national policies and strategies, legal frameworks, regulations, and institutional arrangements.

Credit: Bangkok Principles for the implementation of the health aspects of the Sendai Framework for Disaster Risk Reduction 2015–2030, recommended by The International Conference on the Implementation of the Health Aspect of the Sendai Framework for Disaster Risk Reduction 2015–2030, held on 10–11 March 2016, in Bangkok, Thailand.

The biennial Global Platform for Disaster Risk Reduction is the most prominent global platform for international exchanges and deliberations on the latest development of disaster risk reduction. The 2017 discussion held in Cancun, Mexico focused on the implementation of the Sendai Framework. While the Cancun deliberation did call for the operationalization of the Bangkok Principles, it highlighted more the impact of health disasters rather than the role of health in disaster risk reduction and the general health outcomes of disasters (Global Platform for Disaster Risk Reduction 2017).

The role of health in disaster risk reduction has been recognized beyond doubt by the Sendia Framework. However, it requires further efforts for it to be operationalized and for sustained relevance to global discussions. To facilitate this process, the chapter will illustrate how health and public health principles and practices intersect with disaster risk management under the four priorities of the Sendai Framework. This will be preceded by a discussion on some definitions in health and disaster management to lay the foundation.

5.2 Health and Its Meaning in the Disaster Management

Health is '*a state of complete physical, mental and social well-being and not merely the absence of disease or infirmity*' (International Health Conference 1946). Therefore, to consider the role and impact of health in disaster risk reduction, we should have in mind the physical, mental and social well-being of an individual. Looking at the most widely accepted definition of public health, the powerful resonance with the concept of disaster risk management is apparent: '*the science and art of preventing disease, prolonging life and promoting health through organized efforts of society*' (Acheson 1988). Public health in practice generally includes health protection, health improvement and health services, (Griffiths et al. 2005) and all these domains have a role to play in disaster management.

Disaster management can be divided into three phases: (a) pre-impact phase, (b) impact phase and (c) post-impact phase (Chan 2017). The pre-impact stage is concerned mostly with disaster preparedness; disaster response is the key during the impact phase; while the post-impact phase focuses on disaster recovery and rehabilitation. Public health practices and measures have a critical role to play in all three phases for reducing the negative health impact of disasters. The pre-impact phase in particular, has the greatest potential for improving the health outcome of disasters; again, reflecting the fit between public health and disaster risk reduction.

The cross-over of health and disaster risk reduction gave rise to the relatively new discipline of health emergency and disaster risk management (Health-EDRM). Health-EDRM refers to '*the systematic analysis and management of health risks, posed by hazardous events, including emergencies and disasters, through a combination of hazard, exposure and vulnerability reduction to prevent and mitigate risks, preparedness, response, and recover*' (World Health Organization and Public Health England 2017).

It is increasingly apparent that Health-EDRM could be considered as a core strategy disaster risk reduction and for sustainable development and thus part of the development process involving individuals, communities, governments at local, national and international levels. As there are a wide spectrum of determinants of health in the disaster context, e.g. basic physical needs of water, sanitation, food; social support; security; infrastructure; communications; etc. Health-EDRM requires multisectoral action anchored around the wider health system.

Health-EDRM emphasises in particular the role of health not only during the impact phase but also the pre-impact stage, i.e. on reducing disaster risk and its impact on health. Disaster risk management therefore has shifted from 'response-driven' to 'risk-driven' approaches (Aitsi-Selmi and Murray 2016). Morbidity and mortality impact reduction is a natural goal of disaster risk reduction and is included in the Sendai Framework Global Targets and indicators UN member state reporting mechanisms.

This paradigm shift represented by the Health-EDRM discourse means that health is not only a responsive component in disaster response; but plays a major and proactive role in disaster risk reduction. Health-EDRM has been much engaged in this activity since 2011 via the development of a series of WHO Health Emergency and Disaster Risk Management Fact Sheets which were most recently updated in 2017. The Fact Sheets introduces health professionals and partners from other disciplines to the concepts and principles of Health-EDRM and guides the integration of health into wider disaster risk reduction strategies. The Overview explains disaster risk management in the multi-sectoral context and the generic elements of disaster risk management, in particular the concepts of hazards, vulnerabilities and capacities. As all health domains are inter-linked; fact sheet on individual topic should be read as part of the entire set and in conjunction with the overview. The fact sheets address the topics of chemical emergencies; child health; climate risk management; communicable diseases; ethics; injury prevention and trauma care; mass gatherings; mass fatalities/dead bodies; mental health and psychosocial support; natural hazards; non-communicable diseases; nutrition; people with disabilities and older people; radiation emergencies; safe hospitals: prepared for emergencies and disasters; sexual and reproductive health and water, sanitation and hygiene.

In order to capture the breadth of the subject and for health and non-health actors to better understand its scope and principles, a comprehensive set of Fact Sheets have been published and regularly updated by WHO and its partners. Part of this work falls into WHO's Technical Support and Expert Networks on Health Emergency and Disaster Risk Management (Health-EDRM).

In summary the Platform was launched in 2009 by WHO and UNISDR as a dedicated global mechanism for protecting public health against disasters, and in support of the Hyogo Framework for Action 2005–2015. The Platform has since then facilitated multidisciplinary and multisectoral action furthering the development of Health-EDRM, brought public health to the centre stage of the Sendai Framework, and now focusing on the implementation of the Health-EDRM principles enshrined in the Sendai Framework. A key goal of the Platform is to improve the health and well-being of millions of people at risk through enhanced health emergency and disaster risk management, embracing prevention, preparedness, response and recovery measures and enabling capacities in health and other sectors at all levels of society.

The WHO Thematic Platform on Health Emergency and Disaster Risk Management reviewed the in 2016 highlighted the major challenges facing the academic and research community in Health-EDRM: overlap in research activities, lack of strategic research agenda, lack of coordination between key stakeholders, and lack of resources. The need to develop a research network to strengthen the evidence base of Health-EDRM policies and practices was recognised in the WHO Official Statement to the Global Platform for Disaster Risk Reduction in Cancun, Mexico, in May 2017. The TPRN was subsequently established by WHO as a subgroup to the Platform, with the aim to serve as an international multi-stakeholder and inter-disciplinary platform to exchange information, share views and advise WHO in the area of Health-EDRM research and evidence-related activities.

5.3 Sendai Framework and Health

Noting the interplay between health and disaster risk reduction, and the essence of Health-EDRM, is summarised in through the four priority areas of the Sendai Framework

Priority 1: Understanding disaster risk

Priority 1 emphasizes that an understanding of disaster risk is the foundation for developing disaster risk management policies and practices. Disaster risk should be well understood from different aspects: vulnerability, capacity, exposure, hazard and environment. In practice, it requires reliable disaster data to be collected, analysed, managed and used; from developing risk maps, evaluating health impact of disasters, to public education of disaster knowledge. A specific reference to health under this priority area is:

> To systematically evaluate, record, share and publicly account for disaster losses and understand the economic, social, health, education, environmental and cultural heritage impacts, as appropriate, in the context of event-specific hazard-exposure and vulnerability information. (paragraph 24(b))

This recent development of disaster risk management mirrored that of health: disaster risk management is now concerned not only with on-the-ground rescue of victims but also risk assessment to address vulnerabilities; as like heath which emphasizes more and more on prevention, and not only removal of disease for individual patient. There has been call for disaster risk management to address more the underlying vulnerability from social, economic and environmental factors, including poverty, land use, inequity (Aitsi-Selmi and Murray 2015). This represents unequivocal recognition of the role of upstream factors in both fields: social, economic, political and environmental determinants over health outcomes and disaster vulnerabilities; since it has been established that disaster risks and poor health outcomes were socially patterned and affecting the disadvantaged disproportionately (Phibbs et al. 2016). Epidemiology, the core methodological tool in public health to understand the distribution and determinants of diseases, has huge potential to contribute to the understanding of disaster risk. Disaster epidemiology has a role to play in all phases of a disaster cycle in terms of assessing needs and understanding risk, and its application to the development of the Interagency Emergency Health Kit and The Sphere Project is a case in point (Chan 2017). Epidemiological methods however have not been routinely adopted in disaster risk management programme and policies globally or nationally (Malilay et al. 2014).

Availability, accessibility, quality and applicability of reliable data, for both baselines and measuring disaster loss, are critical to understanding disaster risks. Data availability has been one of the themes dominating the discussion on Sendai Framework implementation. Concerns on health-related data in disaster setting have been raised, including non-standardized recoding of health data due to the absence of universal definition and classification of disasters, temporality and attribution issues in linking mortality and morbidity to disasters in particular for slow-onset disasters like drought, the high cost of building historic datasets (Maini et al. 2017). The Sendai Framework Data Readiness Review has been conducted to ascertain how ready countries were to report against the global targets under the Sendai Framework. The Review found that in terms of data availability, data on economic loss and damage to critical infrastructure was very limited, although the picture is slightly better for mortality and morbidity data. In terms of data quality, which covers the issues of relevance, accuracy, timeliness, punctuality, accessibility, clarity, and comparability, the Review found that data quality varied significantly between countries, with different measurement, data hosting systems and aggregate data used. These findings provide useful insights on the availability of disaster data globally and their application in understanding disaster risk.

Priority 2: Strengthening disaster risk governance to manage disaster risk

Priority 2 focuses on the significance of governance within and across relevant sectors at local, national, regional and global levels for effective disaster risk reduction, which requires leadership, collaboration and participation of stakeholders during all phases of a disaster. Disaster risk reduction should be mainstreamed into laws, regulations, policies, administrative structures and budgets. The specific reference to health in under this priority is:

> To promote transboundary cooperation to enable policy and planning … to build resilience and reduce disaster risk, including epidemic and displacement risk. (paragraph 28(d))

While the explicit reference to health under this priority area is limited to epidemic risk, a much wider health perspective is recommended in terms of disaster risk governance. Health perspective should be integrated in national disaster risk reduction strategies and plans, relevant laws, regulations and policies. Structures, roles and responsibilities within the governments in relation to Health-EDRM should be defined; finances and manpower requirements should be planned; and coordination mechanism for the operation of Health-EDRM should be in place (World Health Organization and Public Health England 2017).

It is one of the global targets under the Sendai Framework to substantially increase the number of countries with national and local disaster risk reduction strategies. Although Health-EDRM is not referred to in this target and the corresponding indicators, this target presents a window to lobby for the mainstreaming of health perspective when the disaster risk reduction strategy is being developed or revised. This process would need to be supported by strong health science and evidence in disaster context. The research linking health and disaster risk management had been considered as fragmented and piecemeal in the last decade (Tekeli-Yeşil 2006). While the Sendai Framework has improved the process and provided a structure for relevant research, Health-EDRM research should be planned with a view to influence disaster policy and practice, and that past research should be reviewed to draw up practical policy recommendations (Lo et al. 2017).

Priority 3: Investing in disaster risk reduction for resilience

Priority 3 is about building resilience through cost-effective structural and non-structural measures to prevent and reduce disaster risk. Economic, social, health and cultural resilience of persons, communities, countries, and environment are all covered by the Sendai Framework. Resilience is defined as "the ability of a system, community or society exposed to hazards to resist, absorb, accommodate to and recover from the effects of a hazard in a timely and efficient manner, including through the preservation and restoration of its essential basic structures and functions" (United Nations Office for Disaster Risk Reduction [UNISDR] 2009). This is the priority area under which health is featured most prominently, with the intersection of health and disaster risk reduction at the crux, highlighting in particular the role of the health system. Specific references to health under this priority area are:

Safe hospitals

> To strengthen, as appropriate, disaster-resilient public and private investments… in particular schools and hospitals and physical infrastructures…. (paragraph 30(c))

Health system resilience

> To enhance the resilience of national health systems, by
>
> - integrating disaster risk management into primary, secondary and tertiary health care…
> - developing the capacity of health workers in understanding disaster risk and applying and implementing disaster risk reduction approaches in health work
> - promoting and enhancing the training capacities in the field of disaster medicine
> - supporting and training community health groups in disaster risk reduction approaches in health programmes, in collaboration with other sectors
> - the implementation of the International Health Regulations (2005) … (paragraph 30(i))

Access to basic services

> To strengthen the design and implementation of inclusive policies… including … access to basic health-care services, including maternal, newborn and child health, sexual and reproductive health, food security and nutrition. (paragraph 30(j))

Chronic diseases

> People with life-threatening and chronic disease, due to their particular needs, should be included in the design of policies and plans to manage their risks before, during and after disasters, including having access to life-saving services. (paragraph 30(k))

The above references in the Sendai Framework make it clear that the significant role played by resilient health systems in reducing disaster risk cannot be overestimated, including safe hospitals, clinics and related structures and facilitates, access to health care services. It is worth noting in particular the highlight of the needs of chronic disease patients, which have not commanded much attention in disaster management until recent years.

Safe health facilities ensure that disruption to health services following a disaster is kept to the minimal. According to the WHO, damage to hospital and inadequate preparedness are the main reasons that prevent people from receiving essential health services. The Comprehensive Safe Hospital Framework, together with the Safe Hospital Index, issued by the WHO in 2015, provide practical and detailed guidance for national governments to implement the relevant action recommended in the Sendai Framework (Box 3). The Comprehensive Safe Hospital Framework covers different aspects of hospital management, including policy, norms and legislation; coordination and service delivery; resources management; and knowledge and information management (World Health Organization 2015a).

Health system resilience is however much wider than safe health facilities. The WHO framework that describes health systems in terms of six "building blocks" is illuminating for national and local authorities to draw up plans in building resilient health system for disaster risk reduction. The six "building blocks" are (a) service delivery, (b) health workforce, (c) health information systems, (d) access to essential medicines, (e) financing, and (f) leadership/governance (World Health Organization 2007). A set of indicators and measurement strategies have also been published around the six aspects to track progress and evaluate performance (World Health Organization 2010). The WHO health system building blocks have already been applied in devising a climate change resilient health system operational framework, and the same could be considered for general disaster risk reduction (World Health Organization 2015b).

Box 3 Safe Hospitals Initiative

To ensure continued operation of health facilities during disasters, WHO has been promoting safe hospital programmes for over 20 years. "Safe hospitals" refers to all types of health facilities. This concept of ensuring the safety and minimizing the disruption of health facilities during a disaster has been recognized unequivocally as an important disaster risk reduction strategy in the last decade, with specific reference made in the Hyogo Framework for Action (2005–2015):

> … promote the goal of "hospitals safe from disaster" by ensuring that all new hospitals are built with a level of resilience that strengthens their capacity to remain functional in disaster situations …

The critical role played by safe hospitals in disaster risk management was reaffirmed in its succeeding document, the Sendai Framework for Disaster Risk Reduction 2015–2030:

> To strengthen, as appropriate, disaster-resilient public and private investments… in particular schools and hospitals and physical infrastructures…

To promote the concept of safe hospital as a disaster risk reduction strategy under the Sendai Framework, the WHO launched the **Comprehensive Safe Hospital Framework** at the same period when the Sendai Framework was adopted. The Comprehensive Safe Hospital Framework acts as a guiding document for governments and relevant authorities to improve safety and resilience of health facilities at national and subnational levels. At the operational level, the WHO has also published the **Hospital Safety Index**, which is an assessment tool to evaluate the safety and disaster preparedness of hospitals. To ensure that the Hospital Safety Index is applied appropriately, WHO also provides training to at country level.

Credit: Comprehensive Safe Hospital Framework, World Health Organization (2015) Hospital Safety Index: Guide for Evaluators, Second Edition, World Health Organization and Pan American Health Organization (2015).

Following the increasing burden of chronic diseases globally and aging population in many parts of the world, chronic disease management has become an obvious health needs following a disaster (Chan and Kim 2011) (Aitsi-Selmi and Murray 2016). Chronic conditions could be exacerbated, or new conditions may arise during disasters due to physical injuries, change in living conditions or interruption of care (UN Interagency Taskforce on NCDs and World Health Organization 2016). The most important health need of people with chronic disease in a disaster is uninterrupted access to the medication and care required, e.g. supply of insulin for some diabetic patients, dialysis service for patients with severe kidney diseases. Chronic disease management in disaster and emergency settings however has yet to be integrated into the disaster risk reduction plans in most countries. The clear and specific reference to chronic disease management before, during and after a disaster in the Sendai Framework hopefully will encourage more research in this area which is urgently needed.

Priority 4: Enhancing disaster preparedness for effective response, and to "Build Back Better" in recovery, rehabilitation and reconstruction

Priority 4 calls for disaster preparedness in response, recovery, rehabilitation and reconstruction through integrating disaster risk reduction into development measures. Participation and empowerment of women and people with disabilities are highlighted. The direct references to health under this priority are:

> To promote the resilience of new and existing critical infrastructure, including … hospitals and other health facilities… in order to provide life-saving and essential services. (paragraph 33(c))
>
> To establish a mechanism of case registry and a database of mortality caused by disaster in order to improve the prevention or morbidity and mortality. (paragraph 33(n))
>
> To enhance recovery schemes to provide psychosocial support and mental health services for all people in need. (paragraph 33(o))

The importance of resilient health facilities is reinforced under this priority area, for its critical influence over both the impact and post-impact phases. The establishment of case registry and database of disaster mortality is closely related to the discussion on disaster data under Priority 1. The difficulties surrounding disaster data collection have to be addressed and resolved at least to some extent for national and international authorities to be able to build up such mechanism.

Mental health requires special consideration in emergency settings because of increased rates of mental health cases, damaged mental health infrastructure, and difficulties in coordinating relevant agencies in the provision of service (World Health Organization 2013). It is encouraging and timely for the Sendai Framework to call for preparedness in providing support for psychosocial and mental health during the recovery stage. This reference echoes research findings on mental issues related to disaster and the benefits brought about by psychosocial support post-disaster. A number of mental issues, e.g. stress, depression, anxiety, post-traumatic stress disorders, had been found to be associated with population affected by certain weather-related disasters, e.g. hurricane (Neria and Shultz 2012), flooding (Stanke et al. 2012). Social and psychological support and interventions, ranging from basic service and security, community and family supports, focused and non-specialized support, to specialized services, are recommended to improve the mental health outcomes in emergencies (World Health Organization 2011). Particularly on recovery phase, case studies have been developed to demonstrate how mental health care service could be reformed and improved during post-emergency period (World Health Organization 2013).

5.4 Challenges Ahead

From the detailed analysis on the health elements of Sendai Framework above, it is clear that health resilience is now beyond doubt a critical part of the global disaster risk reduction strategy. However, for Health-EDRM to fulfil its aspiration to bring about systematic analysis and management of health-related risks at every stage of a disaster cycle, two major challenges, that are inter-related and particularly relevant for the research community, remained: data quality and research methodology.

Availability and quality of data has been a long-standing issue for the disaster risk reduction community, for understanding risk, conducting research, developing indicators, and evaluating programmes and policies. While the recent data readiness review conducted under the Sendai Framework revealed that basic disaster loss data, including baselines, were available and accessible in some countries, data quality remains an issue and even data accessibility is highly limited in other countries, in particular developing countries (Osuteye et al. 2017). Data limitations mean difficulties in the deployment of indicators, the reporting of which should ideally be integrated with national data collection systems (Chan and Murray 2017). The data issue also represents a major constraint for the development of science-based Health-EDRM policy.

Well-coordinated multisectoral effort across policy areas is essential for the formulation and execution of disaster risk management policy (Aitsi-Selmi et al. 2016), and health authorities, unfortunately, have not traditionally been the leaders in this area. For Health-EDRM to be mainstreamed, it must be demonstrated that health perspective will improve the overall outcome, and that the related policies and measures are cost-effective. Availability of quality data and meaningful indicators are critical in the process. Due to the unique nature and context of every disaster, prospective study designs are by nature more difficult to carry out. At the same time, outcome-based research (Abramson et al. 2007), intervention studies, in particular community intervention trials (Chan and Murray 2017), and an overarching research ethics framework and supporting mechanism (Chan et al. 2019) are much needed in the field for Health-EDRM practices and recommendations to be more evidence-based.

The UNISDR Scientific and Technical Advisory Group (STAG) have been set up to promote scientific disaster research in disaster risk reduction. Although STAG does not focus on health specifically, its mission to provide substantive technical advice and support in forming and implementing disaster risk reduction means that it could be a valuable resource for the Health-EDRM community to leverage on, for facilitating cross-disciplinary collaborative scientific research.

Other disciplines involved in disaster risk management should also be insightful for Health-EDRM, as illustrated in a recent review by the European Union which showcased the use of scientific knowledge and scientific evidence base in disaster risk management actions in Europe. The review suggested that health sciences should play a more active role in disaster risk management, not only through contribution to the health aspects, but also to data collection (Karmen et al. 2017).

5.5 Conclusion

This chapter illustrated how health was interwoven with disaster risk reduction through analysing the four priority areas of the Sendai Framework. Health issues highlighted in the Sendai Framework included health system resilience, safe hospitals/health facilities, access to basic health care services, epidemics and pandemics, needs of chronic disease patients, mental health services, health data management. Our analysis shown that health was a cross-cutting theme across all priority areas, with resilience of health system forming the core. The unequivocal status of Health -EDRM under the Sendai Framework laid strong foundation for its further development; it is however only the first step. As pointed out, many of the existing health research and frameworks/guidelines developed by the WHO could facilitate the implementation of health-related recommendations under the Sendai Framework by relevant stakeholders. The strength of Health-EDRM lies in its twin emphases on systematic management of health-related disaster risks and the role of health in every stage of a disaster, from prevention, response to recovery. The systematic and comprehensive nature of this discipline has strong potential for generating evidence to inform disaster policies at national and global level. Some major challenges nonetheless remain for the operationalization of Health-EDRM principles and for it to be a well-recognized discipline of enquiry, in particular the quality of disaster data and research design of Health-EDRM studies.

There is clear synergy between health concepts and the Sendai Framework, and the health perspective can serve as a conceptual base for the implementation of the relevant sections of the Sendai Framework (Phibbs et al. 2016). The further development of Health-EDRM research in the near future should aim to facilitate the implementation of the health-related sections of the Sendai Framework, and solidify the centrality of health in disaster risk reduction.

References

Abramson DM, Morse SS, Garrett AL, Redlener I (2007) Public health disaster research: surveying the field, defining its future. Disaster Med Public Health Prep 1:57–62. https://doi.org/10.1097/DMP.0b013e318065b7e1

Acheson D (1988) Public health in england: the report of the committee of inquiry into the future development to the public health function. London, UK

Aitsi-Selmi A, Murray V (2015) The Sendai framework: disaster risk reduction through a health lens. Bull World Health Organ 93:362. https://doi.org/10.2471/BLT.15.157362

Aitsi-Selmi A, Murray V (2016) Protecting the health and well-being of populations from disasters: health and health care in the sendai framework for disaster risk reduction 2015–2030. Prehosp Disaster Med 31:74–78. https://doi.org/10.1017/S1049023X15005531

Aitsi-Selmi A, Murray V, Heymann D, McCloskey B, Azhar EI, Petersen E, Zumla A, Dar O (2016) Reducing risks to health and wellbeing at mass gatherings: the role of the Sendai Framework for Disaster Risk Reduction. Int J Infect Dis 47:101–104. https://doi.org/10.1016/j.ijid.2016.04.006

Chan EYY (2017) Public health humanitarian responses to natural disasters, first. Routledge, London

Chan EYY, Kim J (2011) Chronic health needs immediately after natural disasters in middle-income countries: the case of the 2008 Sichuan, China earthquake. Eur J Emerg Med 18:111–114. https://doi.org/10.1097/MEJ.0b013e32833dba19

Chan EYY, Murray V (2017) What are the health research needs for the Sendai Framework? Lancet 390:e35–e36. https://doi.org/10.1016/S0140-6736(17)31670-7

Chan EYY, Wright K, Parker M (2019) Health-emergency disaster risk management and research ethics. Lancet 393:112–113

Global Platform for Disaster Risk Reduction (2017) Global platform for disaster risk reduction 2017—Chair's summary: from commitment to action. Cancun

Griffiths S, Jewell T, Donnelly P (2005) Public health in practice: the three domains of public health. Public Health 119:907–913. https://doi.org/10.1016/j.puhe.2005.01.010

International Conference on the Implementation of the Health Aspect of the Sendai Framework for Disaster Risk Reduction 2015–2030 (2016) Bangkok principles for the implementation of the health aspects of the sendai framework for disaster risk reduction 2015–2030. Bangkok

International Health Conference (1946) Constitution of the world health organization

Karmen P, Montserrat MF, Tom DG, Ian C (2017) Science for disaster risk management 2017: knowing better and losing less. Publications Office of the European Union

Lo STT, Chan EYY, Chan GKW, Murray V, Abrahams J, Ardalan A, Kayano R, Yau JCW (2017) Health emergency and disaster risk management (Health-EDRM): developing the research field within the Sendai Framework paradigm. Int J Disaster Risk Sci 8:145–149. https://doi.org/10.1007/s13753-017-0122-0

Maini R, Clarke L, Blanchard K, Murray V (2017) The Sendai Framework for disaster risk reduction and its indicators—where does health fit in? Int J Disaster Risk Sci 8:150–155. https://doi.org/10.1007/s13753-017-0120-2

Malilay J, Heumann M, Perrotta D, Wolkin AF, Schnall AH, Podgornik MN, Cruz MA, Horney JA, Zane D, Roisman R, Greenspan JR, Thoroughman D, Anderson HA, Wells EV, Simms EF (2014) The role of applied epidemiology methods in the disaster management cycle. Am J Public Health 104:2092–2102. https://doi.org/10.2105/AJPH.2014.302010

Neria Y, Shultz JM (2012) Mental health effects of hurricane sandy: characteristics, potential aftermath, and response. JAMA 308:2571–2572. https://doi.org/10.1001/jama.2012.110700

Osuteye E, Johnson C, Brown D (2017) The data gap: an analysis of data availability on disaster losses in sub-Saharan African cities. Int J Disaster Risk Reduct 26:24–33. https://doi.org/10.1016/j.ijdrr.2017.09.026

Phibbs S, Kenney C, Severinsen C, Mitchell J, Hughes R (2016) Synergising public health concepts with the Sendai Framework for disaster risk reduction: a conceptual glossary. Int J Environ Res Public Health 13. https://doi.org/10.3390/ijerph13121241

Stanke C, Murray V, Amlôt R, Nurse J, Williams R (2012) The effects of flooding on mental health: outcomes and recommendations from a review of the literature. PLoS Curr. https://doi.org/10.1371/4f9f1fa9c3cae

Tekeli-Yeşil S (2006) Public health and natural disasters: disaster preparedness and response in health systems. J Public Health (Bangkok) 14:317–324. https://doi.org/10.1007/s10389-006-0043-7

UN Interagency Taskforce on NCDs and World Health Organization (2016) Noncommunicable diseases in emergencies

United Nations Office for Disaster Risk Reduction (UNISDR) (2009) 2009 UNISDR Terminology on disaster risk reduction. Geneva

United Nations Office for Disaster Risk Reduction (2015) Sendai framework for disaster risk reduction. https://www.unisdr.org/we/coordinate/sendai-framework. Accessed 1 Jun 2018

World Health Organization (2007) Everybody business: strengthening health systems to improve health outcomes : WHO's framework for action

World Health Organization (2010) Monitoring the building blocks of health systems: a handbook of indicators and their measurement strategies

World Health Organization (2011) Disaster risk management for health—mental health and psychosocial support

World Health Organization (2013) Building back better: sustainable mental health care after emergencies

World Health Organization (2015a) Comprehensive safe hospital framework

World Health Organization (2015b) Operational framework for building climate resilient health systems

World Health Organization and Public Health England (2017) Health emergency and disaster risk management: overview. http://www.who.int/hac/techguidance/preparedness/who-factsheet-overview-december2017.pdf?ua=1. Accessed 1 Jun 2018

Chapter 6
Health-EDRM in International Policy Agenda II: Paris Climate Agreement

Emily Ying Yang Chan, Heidi Hung and Rajib Shaw

Abstract In December 2015, the twenty-first Session of the Conference of the Parties (COP21) convened in Paris, France and the Paris Agreement was adopted on 12 December. It is the latest step in the evolution of the UN climate change framework and it builds on the work undertaken under the United Nations Framework Convention on Climate Change (UNFCCC) since its adoption in 1992. The central aim of the Paris climate agreement, which was agreed by 197 countries/parties, is to keep a global temperature rise for this century well below 2 °C above pre-industrial levels and to drive efforts to limit the temperature increase even further to 1.5 °C. It also aims to strengthen the ability of countries to deal with climate change impacts. The mobilization of US$100 billion a year to help developing countries cope with climate change is to extend beyond 2020 and the figure will be considered a floor until 2025. A higher financial goal is to be adopted before 2025. Under the agreement, each country has to set its own national cap for emissions, known as "nationally determined contributions" (NDCs). The pledges made by the countries are voluntary and not legally binding, but their progress will be reviewed every five years. This chapter will discuss the health implications of climate change-related disasters and how the reduction of carbon emission can work together with health-promoting behaviours to achieve health and environmental co-benefits (HEC).

Keywords Climate change · Climate-related disaster · Health Emergency and Disaster Risk Management (Health-EDRM) · Paris Agreement

E. Y. Y. Chan (✉) · H. Hung
JC School of Public Health and Primary Care, The Chinese University of Hong Kong, Shatin, Hong Kong
e-mail: emily.chan@cuhk.edu.hk

R. Shaw
Keio University, Fujisawa, Japan

E. Ying Yang Chan and R. Shaw (eds.), *Public Health and Disasters*, Disaster Risk Reduction, https://doi.org/10.1007/978-981-15-0924-7_6

6.1 Introduction

Climate change means "a change of climate which is attributed directly or indirectly to human activity that alters the composition of the global atmosphere and which is in addition to natural climate variability observed over comparable time periods" (Article 1 United Nations Framework Convention on Climate Change 1992). There are in general four climate change phenomena, namely (1) raised average and extreme temperature, (2) altered rainfall pattern, (3) sea-level rise and (4) extreme weather events (Watts et al. 2015). With the debate on the existence of climate change long settled, the focus has shifted to the prediction of future changes; the impact of these changes on the environment, ecosystems and human beings, including human health; and the formulation of mitigation and adaptation measures. Mitigation, i.e. reduction of greenhouse gas (GHG) emission, has been the main global and national efforts to combat climate change over the years. GHG are gases that trap heat in the atmosphere, including in particular carbon dioxide, methane, nitrous oxide, fluorinated gases; and the main sources are the use of fossil fuels in transportation, electricity generation, industrial, commercial and household settings, emission from husbandry (United States Environmental Protection Agency n.d.). The multifaceted impact of climate change is now better understood with the advancement of scientific knowledge, and hence adaptation, i.e. the reduction of damage caused by climate change, has been gaining importance as the other pillar to cope with climate change. In particular, the impacts of climate change on human health have been drawing increasing attention in recent years, with those associated with climate-related disasters forming the core.

WHO estimates that climate change will lead to an additional 250,000 potential deaths per year between 2030 and 2050 due to malaria, malnutrition, diarrhea and heat stress and calls climate change "the defining issue for health systems in the twenty-first century" (World Health Organization 2016a). In 2015, international experts from a multidisciplinary background reached the conclusion that the effects of climate change "represent an unacceptably high and potentially catastrophic risk to human health" (Watts et al. 2015). Climate change could affect human health in the following ways (Smith et al. 2014; Watts et al. 2015):

- Direct impacts: effects from changes in climate, including frequency of extreme temperature and rainfall; and
- Indirect impacts: (a) effects mediated through natural systems, e.g. vector-borne diseases, waterborne diseases, foodborne diseases, diseases related to air pollution and (b) effects mediated by human systems, e.g. undernutrition, occupational impacts, disruption to health system, mental stress, violence and conflict.

Climate change affects the intensity and frequency of climate-related disasters, e.g. flooding, heat waves and occurrence of climate-related disasters could cause tremendous damage and disruptions and will lead to surge capacity needs of health systems and increase the burden on urban infrastructures. This chapter will explain the impacts of climate-related disasters on human health and outline the environment and health co-benefits (HEC) of mitigation measures. Before going into the details, a general introduction on the major international agreements on climate change and the role of health under these instruments will be set out.

6.2 Role of Health in Global Climate Change Efforts: From UNFCCC to the Paris Agreement

The United Nations Framework Convention on Climate Change (UNFCCC), adopted in 1992 with a close-to-universal membership of 197 parties, is the single most important international treaty to combat climate change and has laid a solid foundation for future global development in this area. Yearly meetings are held under the UNFCC, known as the Conference of the Parties (COP). The Kyoto Protocol, linked to UNFCCC, was adopted in Kyoto, Japan in 1997. The Kyoto Protocol set internationally binding emission targets for its parties under the "common but differentiated responsibilities" principle (Article 10), with heavier burden on developed nations. By 2015, historical temperature goal was reached at COP 21, which resulted in the Paris Agreement. The Paris Agreement built upon the UNFCCC and aimed to hold the increase in global average temperature to well below 2 °C above pre-industrial levels and to pursue efforts to limit the temperature increase to 1.5 °C (Article 2(a)). Paris Agreement also emphasized the building up of parties' capacities to adapt to the impacts of climate change and the fostering of climate resilience (Article 2(b)). Parties had been requested to put forward their "nationally determined contributions" (NDCs) (Article 3), which includes regular reporting on their emissions and implementation efforts. The global stock-take will be conducted every five years. In 2018, parties will report their preparations for NDCs and conduct a preliminary reporting of efforts (UN Climate Change n.d.).

While UNFCCC focuses primarily on the responsibility of the developed nations to stabilize GHG concentration, Article 1 clearly spells out that "adverse effects of climate change" include significant deleterious effects on human health, and Article 7 commits all parties to employ appropriate methods to minimize effects of climate change on public health. As the early days of the climate change discourse focused mainly on mitigation, there was no specific reference to impact of climate change and related disasters on human health in the Kyoto Protocol. As for the Paris Agreement, adaptation efforts, which are highly relevant for reducing health impact of climate-related disasters, featured prominently. Following the adoption of the Paris Agreement, the Second Global Conference on Health and Climate was hosted by the COP 21 presidency of France and the World Health Organization (WHO) in July 2016. The conference proposed key actions to reduce climate-related health risks as adaptation measures under the Paris Agreement, including investment in protecting health from risks linked to extreme weather events, more sustainable food production and healthier diets, more sustainable transport systems and urban planning that reduce major health risks (World Health Organization 2016b). In fact, WHO has been actively trying to mainstream health into global discussions over both climate change and related disaster risk reduction over the years. WHO has in 2009 developed a global Work Plan to support member states in climate change and health protection (updated in 2014). The four objectives of the latest Work Plan on Climate Change and Health 2014–2019 are to (a) advocate and raise awareness; (b) strengthen partnerships; (c) enhance scientific evidence and (d) strengthen health system (World Health Organization 2015a). WHO not only works closely with UNFCCC to put health at the centre of climate agreements; it has strengthened collaboration with the World Meteorological Organization (WMO) through conducting joint assessments and establishing a joint office to improve the use of climate information to protect public health (World Health Organization n.d.a).

Strong scientific efforts have underpinned the development of global climate change instruments since the 1980s. The Inter-governmental Panel on Climate Change (IPCC) was created by the WMO and United Nations Environment Programme in 1988 to assess the science related to climate change. Every five to six years, IPCC publishes an assessment report after reviewing and assessing the most recent scientific, technical and socio-economic information relevant to the understanding of climate change and related disasters, involving thousands of scientists from all over the world (IPCC). Impacts of climate disasters on human health have been gaining coverage in the assessment reports since mid-1990s. IPCC predicts that climate change will exacerbate existing health problems, increase the number of ill-health in many regions, especially in developing countries, and by 2100, even common human activities will be compromised, including growing of food and working outdoors (Pachauri and Meyer 2014).

6.3 Climate Change and Disaster Risk Reduction

It is worth noting that climate change, including its effect on climate-related disasters, is the focus not only under the climate change international agreements—it features prominently also in a few other instruments guiding global developments: The Sustainable Development Goals, the Sendai Framework and the New Urban Agenda, indicating the cross-cutting effect of climate change in human development in different settings. The 2030 Agenda for Sustainable Development, also known as Sustainable Development Goals (SDGs) was adopted in 2015 to end poverty, protect the planet and ensure prosperity for all as part of a new sustainable development agenda. Goal 13 commits all countries to take urgent action to combat climate change and its impacts and Goal 3 calls for the strengthening of national capacity for early warning, risk reduction and management of national and global health risks, which are highly relevant to climate-related disasters. Given the far-reaching health impacts of climate change, its relevance cuts across many other goals, e.g. Goal 2 on improving food security and nutrition (given that extreme temperatures and rainfall could result in crop failure and food shortage), Goal 6 on water and sanitation (given that climate-related disaster like flooding could result in water contamination). The New Urban Agenda adopted in 2016 set out global standards for sustainable urban development through reforming the way that cities are built and managed. The New Urban Agenda recognized that unplanned urbanization caused human settlement to be vulnerable to the impacts of climate change and climate-related disasters and called the adoption of climate change mitigation and adaptation efforts into urban development and planning process.

The Sendai Framework for Disaster Risk Reduction 2015–2030 (Sendai Framework) was adopted in 2015 charting the global direction in disaster risk reduction for the next decade. For the purpose of this chapter, the relationship between climate change and the Sendai Framework is the most relevant. As climate change is playing an increasing role in inducing and exacerbating natural disasters, e.g. flood, storm, drought, climate change is considered as one of the major drivers of disaster risk in the Sendai Framework, and countries are urged to include climate change scenarios in their disaster risk assessments. With "climate change" mentioned 15 times in the Sendai Framework, mostly as a risk driver, it has been considered as well-represented (Kelman 2015). As a major disaster risk driver, research indicates that climate change discourse should be placed within the wider disaster risk reduction context. The link between climate change and disaster risk is the most apparent through climate-related disasters.

6.4 Relationship Between Climate Change and Climate-Related Disasters

Climate-related disasters or extreme weather events refer to "the occurrence of a value of a weather or climate variable above (or below) a threshold value near the upper (or lower) ends of the range of observed values of the variable" (Field et al. 2012). Climate-related disasters are generally classified into three categories depending on the nature of the hazards involved (Centre for Research on Epidemiology of Disasters [CRED] 2018):

(a) Meteorological disasters: e.g. storms/tropical cyclones, extreme temperatures (heat waves, cold air outbreaks);
(b) Hydrological disasters, e.g. flooding, landslide, wave actions and
(c) Climatological disasters, e.g. drought, wildfire, glacial lake outburst.

The frequency of natural disasters recorded in the Emergency Events Database (EM-DAT n.d.) has increased almost three-fold between 2005 and 2014, with the number of hydrological and meteorological events increased sharply during this period (Thomas and Ramón 2015). Over the last decade, which saw an annual average of close to 380 natural disasters globally, climate-related disasters accounted for over 90% of natural disasters, with hydrological disasters constituted over 50%, meteorological disasters about 30% and climatological disasters about

10%, with the remaining attributed to geophysical disasters, e.g. earthquake (less than 10%). The annual average number of deaths caused by natural disasters in the last decade was up to 69,827. In terms of location, Asia continues to take up the largest share of disaster, with close to 50% of all disasters worldwide; and China, the United States, India, Indonesia and the Philippines are the top five countries which had experienced the most number of natural disasters in the past 10 years (Guha-Sapir et al. 2016). In the first half of 2018, there were 56 recorded incidents of climate-related disasters globally involving 460 deaths (EM-DAT n.d.).

Box 1 Number of Natural Disasters, People Killed and People Affected by Natural Disasters in 2016 (*Credit* Guha-Sapir et al. 2016)

Number of natural disasters: 342

Hydrological disasters: 177 (51.8%)
Meteorological disasters: 96 (28.1%)
Climatological disasters: 38 (11.1%)
Geophysical disasters: 31 (9.0%)

Number of people killed by natural disasters: 8733

Flood: 4731 (54.2%)
Storm: 1797 (20.6%)
Earthquake: 1315 (15.1%)
Extreme temperatures: 490 (5.6%)
Landslide: 361 (4.1%)

Number of people affected by natural disasters: 569 million

Climatological disasters: 393 million (over 99.9% due to drought)
Meteorological disasters: 95.8 million (98% due to storm)
Hydrological disasters: 78.1 million (99.7% due to flood)
Geophysical disasters: 2.2 million (94% due to earthquake)

While climate change, in general, has not been considered an essential condition for climate-related disasters to occur, a changing climate leads to changes in the frequency, intensity, spatial extent, duration and timing of these events, including rising speed of tropical cyclones, more heat waves and increased intensity of drought (Field et al. 2012). Apart from an increasing amount of evidence indicating that climate change is a significant driver of these events, we are beginning to see research that finds the effect of climate change alone strong enough to cause extreme events to take place beyond natural variability (Herring et al. 2018). In fact, the National Academies of Science in 2016 concluded that it would no longer be an unqualified blanket statement to claim that individual climate events could not be attributable to climate change (National Academies of Sciences Engineer and

Medicine 2016). On the other hand, understanding of the health impact pathways of climate-related disasters and climate change is therefore crucial for developing strategies to reduce the health risks involved (Banwell et al. 2018) and to manage the simultaneous health impacts of climate change and natural disasters (Phalkey and Louis 2016). As such, the concept of health emergency and disaster risk management (Health-EDRM) is highly relevant in the climate-related disaster setting. Health-EDRM refers to "the systematic analysis and management of health risks, posed by hazardous events, including emergencies and disasters, through a combination of hazard, exposure and vulnerability reduction to prevent and mitigate risks, preparedness, response, and recovery" (World Health Organization and Public Health England 2017). Health-EDRM is now a core component of general disaster risk reduction and is applicable to climate-related disasters specifically.

6.5 Impacts of Climate-Related Disasters on Human Health

Various health impacts are found to be associated with different types of climate-related disasters. This section will first elaborate on the specific health outcomes of the three types of climate-related disasters that caused the greatest number of deaths on average, namely flood, storm and extreme temperatures.

6.5.1 Flood

Flood is a form of hydrological disaster. It is the most frequently occurred natural disaster, accounting for over 50% of natural disasters worldwide, and contributes significantly to disaster-related deaths. In East Asia, more than 62 million people suffered from floods in 2017, which is 1.3 times the 2006–2015 annual average, and a major reason is a flood in southern and central China which alone affected 60 million people. Different health impacts of floods often surface at different stages after the event strikes. The most immediate health impacts of flood are drowning (including driving on flooded roads), injuries from debris, electrical injuries, burns and explosions following disruption to fuel storage tanks and oil leakage and hypothermia (Du et al. 2010). At the early stage of the post-event period, exposure to floodwaters could lead to higher risk of infectious diseases, in particular skin infections, respiratory infections, gastrointestinal diseases, zoonotic diseases (e.g. leptospirosis), vector-borne diseases (e.g. dengue fever) (Paterson et al. 2018). Given the significant health impact of flood, health outcomes of floods are the most well-researched among extreme weather events in China, focusing mostly on infectious diseases, in particular infectious diarrhea. For Southeast Asia, dengue is spreading geographically, and was reported for the first time in mountainous countries of Nepal and Bhutan (Kumaresan et al. 2011). At the later stage, issues related to non-communicable diseases, including mental health events, become increasingly apparent. The main issue related to non-communicable diseases is the disrupted access to required medication, in particular medication non-compliance. Research has found deteriorated glycemic control among insulin users after flooding in Britain (Ng et al. 2011), increased mortality rates in patients with cardiovascular disease and diabetes (Ryan et al. 2015), increase in acute events after natural disasters among patients with chronic respiratory diseases (Robinson et al. 2011). Climate-related disasters, including floods, easily put extra stress on those who are suffering from mental illness and bring about new cases of anxiety, disaster-related psychiatric trauma, depression, aggression, which have both short-term and long-term impacts (Smith et al. 2014). Review found that flooding exacerbates or provokes mental health problems among people of all ages, and most existing literature on the mental health effect of flood focuses on post-traumatic stress disorder (Stanke et al. 2012). Relocation and household income have been identified as predictive factors for reported depression and anxiety among flood victims in England (Lamond et al. 2015).

A climate change phenomenon closely related to flooding is sea-level rise. IPCC projected that under the high greenhouse gas concentration scenario, global mean sea level toward the end of the twenty-first century (2081–2100) is likely to rise by 0.45–0.82 m relative to the average of 1986–2005 (Stocker et al. 2013). As the rise of sea level does not come drastically, its impact on human health is often neglected. Sea-level rise is associated with not only to flooding, but affects human health in a number of ways, in particular in the coastal areas. In addition to flooding, rising sea level could also disrupt freshwater and related food supply causing malnutrition, affect the spread of infectious diseases associated with mosquitos and water (e.g. malaria, cholera).

6.5.2 Storm or Strong Wind Event

Storm is a form of meteorological disaster. Storms include extra-tropical storm, tropical storm and convective storm (e.g. hail, sand/dust storm, winter storm/blizzard) (Centre for Research on Epidemiology of Disasters [CRED]). Southeast Asia, the Western Pacific and America are the regions most hard hit by storms/cyclones/typhoons in the past decades. Research on health impacts of storms focuses on tropical cyclones and hurricanes, in particular, those that had hit the Americas. The risk of tropical cyclones is progressively increasing with warmer oceans and rising sea levels, resulting in storms with higher intensity, more precipitation and higher frequency of intense cyclone; coupled with increasing vulnerabilities due to population growth and urbanization (Shultz et al. 2018). It should be noted that modelling and attribution of health impacts of cyclones have been considered a major challenge as their impacts were modified strongly by local circumstances.

Immediate physical health consequences of cyclones generally include drowning (including deaths in submerged vehicles and structures), physical injuries (including fractures, abrasions) (Shultz and Galea 2017). Increased rainfall from more intense storm also affects patterns of vector-borne diseases, which are transmitted by the bite of blood-sucking arthropods such as mosquitoes, ticks, mites, sandflies and black-flies. According to WHO, vector-borne diseases account for over 17% of all infectious diseases globally and there are over 700,000 deaths arising from diseases such as malaria, dengue, schistosomiasis, human African trypanosomiasis, leishmaniasis, Chagas disease, yellow fever, Japanese encephalitis and onchocerciasis (World Health Organization 2017a). The exact influence of climate change varies with vector species; but in general, higher temperatures and heavier rainfall create favourable conditions for the vector to survive and transmit diseases. In the case of mosquito, for example, increase in temperature induces faster larval development, faster blood digestion, shorter intervals between generations, increased egg production and more generations per year (Vardoulakis and Heaviside 2012), hence more intense and prolonged biting season. More frequent and severe cyclones bring about more extreme rainfall, high humidity and water accumulation, which increase breeding sites for certain vector species. Based on the latest projection on climate change, IPCC concludes that the locations around the world that are climatically suitable for dengue fever to occur will increase, although it is difficult to project the exact impact on disease burden (Smith et al. 2014). Other issues related to infectious diseases and non-communicable diseases are generally similar to those occurring in the context of floods. As for mental health consequences, increases in both suicidal thoughts (from 2.8 to 6.4%) and actual suicidal plans (from 1.0 to 2.5%) were observed in residents 18 months after Hurricane Katrina in the US (Crimmins et al. 2016). According to a literature review on the human impact of tropical cyclones in the past 30 years, the primary cause of cyclone-related mortality is drowning, and the overall mortality and injury burden was concentrated in the less developed countries in Southeast Asia and the Western Pacific (Doocy et al. 2013).

6.5.3 Extreme Temperatures

Extreme temperature events include heat waves, cold waves and severe winter conditions (e.g. snow/ice, frost) [Centre for Research on Epidemiology of Disasters (CRED)]. Elevated temperatures and more frequent and prolonged hot days can induce health complications, e.g. heatstroke, heat exhaustion, worsened cardiovascular conditions; and cold exposures are associated with respiratory infections (World Health Organization n.d.b). Temperature-related deaths are seldom the direct results of hyperthermia or hypothermia, but due to the effects of temperature on cardiovascular and respiratory diseases (Vardoulakis and Heaviside 2012). A large number of epidemiological studies have been conducted on the association between extreme temperatures and different health outcomes, in particular, mortality (Baccini et al. 2008; Wang et al. 2016; Goggins et al. 2015) and cardiorespiratory diseases (Michelozzi et al. 2009; Wang et al. 2015), and IPCC has concluded that the association between hot days and increases in mortality was very robust (Smith et al. 2014). In addition, a recent multi-country observational study involving 384 locations across the world concluded that most of the temperature-related mortality burden was attributable to the contribution of cold weather (Gasparrini et al. 2015).

6.6 Health-EDRM in Climate-Related Disasters

The health outcomes of climate-related disasters explained in the last section pose a significant burden on the health system. A health system that is not able to absorb and cope with the impact of the disasters means exacerbation of the negative health outcomes. This section will illustrate how Health-EDRM can be applied in climate-related disaster contexts to minimize the negative health impact, through developing (1) health system resilience, (3) community resilience and (3) effective early warning system, (World Health Organization 2017b). All these strategies are espoused in the Sendai Framework. There was, in fact, call for a regional framework to reduce health impact of climate change from the WHO Regional Offices of South-East Asia and the Western Pacific regions as early as in 2007 (WHO Regional Office for South-East Asia 2008).

6.6.1 Health System Resilience

Resilience is defined as: "The ability of a system, community or society exposed to hazards to resist, absorb, accommodate to and recover from the effects of a hazard in a timely and efficient manner, including through the preservation and restoration of its essential basic structures and functions" (United Nations Office for Disaster Risk Reduction [UNISDR] 2009). Following a climate-related disaster, demand for emergency rooms and many forms of services will increase markedly, especially among homeless and elderly. Many climate-related disasters could also lead to different forms of losses of the healthcare infrastructure, e.g. physical damage of hospitals after flood or storms, loss of paper-based medical records, inaccessibility of electronic medical records, hospital functions interrupted due to disruption to water supply (Paterson et al. 2018). There may also be a reduction on the number of healthcare professionals available to deliver the required services. As such, it is critical to build up a health system that is resilient to the impact of climate-related disasters, which involves health risk assessment, surveillance system, disaster preparedness, surge capacity planning, cross-sector partnership between public and private health actors. On potential physical damage, The Comprehensive Safe Hospital Framework and the Safe Hospital Index, issued by the WHO in 2015, provide detailed guidance. Based on the WHO's six building blocks of health system, the WHO has issued the Operational Framework for Building Climate Resilient Health Systems, which provides "guidance on how the health sector and its operational basis in health systems can systematically and effectively address the challenges increasingly presented by climate variability and change" (World Health Organization 2015b).

6.6.2 Community Resilience

The definition of resilience has made it clear that it has much wider application than to a specific system—community disaster resilience has been gaining increasing attention in recent years, and is now the cornerstone of disaster risk reduction in many countries (Pfefferbaum et al. 2013). Community resilience covers different sectors in society and involves much cross-sector collaboration, e.g. local emergency management plans, evacuative procedures, community shelters (Shultz et al. 2018). Community resilience is particularly important for improving health outcomes of climate-related disasters as local risk knowledge is utilized, community will be prepared and active in reducing the risk and impact, the affected community can be the first responders before external assistance is available (World Health Organization and Public Health England 2017). A systematic review on the assessment methods and tools of community disaster resilience, including climate-related disasters, identified five domains in community disaster resilience: (1) Social domain (including human capital, lifestyle, risk knowledge), (2) Economic domain, (3) Institutional domain (including governance, emergency response and recovery), (4) Physical domain (including infrastructure, land use) and (5) Natural domain (including ecosystem). The review noted significant disparity in the interpretation of community disaster resilience in existing literature and concluded that there was a need to develop indicators to facilitate the systematic operationalization of community disaster resilience (Ostadtaghizadeh et al. 2015).

For a unique health perspective in community disaster resilience, the development of primary health care, which focuses on basic services to improve health status, is a major anchor. An effective primary health care system is critical to community resilience and is the foundation for emergency response, as it reduces vulnerabilities and improves preparedness. A strong primary health care system also means the capacity to attend to the needs of chronic disease patients which is often not the focus in the acute phase of emergency response but contributing an increasingly large burden (World Health Organization and Public Health England 2017). A valuable but often forgotten lesson is that community resilience must involve empowerment, including participation of the vulnerable population. For example, older people could be active members in community disaster planning, as in the case in Bolivia, where there is a regular association of older people called the "White Brigades" which assist in registering older people, involve in emergency planning, participate in drills, and identify the older people's needs during emergency. In fact, involving the vulnerable population in disaster planning and programming could be an effective way to avoid misconceptions and unfounded assumptions about their needs and capacities (United Nations Population Fund and HelpAge International 2012).

6.6.3 *Effective Early Warning System*

Effective early warning system is a key public health measure against climate-related disasters. The system should target especially at the most at-risk group with effective health content through the appropriate mode of communication. The WHO and WMO had in 2015 jointly issued the heat-health warning system guidelines for reference by national authorities (World Meteorological Organization and World Health Organization 2015). In developing the early warning system, the unique nature of individual health outcomes brought about by climate-related disaster, in terms of the appropriate time-scale involved, must be taken into account: from hours or days for flood or heat wave warnings, to weeks for seasonal epidemics of vector-borne disease, to months for seasonal forecasts of precipitation anomalies, to years for drought and associated food insecurity (World Health Organization 2017b). The Operational Framework for Building Climate Resilient Health Systems also provides guidance on developing early warning system. With financial support from a few developed countries, the World Bank, WMO and UN Office for Disaster Risk Reduction (UNISDR) in 2015 jointed launched The Climate Risk and Early Warning Systems (CREWS) initiative to build up the capacity of Least Developed Countries and Small Island Developing States to develop "effective, impact-based, multihazard, gender-informed early warnings and risk information". CREWS has been supporting projects in 19 countries in Africa and the Pacific against hazards including tropical cyclones and floods (World Meteorological Organization 2018).

6.7 Health and Environmental Co-benefits of Climate Change Mitigation

We have in the previous sections illustrated the strong impact of climate change on human health, the relationship between climate change and climate-related disasters, and the application of Health-EDRM in climate-related disasters setting. This chapter will end with a discussion on a unique concept that links climate change mitigation and health: health and environmental co-benefits (HEC). Since climate change mitigation policies are concerned with the reduction of GHG, the effectiveness of such policies is traditionally measured according to the amount of GHG reduced (Hildén et al. 2014). However, such an approach fails to take into account the wider impact mitigation policies have on other areas, e.g. health, environment, economy, etc. Climate change and human health are so closely linked that many mitigation measures naturally promote public health, and these so-called health co-benefits include "health gains from strategies that are directed primarily at climate change, and mitigation of climate change from well-chosen policies for health advancement" (Smith et al. 2014). The subject of health co-benefits of climate mitigation now commands much attention at the global level (World Health Organization 2015a). According to IPCC, co-benefits associated with climate change mitigation measures can be categorized in general as follows (Smith et al. 2014):

- Reduced pollution: decreased emissions of health-damaging pollutants;
- Dietary change: decreased meat consumption;
- More physical activities: replace transportation with walking; and
- Better land-use planning: increased urban green space.

A landmark series on the public health benefits of climate change mitigation was published in The Lancet in 2009, and presented evidence of such benefits in a range of areas: lower carbon and more active transport reduce GHG emission and improve ischaemic heart disease, cerebrovascular disease, dementia, certain types of cancers, depression, diabetes; lower consumption of animal products reduces agricultural emission from the livestock sector and reduces burden of ischaemic heart disease with lower saturated fat intake; lower carbon fuel in electricity generation improves cardiopulmonary mortality, lung cancer and occupational mortality (Haines et al. 2009). It was emphasized that the public health benefits of mitigation should be more prominent in both international negotiations and domestic policymaking (Ganten et al. 2010). An increasing amount of research has been devoted to examine the wider benefits of mitigation action, e.g. air quality co-benefits (Nemet et al. 2010), expected reduction in human mortality and reduction of acidification and eutrophication over ecosystems (Rafaj et al. 2013). A study based on global atmospheric model and future GHG emission scenarios found the global average marginal co-benefits of avoided mortality to be between US$50–380 per tonne of carbon dioxide, which exceed marginal abatement costs in 2030 and 2050, and are within the low range of costs in 2100 (West et al. 2013). A systematic review examined the wider impacts of climate mitigation actions in the UK on HEC, including improvements in air pollution, noise, fuel extraction and the lifestyle benefits from policies over power generation and energy use, concluded that HEC of climate action significantly outweigh the negative impacts, with a net present value of more than £85 billion from 2008 to 2030, taking into account the monetized benefits alone (Hildén et al. 2014). A 2016 systematic review found that sustainable dietary patterns could lead to 70–80% reduction of GHG emissions (Aleksandrowicz et al. 2016).

In terms of HEC at individual level, a study series in Hong Kong, a densely populated Asian metropolis, found that among the common co-benefit behaviours, using less packaging and disposable shopping bags were practiced by over 70% of the respondents. Bringing personal eating utensils when eating out, showering less than five minutes, having one vegetarian meal a week, and buying more organic food were practiced by less than half of the respondents (Chan et al. 2017). The second study in the series was conducted in 2018 which covers an extended number of HEC behaviours. Examples of HEC behaviours which were considered as "habits" by majority of the respondents are carrying their own water bottle or cup, switching off standby mode and separating household waste (Collaborating Centre for Oxford University and CUHK for Disaster and Medical Humanitarian Response [CCOUC] 2018). Overall speaking, the practice of HEC behaviours of the Hong Kong population is steady over the two years period, with improvements shown in certain specific behaviours, in particular the purchase/consumption of organic food. Gender and age were consistently associated with the frequency of practicing HEC behaviour in both studies, with female and older age groups being more engaged.

Apart from a technical understanding of the overall impact of climate change mitigation policy, the HEC discourse has major practical implications: it lends additional justifications to climate change mitigation policies and possibly catalyzes the policy process (Haines 2017). The real cost of mitigation policies will be much lower when the full range of their benefits are considered, over economy, health and environment. National governments should adopt a wider approach in conducting impact assessment of their mitigation measures.

6.8 Conclusion

Climate change is no longer a subject solely about the environment, but a cross-cutting issue with significant impact on disaster management and wider sustainable development, which explains its prominence in international agreements covering a wide of subjects: The Sendai Framework, SDGs, the New Urban Agenda. This chapter illustrated that health was an essential component of climate-related disaster management and Health-EDRM provided a comprehensive framework to improve health outcomes. With the far-reaching impact of climate change over different aspects of our life and development, climate change adaptation now goes hand in hand with mitigation, which was traditionally the major focus of climate change policy. Flood, storm, drought and extreme temperatures are the most significant climate-related disasters that affect the most number of people and cause preventable deaths. Health-EDRM is applicable in the climate-disaster context and calls for a range of actions to improve the health outcomes, in particular the building of health system and community resilience and development of effective early warning systems, which are all advocated in the Sendai Framework. An understanding of the role of health in climate change and climate-related disasters naturally brings us to the concept of HEC, which provides additional justifications for climate change mitigation policies, and has the potential to galvanize much stronger public support for climate actions across the globe.

References

Aleksandrowicz L, Green R, Joy EJM, Smith P, Haines A (2016) The impacts of dietary change on greenhouse gas emissions, land use, water use, and health: a systematic review. PLoS ONE 11: e0165797. https://doi.org/10.1371/journal.pone.0165797

Article 1 United Nations Framework Convention on Climate Change (1992) United nations framework convention on climate change

Baccini M, Biggeri A, Accetta G, Kosatsky T, Katsouyanni K, Analitis A, Anderson HR, Bisanti L, D'Ippoliti D, Danova J, Forsberg B, Medina S, Paldy A, Rabczenko D, Schindler C, Michelozzi P (2008) Heat effects on mortality in 15 European cities. Epidemiology 19:711–719. https://doi.org/10.1097/EDE.0b013e318176bfcd

Banwell N, Rutherford S, Mackey B, Street R, Chu C (2018) Commonalities between disaster and climate change risks for health: a theoretical framework. Int J Environ Res Public Health 15:538. https://doi.org/10.3390/ijerph15030538

Centre for Research on Epidemiology of Disasters (CRED) (2018) The International Disaster Database (EM-DAT). https://www.emdat.be/classification. Accessed 12 Jun 2018

Chan EYY, Wang SS, Ho JY-E, Huang Z, Liu S, Guo C (2017) Socio-demographic predictors of health and environmental co-benefit behaviours for climate change mitigation in urban China. PLoS ONE 12:e0188661. https://doi.org/10.1371/journal.pone.0188661

Collaborating Centre for Oxford University and CUHK for Disaster and Medical Humanitarian Response (CCOUC) (2018) Final report for a study on health and environmental co-benefit behaviours for climate change mitigation in Hong Kong. Hong Kong

Crimmins A, Balbus J, Gamble J, Beard C, Bell J, Dodgen D, Eisen R, Fann N, Hawkins M, Herring S, Jantarasami L, Mills D, Saha S, Sarofim M, Trtanj J, Ziska L (2016) USGCRP, 2016: the impacts of climate change on human health in the United States: a scientific assessment. Washington, D.C.

Doocy S, Dick A, Daniels A, Kirsch TD (2013) The human impact of tropical cyclones: a historical review of events 1980–2009 and systematic literature review. PLoS Curr. https://doi.org/10.1371/currents.dis.2664354a5571512063ed29d25ffbce74

Du W, FitzGerald GJ, Clark M, Hou X-Y (2010) Health impacts of floods. Prehosp Disaster Med 25:265–72

EM-DAT (n.d.). The emergency events database. Brussels

Field C, Barros V, Stocker T, Qin D, Dokken D, Ebi K, Mastrandrea M, Mach K, Plattner G-K, Allen S, Tignor M, Midgley P (2012) IPCC, 2012: managing the risks of extreme events and disasters to advance climate change adaptation. A special report of working groups I and II of the intergovernmental panel on climate change. Cambridge

Ganten D, Haines A, Souhami R (2010) Health co-benefits of policies to tackle climate change. Lancet (London, England) 376:1802–1804. https://doi.org/10.1016/S0140-6736(10)62139-3

Gasparrini A, Guo Y, Hashizume M, Lavigne E, Zanobetti A, Schwartz J, Tobias A, Tong S, Rocklöv J, Forsberg B, Leone M, De Sario M, Bell ML, Guo Y-LL, Wu C, Kan H, Yi S-M, de Sousa Zanotti Stagliorio Coelho M, Saldiva PHN, Honda Y, Kim H, Armstrong B (2015) Mortality risk attributable to high and low ambient temperature: a multicountry observational study. Lancet (London, England) 386:369–375. https://doi.org/10.1016/S0140-6736(14)62114-0

Goggins WB, Yang C, Hokama T, Law LSK, Chan EYY (2015) Using annual data to estimate the public health impact of extreme temperatures. Am J Epidemiol 182:80–87. https://doi.org/10.1093/aje/kwv013

Guha-Sapir D, Hoyois P, Below R (2016) Annual disaster statistical review 2016: The numbers and trends. Brussels

Haines A (2017) Health co-benefits of climate action. Lancet Planet Heal 1:e4–e5. https://doi.org/10.1016/S2542-5196(17)30003-7

Haines A, McMichael AJ, Smith KR, Roberts I, Woodcock J, Markandya A, Armstrong BG, Campbell-Lendrum D, Dangour AD, Davies M, Bruce N, Tonne C, Barrett M, Wilkinson P

(2009) Public health benefits of strategies to reduce greenhouse-gas emissions: overview and implications for policy makers. Lancet (London, England) 374:2104–2114. https://doi.org/10.1016/S0140-6736(09)61759-1

Herring S, Christidis N, Hoell A, Kossin J, Schreck J III, Stott P (2018) 2018: explaining extreme events of 2016 from a climate perspective. Bull Amer Meteor Soc 99:S1–S157

Hildén M, Jordan A, Rayner T (2014) Climate policy innovation: developing an evaluation perspective. Env Polit 23:884–905. https://doi.org/10.1080/09644016.2014.924205

Kelman I (2015) Climate change and the sendai framework for disaster risk reduction. Int J Disaster Risk Sci 6:117–127. https://doi.org/10.1007/s13753-015-0046-5

Kumaresan J, Narain JP, Sathiakumar N (2011) Climate change and health in South East Asia. Int J Clim Chang Strateg Manag 3:200–208. https://doi.org/10.1108/17568691111129020

Lamond JE, Joseph RD, Proverbs DG (2015) An exploration of factors affecting the long term psychological impact and deterioration of mental health in flooded households. Environ Res 140:325–334. https://doi.org/10.1016/j.envres.2015.04.008

Michelozzi P, Accetta G, De Sario M, D'Ippoliti D, Marino C, Baccini M, Biggeri A, Anderson HR, Katsouyanni K, Ballester F, Bisanti L, Cadum E, Forsberg B, Forastiere F, Goodman PG, Hojs A, Kirchmayer U, Medina S, Paldy A, Schindler C, Sunyer J, Perucci CA, PHEWE Collaborative Group (2009) High temperature and hospitalizations for cardiovascular and respiratory causes in 12 European cities. Am J Respir Crit Care Med 179:383–9. https://doi.org/10.1164/rccm.200802-217oc

National Academies of Sciences Engineer and Medicine (2016) Attribution of extreme weather events in the context of climate change. National Academies Press, Washington, D.C.

Nemet GF, Holloway T, Meier P (2010) Implications of incorporating air-quality co-benefits into climate change policymaking. Environ Res Lett 5:014007. https://doi.org/10.1088/1748-9326/5/1/014007

Ng J, Atkin SL, Rigby AS, Walton C, Kilpatrick ES (2011) The effect of extensive flooding in Hull on the glycaemic control of patients with diabetes. Diabet Med 28:519–524. https://doi.org/10.1111/j.1464-5491.2011.03228.x

Ostadtaghizadeh A, Ardalan A, Paton D, Jabbari H, Khankeh HR (2015) Community disaster resilience: a systematic review on assessment models and tools. PLoS Curr 7. https://doi.org/10.1371/currents.dis.f224ef8efbdfcf1d508dd0de4d8210ed

Pachauri R, Meyer L (2014) IPCC, 2014: climate change 2014: synthesis report. Contribution of working groups I, II and III to the fifth assessment report of the intergovernmental panel on climate change. Geneva

Paterson DL, Wright H, Harris PNA (2018) Health risks of flood disasters. Clin Infect Dis. https://doi.org/10.1093/cid/ciy227

Pfefferbaum RL, Pfefferbaum B, Van Horn RL, Klomp RW, Norris FH, Reissman DB (2013) The Communities Advancing Resilience Toolkit (CART): an intervention to build community resilience to disasters. J Public Health Manag Pract 19:250–8. https://doi.org/10.1097/phh.0b013e318268aed8

Phalkey RK, Louis VR (2016) Two hot to handle: How do we manage the simultaneous impacts of climate change and natural disasters on human health? Eur Phys J Spec Top 225:443–457. https://doi.org/10.1140/epjst/e2016-60071-y

Rafaj P, Schöpp W, Russ P, Heyes C, Amann M (2013) Co-benefits of post-2012 global climate mitigation policies. Mitig Adapt Strateg Glob Chang 18:801–824. https://doi.org/10.1007/s11027-012-9390-6

Robinson B, Alatas MF, Robertson A, Steer H (2011) Natural disasters and the lung. Respirology 16:386–395. https://doi.org/10.1111/j.1440-1843.2011.01923.x

Ryan B, Franklin RC, Burkle FM, Aitken P, Smith E, Watt K, Leggat P (2015) Identifying and describing the impact of cyclone, storm and flood related disasters on treatment management, care and exacerbations of non-communicable diseases and the implications for public health. PLoS Curr. https://doi.org/10.1371/currents.dis.62e9286d152de04799644dcca47d9288

Shultz JM, Galea S (2017) Mitigating the mental and physical health consequences of Hurricane Harvey. JAMA 318:1437. https://doi.org/10.1001/jama.2017.14618

Shultz JM, Shepherd JM, Kelman I, Rechkemmer A, Galea S (2018) Mitigating tropical cyclone risks and health consequences: urgencies and innovations. Lancet Planet Heal 2:e103–e104. https://doi.org/10.1016/S2542-5196(18)30021-4

Smith K, Woodward A, Campbell-Lendrum D, Chadee D, Honda Y, Liu Q, Olwach J, Revich B, Sauerborn R (2014) 2014: human health: impacts, adaptation, and co-benefits. In: Climate change 2014: impacts, adaptation, and vulnerability. Part A: global and sectoral aspects. Contribution of working group II to the fifth assessment report of the IPCC. Cambridge

Stanke C, Murray V, Amlôt R, Nurse J, Williams R (2012) The effects of flooding on mental health: Outcomes and recommendations from a review of the literature. PLoS Curr. https://doi.org/10.1371/4f9f1fa9c3cae

Stocker TF, Qin D, Plattner G-K, Tignor M, Allen SK, Boschung J, Nauels A, Xia Y, Bex V, PMM (eds) (2013) Summary for policymakers. New York

Thomas V, Ramón E (2015) Global increase in climate-related disasters. Manila

UN Climate Change (n.d.) The Paris Agreement. https://unfccc.int/process-and-meetings/the-paris-agreement/the-paris-agreement. Accessed 9 Jun 2018

United Nations Office for Disaster Risk Reduction (UNISDR) (2009) 2009 UNISDR terminology on disaster risk reduction. Geneva

United Nations Population Fund and HelpAge International (2012) Ageing in the twenty-first century: a celebration and a challenge

United States Environmental Protection Agency (n.d.) Greenhouse Gas Emissions. https://www.epa.gov/ghgemissions/overview-greenhouse-gases. Accessed 8 Jun 2018

Vardoulakis S, Heaviside C (2012) Health effects of climate change in the UK 2012: current evidence, recommendations and research gaps. London

Wang X, Li G, Liu L, Westerdahl D, Jin X, Pan X (2015) Effects of extreme temperatures on cause-specific cardiovascular mortality in China. Int J Environ Res Public Health 12:16136–16156. https://doi.org/10.3390/ijerph121215042

Wang Y, Shi L, Zanobetti A, Schwartz JD (2016) Estimating and projecting the effect of cold waves on mortality in 209 US cities. Environ Int 94:141–149. https://doi.org/10.1016/j.envint.2016.05.008

Watts N, Adger WN, Agnolucci P, Blackstock J, Byass P, Cai W, Chaytor S, Colbourn T, Collins M, Cooper A, Cox PM, Depledge J, Drummond P, Ekins P, Galaz V, Grace D, Graham H, Grubb M, Haines A, Hamilton I, Hunter A, Jiang X, Li M, Kelman I, Liang L, Lott M, Lowe R, Luo Y, Mace G, Maslin M, Nilsson M, Oreszczyn T, Pye S, Quinn T, Svensdotter M, Venevsky S, Warner K, Xu B, Yang J, Yin Y, Yu C, Zhang Q, Gong P, Montgomery H, Costello A (2015) Health and climate change: policy responses to protect public health. Lancet (London, England) 386:1861–1914. https://doi.org/10.1016/S0140-6736(15)60854-6

West JJ, Smith SJ, Silva RA, Naik V, Zhang Y, Adelman Z, Fry MM, Anenberg S, Horowitz LW, Lamarque J-F (2013) Co-benefits of global greenhouse gas mitigation for future air quality and human health. Nat Clim Chang 3:885–889. https://doi.org/10.1038/NCLIMATE2009

WHO Regional Office for South-East Asia (2008) Climate change and human health in Asia and the Pacific: from evidence to action

World Health Organization (2015a) WHO workplan on climate change and health

World Health Organization (2015b) Operational framework for building climate resilient health systems

World Health Organization (2016a) WHO message to health ministers. http://www.who.int/globalchange/publications/didyouknow-health-ministers.pdf?ua=1. Accessed 9 Jun 2018

World Health Organization (2016b) Global conference sets health action agenda for the implementation of the Paris Agreement. http://www.who.int/globalchange/conferences/second-global/conclusions/en/. Accessed 9 Jun 2018

World Health Organization (2017a) Vector-borne diseases. http://www.who.int/mediacentre/factsheets/fs387/en/. Accessed 12 Jun 2018

World Health Organization (2017b) Health emergency and disaster risk management: climate risk management

World Health Organization (n.d.a) WMO and WHO Establish Joint Office for Climate and Health. http://www.who.int/globalchange/mediacentre/news/joint-office/en/. Accessed 11 Jun 2018a

World Health Organization (n.d.b) Exposure to Extreme Heat and Cold. http://www.who.int/sustainable-development/housing/health-risks/extreme-heat-cold/en/. Accessed 13 Jun 2018b

World Health Organization and Public Health England (2017) Health emergency and disaster risk management: overview. http://www.who.int/hac/techguidance/preparedness/who-factsheet-overview-december2017.pdf?ua=1. Accessed 1 Jun 2018

World Meteorological Organization (2018) CREWS report series annual report 2017. Geneva

World Meteorological Organization and World Health Organization (2015) Heatwaves and health: guidance on warning-system development

Chapter 7
Health-EDRM in International Policy Agenda III: 2030 Sustainable Development Goals and New Urban Agenda (Habitat III)

Emily Ying Yang Chan, Janice Y. Ho, Chi Shing Wong and Rajib Shaw

Abstract In September 2015, world leaders adopted the 2030 Agenda for Sustainable Development at the United Nations Sustainable Development Summit. The ambitious Agenda includes a set of 17 Sustainable Development Goals (SDGs) with 169 targets to end poverty, fight inequality and injustice, and tackle climate change by 2030. The SDGs have been built on the Millennium Development Goals (MDGs) adopted in 2000, but are broader in scope and go further than the MDGs by addressing the root causes of poverty and calling for action by all countries. The new goals cover the three dimensions of sustainable development: economic growth, social inclusion, and environmental protection, where health is a prominent cross-cutting theme. Other than the goals directly related to health, the SDGs recognize that tackling climate change is essential for sustainable development and poverty eradication. Habitat III, the United Nations Conference on Housing and Sustainable Urban Development, held in Quito of Ecuador in October 2016 reinforces the global commitment to sustainable urbanization and the implementation of a "New Urban Agenda". The conference opened discussions on important urban challenges and questions, such as how to plan and manage cities, towns, and villages for sustainable development. The discussion of these questions, in turn, shapes the implementation of new global development and climate change goals. This chapter will explore how SDGs and New Urban Agenda fit into the framework of Health-EDRM, particularly the cross-cutting goals that coincide with a holistic approach to health in disaster risk reduction.

E. Y. Y. Chan (✉) · J. Y. Ho · C. S. Wong
JC School of Public Health and Primary Care,
The Chinese University of Hong Kong, Shatin, Hong Kong
e-mail: emily.chan@cuhk.edu.hk

R. Shaw
Keio University, Fujisawa, Japan

E. Ying Yang Chan and R. Shaw (eds.), *Public Health and Disasters*,
Disaster Risk Reduction, https://doi.org/10.1007/978-981-15-0924-7_7

Keywords Habitat III · Health Emergency and Disaster Risk Management (Health-EDRM) · New urban agenda · Sustainable Development Goals · Vulnerability

7.1 Introduction to Sustainable Development Goals

The Sustainable Development Goals (SDGs) were adopted in September 2015 at the United Nations Sustainable Development Summit to succeed its predecessor the Millennium Development Goals (MDGs). The MDGs were a global effort between 2000 and 2015 to combat poverty, hunger, and ill-health. Expanding beyond the eight original MDGs, the SDGs seek to address and balance the three dimensions of sustainable development: economic growth, social inclusion, and environmental protection, also known as prosperity, people and the planet. The world leaders and heads of state set an ambitious universal agenda of 17 Sustainable Development Goals with 169 targets that will be guiding global and national action between 2015 and 2030 (United Nations General Assembly 2015). See Fig. 7.1 and Box 7.1 for the full list of Sustainable Development Goals.

Fig. 7.1 Sustainable development goals (SDGs). Reprinted from United Nations Copyright © United Nations

Box 7.1 Sustainable Development Goals (SDGs)

Goal 1	End poverty in all its forms everywhere
Goal 2	End hunger, achieve food security and improved nutrition and promote sustainable agriculture
Goal 3	Ensure healthy lives and promote well-being for all at all ages
Goal 4	Ensure inclusive and equitable quality education and promote lifelong learning opportunities for all
Goal 5	Achieve gender equality and empower all women and girls
Goal 6	Ensure availability and sustainable management of water and sanitation for all
Goal 7	Ensure access to affordable, reliable, sustainable and modern energy for all
Goal 8	Promote sustained, inclusive and sustainable economic growth, full and productive employment and decent work for all
Goal 9	Build resilient infrastructure, promote inclusive and sustainable industrialization and foster innovation
Goal 10	Reduce inequality within and among countries
Goal 11	Make cities and human settlements inclusive, safe, resilient and sustainable
Goal 12	Ensure sustainable consumption and production patterns
Goal 13	Take urgent action to combat climate change and its impacts*
Goal 14	Conserve and sustainably use the oceans, seas and marine resources for sustainable development
Goal 15	Protect, restore and promote sustainable use of terrestrial ecosystems, sustainably manage forests, combat desertification, and halt and reverse land degradation and halt biodiversity loss
Goal 16	Promote peaceful and inclusive societies for sustainable development, provide access to justice for all and build effective, accountable and inclusive institutions at all levels
Goal 17	Strengthen the means of implementation and revitalize the global partnership for sustainable development

* Acknowledging that the United Nations Framework Convention on Climate Change is the primary international, intergovernmental forum for negotiating the global response to climate change.

Source United Nations General Assembly (2015, p. 14)

The SDGs are "integrated and indivisible", aiming to enable sustainable and resilient development of all countries (United Nations General Assembly 2015). Recognizing that different countries face their specific obstacles to sustainable development, the 2030 Agenda for Sustainable Development maintains that the primary responsibility for achieving the SDGs in national development lies within each country, while a global partnership will ensure implementation (United Nations General Assembly 2015). Periodic reviews and follow-ups will be held from national to global levels, and indicator data will be collected to track the progress of attaining these SDGs.

7.2 Sustainable Development Goals and Health-EDRM

Aiming to empower the vulnerable in particular, the Sustainable Development Goals are well aligned with prevention-focused Health-EDRM. On the one hand, disasters are inevitable inhibitors to sustainable development, intermittently and increasingly causing damage and setbacks in communities, which lead not only to mortality and infrastructural losses but also losses in economic productivity, human capital, and health. On the other hand, "better development can reduce the need for emergency relief" (Buchanan-Smith and Maxwell 1994). In order for societies to attain sustainable development, there is a need to increase health resilience and reduce vulnerability to health risks posed by disasters, a key to the prevention approach of Health-EDRM. As a key concept in disaster studies and Health-EDRM, **vulnerability** can be defined as "the characteristics and circumstances of a community, system or asset that make it susceptible to the damaging effects of a hazard" (Matyas and Pelling 2012). This preventive effort is particularly needed among vulnerable populations such as the poor, the older, persons with disabilities and ethnic minorities, who are more susceptible to the negative impact of disasters. The root of human vulnerability in disasters and emergencies come from a complex multi-faceted group of "social, economic, health and cultural factors" (World Health Organization and Public Health England 2017a). As the 17 Sustainable Development Goals aim to address these multi-faceted factors, by meeting these goals, the vulnerability of at-risk populations in disasters will also be reduced. The following section will explore several Sustainable Development Goals as related to Health-EDRM and vulnerability reduction in disasters.

7.2.1 SDG 1: End Poverty in All Its Forms Everywhere

Poverty has long been recognized as a key demographical vulnerability in disasters. Globally, the poorest 20% are 14 times more likely to face childhood mortality than the richest 20% (Steele et al. 2007). The poor might be living under more vulnerable conditions, potentially with greater pre-existing health risks due to the limited access to needed medical services (Nicogossian et al. 2012). There is an uneven distribution of risk, as poverty reduces people's choice (DFID 2006; World Health Organization and Public Health England 2017a). Those with lower socioeconomic status may be forced to live in more hazard-prone areas, have poorer housing infrastructure, have less access to resources and comprehensive warning systems, and focus more on basic survival priorities rather than long-term planning or disaster preparation (DFID 2006; UNISDR 2008). Poverty is also related to greater risk of mental health issues, as the poor may have to cope with chronic life stressors, which can become further exacerbated when a disaster strikes (Goldmann and Galea 2014; Elliott 2016).

Poorer people may also experience poverty traps as a result of disasters, whereby a downward spiral of poverty is perpetuating (Rentschler 2013). Facing relatively high vulnerability in disasters, the poor are disproportionately affected by the damaging effect when a disaster strikes. Although the economic losses of the poor in absolute terms are usually less than those of the rich, they may lose a greater proportion of assets or even potential income sources in every disaster (Rentschler 2013). For example, farmers dependent on a crop produce for the coming year's income can have their income source reduced in a singular disaster event such as a flood (CARE 2011). These have implications for their ability to access resources or health services in order to maintain health post-disaster.

The disaster impact would, in turn, reduce their already-limited assets, further deepening their poverty and subsequently increasing their vulnerability in the next disaster.

Target 1.5 of SDG 1 specifically pledges to "build the resilience of the poor and those in vulnerable situations and reduce their exposure and vulnerability to climate-related extreme events and other economic, social and environmental shocks and disasters [by 2030]". By reducing poverty and building the disaster resilience of the poor, this would prevent adverse health outcomes when disasters strike.

7.2.2 SDG 2: End Hunger, Achieve Food Security and Improved Nutrition and Promote Sustainable Agriculture

Adequate and appropriate nutrition is essential to health. Malnutrition includes both undernutrition and overnutrition, resulting from inadequate or excessive consumption of macronutrients or micronutrients, which will become health risks to individuals. In 2013, undernutrition was the top cause of under-five mortality globally, accounting for 45% of those deaths, and only 6 out of 20 countries in the Asia-Pacific region were on track to meet the 2025 global goal of reduced stunting (Food and Agriculture Organization of the United Nations 2016). Those who are malnourished are more likely to be vulnerable in disasters. Disasters additionally affect the four pillars of food security, namely food availability, access, utilization, and stability, as defined by the 1996 World Food Summit (EC–FAO Food Security Programme 2008; Garschagen et al. 2015). Food **availability** refers to the supply side of producing adequate food (EC–FAO Food Security Programme 2008). In disasters, such as floods or droughts, crop failure or destruction of food stocks would lead to reduced availability of food. **Access** refers to the ability to acquire the available food (Garschagen et al. 2015). In disaster settings, collapsed infrastructure may inhibit the access to food, as blocked roads or flooded communities would prevent food from reaching those in need. There may also be unintended unequal distribution of food during disaster response to the disadvantage of the most vulnerable, thereby limited their food access. There is also an economic aspect to food access since disasters may inflate food prices or reduce people's ability to pay for food as they have suffered asset losses and have to use their economic resources for other urgent relief needs. **Utilization** refers to the ability to use the food provided, at both the food preparation level and at the individual body level, whereby the body is able to absorb the nutrients from the food (Garschagen et al. 2015). Disasters may reduce the amount of clean water required for food preparation and prevent disaster victims from fully utilizing the food provided to them. Finally, **stability** refers to the maintenance of the aforementioned three dimensions of food security over time (Garschagen et al. 2015). With repeated disaster crises or even the once-in-a-while major disaster, the stability required in food security would be interrupted.

Target 2.4 of SDG 2 aims to "ensure sustainable food production systems and implement resilient agricultural practices that increase productivity and production, that help maintain ecosystems, that strengthen capacity for adaptation to climate change, extreme weather, drought, flooding, and other disasters and that progressively improve land and soil quality [by 2030]".

7.2.3 SDG 3: Ensure Healthy Lives and Promote Well-Being for All at All Ages

Beyond the traditional response to emergencies like search and rescue, health plays a broader role in disaster risk management. The SDGs, particularly SDG 3, are founded on the principle of equity and "leaving no one behind" (World Health Organization 2016). Efforts should be made to minimize the discrepancy between those most vulnerable and least vulnerable to a disaster. An overall healthy population would enable all people of all ages to be more resilient in a disaster situation. At the same time, the provision of robust health services can help improve the health status of communities and reduce people's vulnerabilities. Moreover, a vaccinated population can minimize the potential risk of communicable disease outbreaks after a disaster. A healthy population can also prevent complications of maintaining disease management in a post-disaster scenario, such as management of tuberculosis and diabetes. Furthermore, a reliable disease surveillance system can detect imminent health risks quickly, preventing the potential biological disaster.

Most of the targets in SDG 3 are relevant to Health-EDRM, the full list of which can be found in Box 7.2. Particularly, Target 3.d of "[s]trengthen[ing] the capacity of all countries, in particular developing countries, for early warning, risk reduction and management of national and global health risks" is applicable to all hazards that pose health risks to the population. Additionally, Target 3.8 of maintaining access to quality essential healthcare services and access to essential medicines and vaccinations is even more crucial in a post-disaster setting.

Box 7.2 Targets in SDG 3: Ensure Healthy Lives and Promote Well-Being for All at All Ages

3.1 By 2030, reduce the global maternal mortality ratio to less than 70 per 100,000 live births

3.2 By 2030, end preventable deaths of newborns and children under 5 years of age, with all countries aiming to reduce neonatal mortality to at least as low as 12 per 1000 live births and under-5 mortality to at least as low as 25 per 1000 live births

3.3 By 2030, end the epidemics of AIDS, tuberculosis, malaria and neglected tropical diseases and combat hepatitis, water-borne diseases and other communicable diseases

3.4 By 2030, reduce by one-third premature mortality from non-communicable diseases through prevention and treatment and promote mental health and well-being

3.5 Strengthen the prevention and treatment of substance abuse, including narcotic drug abuse and harmful use of alcohol

3.6 By 2020, halve the number of global deaths and injuries from road traffic accidents

3.7 By 2030, ensure universal access to sexual and reproductive healthcare services, including for family planning, information and education, and the integration of reproductive health into national strategies and programmes

3.8 Achieve universal health coverage, including financial risk protection, access to quality essential healthcare services and access to safe, effective, quality and affordable essential medicines and vaccines for all

3.9 By 2030, substantially reduce the number of deaths and illnesses from hazardous chemicals and air, water, and soil pollution and contamination

3.a Strengthen the implementation of the World Health Organization Framework Convention on Tobacco Control in all countries, as appropriate

3.b Support the research and development of vaccines and medicines for the communicable and non-communicable diseases that primarily affect developing countries, provide access to affordable essential medicines and vaccines, in accordance with the Doha Declaration on the TRIPS Agreement and Public Health, which affirms the right of developing countries to use to the full the provisions in the Agreement on Trade-Related Aspects of Intellectual Property Rights regarding flexibilities to protect public health, and, in particular, provide access to medicines for all

3.c Substantially increase health financing and the recruitment, development, training, and retention of the health workforce in developing countries, especially in the least developed countries and small island developing States

3.d Strengthen the capacity of all countries, in particular developing countries, for early warning, risk reduction and management of national and global health risks

Source United Nations General Assembly (2015, pp. 16–17)

7.2.4 SDG 4: Ensure Inclusive and Equitable Quality Education and Promote Lifelong Learning Opportunities for All

Education that addresses health and disaster may be beneficial to increasing disaster health risk literacy, which may enhance survival and improve life-saving efforts in a disaster setting. Effective Health-EDRM education can promote disaster preparedness among individuals and in the community. This can be delivered through the avenues of formal education, non-formal education, and informal education (Shaw et al. 2011). Beyond increased knowledge and awareness of disaster risk, people need to know what actions to be taken and how to respond – this disaster preparedness can mitigate health risks in times of disasters.

Additionally, the first couple of hours after an emergency are known to be crucial to the survival of affected people. However, this time period is often before external professional help arrives. If the local community can be equipped to adequately respond, including aspects such as first aid training and coordinated emergency responses, these can help save the lives of many people. By incorporating disaster health education into the local curriculum and providing training for the community, these can help reduce adverse disaster outcomes.

Disasters may also directly or indirectly affect education. For example, the destruction of a school building in a disaster, like the 2005 northern Pakistan earthquake which destroyed more than 10,000 school buildings and killed over 17,000 students (Asian Disaster Preparedness Center 2008), would directly interrupt the educational opportunities for the children of the affected community (Rentschler 2013). Depending on the ability of the community and outside help to rebuild, this interruption might have a longer-term impact on the quality of the education. Moreover, a disaster can have an indirect impact on education at either the micro or the macro levels. For example, families or the government may have a reduced ability to pay for education as resources are directed at responding to the disaster situation.

Although disaster education is not explicitly mentioned in the SDGs, Target 4.a aims to "[b]uild and upgrade educational facilities that are child, disability and gender sensitive and provide safe, non-violent, inclusive and effective environment for all", which is consistent with the emphasis of safe facilities and the preventive imperative of Health-EDRM.

7.2.5 SDG 5: Achieve Gender Equality and Empower All Women and Girls

Gender is defined as "socially-constructed roles, behaviors, activities and attributes that a society considers appropriate for a person based on his or her assigned sex at birth" (Ferris et al. 2013). Gender is often a determinant of a person's duties, power, and ability to access and control resources (IASC Inter-Agency Standing Committee 2017). Existing gender inequalities are usually further exacerbated in disaster settings: Women and men experience different vulnerabilities and needs during and after a disaster, which are not only a result of biological and physiological differences but also socioeconomic and power differences (Ferris et al. 2013). Often times, there is a gender differential in disasters, with greater mortality and greater disaster risk among women (Ferris et al. 2013). For example, women in some cultures may be more likely not to have the chance to learn to swim, which poses a risk in flooding situations. Or, they may try to flee while ensuring the safety of other family members, children or older people, thus limiting their survivability as opposed to fleeing alone (United Nations 2009). Additionally, access to life-saving measures such as knowledge of early warning information may be limited for women in some circumstances (Ferris et al. 2013).

After a disaster has occurred or during a prolonged disaster, women can continue facing discrimination and limited security. These can prevent women from accessing needed aid and can also lead to an increased risk of gender-based violence. In the recent Yemen conflict, there had been a 63% increase in reported gender-based violence incidents since the conflict began, with over 10,000 cases reported in 2016 (Rohwerder 2017). Furthermore, menstrual hygiene management and reproductive health interventions are often overlooked in planning for post-disaster recovery (Bradshaw and Fordham 2013; Krishnan and Twigg 2016). Hence, care must be taken to understand the risks women face and meet their needs equally in disasters.

Gender equality refers to the equal enjoyment of rights, opportunities, resources and quality of life for all human beings, not governed or limited by their gender and the societal systems that maintain it (IASC Inter-Agency Standing Committee 2017). Target 5.1 of "[e]nd[ing] all forms of discrimination against all women and girls everywhere" and Target 5.2 of "[e]liminat[ing] all forms of violence against all women and girls in the public and private spheres, including trafficking and sexual and other types of exploitation" are particularly applicable in both pre- and post-disaster contexts. Additionally, Target 5.6 of "[e]nsur[ing] universal access to sexual and reproductive health and reproductive rights" must be considered in planning for emergency disaster responses.

7.2.6 *SDG 6: Ensure Availability and Sustainable Management of Water and Sanitation for All*

Water and sanitation are key determinants of health and also essential in disaster settings. Box 7.3 introduces the aspects of water-related to public health that should be assessed during any disaster stage. In pre-disaster contexts, lack of access to safe and adequate water can heighten a population's health risk and vulnerability in disasters. **Water-related diseases** are "any significant adverse effects on human health, such as death, disability, illness or disorders, caused directly or indirectly by the condition, or changes in the quantity or quality, of any waters" (United Nations Economic and Social Council 2010). These include **waterborne diseases** caused by ingestion of faecally contaminated water, **water-washed diseases** by inadequate water supply and the resulting poor hygiene, **water-based diseases** from contact with water-dwelling parasitic organisms, and **vector-borne diseases** from water-breeding insects, such as mosquito-transmitted malaria (United Nations Economic and Social Council 2010). An estimated 842,000 people globally die each year from diarrhoea resulting from unsafe water sanitation and hygiene, which is largely preventable (World Health Organization 2018).

Taking preventive actions to design water supply and sanitation systems with Health-EDRM in mind can have a huge influence on the health of populations in the event of a disaster. Clean water provision and sanitation are essential and the lack of them in disaster settings can easily lead to water-related disease outbreaks as well as have cascading impacts on water use for healthcare services, food preparation, and rescue services (World Health Organization 2011a). Without preventive measures, the water systems may instead become a cause of contamination for the outer environment in severe disasters (World Health Organization 2011a). Effective water management should take into account the resiliency of the various water systems and make appropriate modifications such that the systems would not as easily be physically damaged by earthquakes or other local disaster risks (World Health Organization 2011a). Systems should be designed to consider runoff from heavy rainfall and flooding, which might contaminate the drinking water source or overflow the sanitation and drainage systems. Robust water treatment systems should also be designed to account for potential changes in incoming water quality due to disasters.

Particularly, Target 6.1 of "achiev[ing] universal and equitable access to safe and affordable drinking water for all [by 2030]" and Target 6.2 of "achiev[ing] access to adequate and equitable sanitation and hygiene for all and end open defecation, paying special attention to the needs of women and girls and those in vulnerable situations [by 2030]" need to be upheld in both pre- and post-disaster settings and be robust against disaster risks to maintain the health of populations.

Box 7.3 Assessment of Water Supply from a Public Health Framework

The following five criteria should be examined by the water supply at any disaster stage.

Quality: the water supply should be free of hazardous substances that might be harmful to human health. This can be assessed by looking at the location and quality of water at source; through examining the local patterns and habits of water obtainment, storage and utilization; and through surveillance of diarrhoea and waterborne diseases.
Quantity: the water supply should be enough to ensure hygiene practice is not compromised, which is measured by service level or how much each person can have regular access to. The minimum standard of water usage is 7.5 – 15 L daily per person among those affected by disasters, and at least 25 L per person per day under normal conditions.
Reliability: the water supply should be regular and reliable, measured as the percentage of time where drinking water is available.
Accessibility: the water supply should be easily acquired by the population, measured as the time required to access and the percentage of population who can acquire water with ease. Particularly among rural areas with decentralized water systems, each household may have different accessibility and vulnerable subgroups may encounter barriers to access.
Affordability: the water supply should be affordable for all and the cost of water should not hinder people's access to it, measured as the price paid by consumers to obtain their water supply.

Sources World Health Organization (2011b), Chan (2018), Chan and Ho (2019)

7.2.7 SDG 9: Build Resilient Infrastructure, Promote Inclusive and Sustainable Industrialization and Foster Innovation

Critical infrastructure, such as electricity supplies, information, and communications technology (ICT) systems, roads and transport systems, needs to be built to withstand the disaster risks based on the local context. Ensuring the operationality or quick recovery of these systems in a disaster can be critical to maintain the health of those affected. For example, a massive power outage in New York City in 2003 left over 9 million people without power for up to two days and led to a significant increase in diarrhoeal illnesses (Marx et al. 2006). Poorly designed underground subway systems may put many people's lives at risk in the case of floods. Collapsed roads may prevent timely access of outside assistance and life-saving supplies (Independent Evaluation Department 2013). Specifically, a resilient healthcare system that can withstand the hazards and remain operational (World Health Organization and Public Health England 2017a) is essential for reducing the health risk in the case of disasters. The Comprehensive Safe Hospital Framework mentioned earlier in Chap. 5, is an example of a global intervention aiming to attain this SDG. Such a resilient healthcare system would also support tertiary prevention measures - while a disaster has already happened, steps can be taken to reduce survivors' health risk post-disaster. Disruption of health services after a disaster would cause even more damage than the disaster itself. Not only would there be a loss of emergency services for those affected by the disaster, but also the loss of the community's acute and chronic healthcare services (World Health Organization and Public Health England 2017b). Existing patients and life-sustaining services could also be affected in the case of an evacuation or interruption of the power supply, particularly refrigerated medical stock and powered medical devices (such as monitors and ventilators) (U.S. Food and Drug Administration 2018). Damaged health facilities may limit people's post-disaster access to healthcare, particularly chronic disease patients who may need regular check-ups, treatments or medications (Redlener and Reilly 2012). All these make disaster preparedness essential for the healthcare system prior to disaster.

Although SDG 9 fails to explicitly mention the necessity of resilient healthcare facilities, Target 9.1 of "[d]eveloping quality, reliable, sustainable and resilient infrastructure, including regional and transborder infrastructure, to support economic development and human well-being, with a focus on affordable and equitable access for all" is relevant to Health-EDRM. Strengthening critical infrastructure of not only hospitals but also the water and energy systems hospitals dependent on should also be upheld.

7.2.8 SDG 11: Make Cities and Human Settlements Inclusive, Safe, Resilient and Sustainable

Among the world population in 2018, over 55% are urban dwellers (UNDESA 2018). The urban population is further projected to rise to 68% of the world's population by 2015. Additionally, 33 megacities globally are holding 10 million people or more and these numbers are growing. As a result, disasters that occur in these urban areas could potentially affect large populations simultaneously.

Particularly, those dwelling in informal settlements and slum areas of the city are at a higher risk of adverse outcomes in disasters. Without adequate basic services, hazards that would not make a huge impact in formal settlements, such as a small fire or heavy rainfall, could develop into a disaster in these overcrowded living conditions (Baker 2011). Urban planning and management must be improved to reduce health and disaster risks in urban settings, particularly among the urban poor

(Baker 2011). This is aligned with the New Urban Agenda and more will be discussed in the later section on the linkage between New Urban Agenda and Health-EDRM.

Target 11.1 of "ensur[ing] access for all to adequate, safe and affordable housing and basic services and upgrad[ing] slums [by 2030]" would be essential to prevent disasters among the urban poor. In addition, the following two targets are disaster-focused and provide a linkage with Sendai Framework and Health-EDRM.

> Target 11.5: "By 2030, significantly reduce the number of deaths and the number of people affected and substantially decrease the direct economic losses relative to global gross domestic product caused by disasters, including water-related disasters, with a focus on protecting the poor and people in vulnerable situations"
>
> Target 11.b: "By 2020, substantially increase the number of cities and human settlements adopting and implementing integrated policies and plans towards inclusion, resource efficiency, mitigation and adaptation to climate change, resilience to disasters, and develop and implement, in line with the Sendai Framework for Disaster Risk Reduction 2015–2030, holistic disaster risk management at all levels"

7.2.9 SDG 13: Take Urgent Action to Combat Climate Change and Its Impacts

As discussed in Chap. 6 regarding the linkage between the Paris climate agreement and Health-EDRM, climate change is also an essential factor affecting sustainable development, as acknowledged by SDG 13. Climate change is one of the major drivers that are increasing the frequency and intensity of hazards, such as temperature extremes, possible rainfall variability (leading to increased flooding or droughts), sea-level rise, extreme weather events (e.g. typhoons and storms) and climate-sensitive diseases (e.g. dengue and malaria) (Kelman 2017; Banwell et al. 2018a). These hazards may range from acute events to slow onsets; however they all commonly have direct impacts on health, such as mortality, morbidities, injuries, mental health and other health effects (Banwell et al. 2018a). Health risks and interventional strategies in the overlapping spheres of development, climate change, and disasters should be considered and assessed holistically. This coherency would maximize the use of resources, as well as prevent contradictory efforts (Kelman 2017; Banwell et al. 2018b).

Under this SDG, Target 13.1 and its indicators are particularly relevant to Health-EDRM.

> Target 13.1: "Strengthen resilience and adaptive capacity to climate-related hazards and natural disasters in all countries"
>
> 13.1.1 Number of deaths, missing persons and persons affected by disaster per 100,000 people
>
> 13.1.2 Number of countries with national and local disaster risk reduction strategies
>
> 13.1.3 Proportion of local governments that adopt and implement local disaster risk reduction strategies in line with national disaster risk reduction strategies

In addition to acknowledging this important interlinkage between climate change and sustainable development, this SDG also acknowledges "that the United Nations Framework Convention on Climate Change (UNFCCC) is the primary international, intergovernmental forum for negotiating the global response to climate change" (United Nations General Assembly 2015).

7.2.10 SDG 16: Promote Peaceful and Inclusive Societies for Sustainable Development, Provide Access to Justice for All and Build Effective, Accountable and Inclusive Institutions at All Levels

Conflicts, wars, violence, and terrorism are direct threats to life (World Health Organization 2000). Health and human security decline in conflict-filled communities and fragile states since basic services and provisions for basic needs such as food and water are reduced. Large-scale violence and conflict lead to population displacements as people fear for their lives, and survival becomes the principal goal of existence (World Health Organization 2000). Furthermore, conflicts have not only acute effects on violence-related mortality and injuries, but also broader and lingering effects on communicable diseases, malnourishment, non-communicable disease treatments, and mental health. Research evidence from the Darfur conflict shows there was a lingering heightened level of mortality one year post-conflict as a result of diarrhoea-related mortality, particularly among displaced populations (Degomme and Guha-Sapir 2010).

In large-scale conflict settings, it can become increasingly difficult to provide basic standards of care, as the insecurity inhibits the provision of healthcare services (Checchi 2010). Healthcare infrastructure may be destroyed and delivery of medical supplies, medications and vaccines may be disrupted and delayed. Existing chronic conditions such as tuberculosis, HIV, and diabetes may be exacerbated as access to healthcare and routine treatment is inhibited. Conflicts can also displace the healthcare providers or inflict causalities upon them, not only resulting in loss of life but also weakening the healthcare system. Protracted conflicts can reduce immunization coverage and put entire population segments at risk of communicable disease outbreaks (Quinn et al. 2017). Exploratory evidence has also shown that long-term conflict impacts the public health of populations outside of the conflict zone, measured by infant and maternal mortality rates (Johnson 2017). The ability for the international humanitarian community to provide assistance is also narrowed in conflict-filled settings, particularly since the nature of conflict has been changing in the recent years to compromise the security of humanitarian assistance (Spiegel et al. 2010).

Violence, exploitation, and trafficking, at any scale, are associated with increased health risks – including but not limited to psychological trauma, injuries from violence, HIV/AIDS, substance misuse, sexually transmitted infections and other adverse reproductive health outcomes (Zimmerman et al. 2003). Those who are trafficked often experience not only reduced access to health services and social care, but also exploitative and unsafe working conditions. They may not have legal protections but are often criminalized. Anti-trafficking interventions need to take into account the context of those at risk, and policies should be devised to minimize the harm and exploitation against them (Busza et al. 2004).

Target 16.1 aims to "[s]ignificantly reduce all forms of violence and related death rates everywhere", particularly conflict-related deaths as mentioned in indicator 16.1.2. Target 16.2 of "[e]nding abuse, exploitation, trafficking and all forms of violence against and torture of children" is also essential for better health. Finally, Target 16.A of "[s]trengthening relevant national institutions, including through international cooperation, for building capacity at all levels, in particular in developing countries, to prevent violence and combat terrorism and crime" is critical in the coming decades since societal disasters have become increasingly frequent.

7.2.11 Connections and Interlinkages with Other SDGs

Other Sustainable Development Goals also have linkages and connections to Health-EDRM, albeit not as direct. However, if these other SDGs are not achieved, it will be hard to meet the goal of Health and Emergency Disaster Risk Management. These include economic growth-related goals (SDG 7, SDG 8, SDG 10), environmental protection-related goals (SDG 12, SDG 14, SDG 15), and the goal to enhance global partnerships (SDG 17). Taking the example of SDG 10: Reduce inequality within and among countries. Low-income countries experience greater fatality losses than higher income countries (Rentschler 2013). Lower income countries also experience higher proportional economic losses compared to their overall GDP, close to a quarter of their GDP (Rentschler 2013). This affects how well they are able to recover from disasters as well as "build back better". Reducing inequalities within and among countries will provide countries with more equal opportunities to reduce their populations' vulnerabilities to disasters. Additionally, the protection of terrestrial ecosystems (SDG 15) can help reduce hazard and vulnerability risks. Damaged ecosystems, caused by deforestation, overgrazing and other human activities, have led to increased disaster risks. As ecosystems are restored, they can serve as buffers for hazards.

Furthermore, these SDGs as well as their Health-EDRM connections are more intertwined than they appear. For example, poverty conditions, which SDG 1 aims to eradicate, further exacerbates the access and availability of the resources mentioned in other goals. It is connected to SDG 2 of ending hunger since poor households are more vulnerable to malnutrition in disasters. Poorer communities are also more likely to have their education infrastructure destroyed by disasters, affecting the educational achievements as aimed by SDG 4. These interlinkages highlight the synergies available among attaining SDGs and Health-EDRM goals holistically, and they should be considered in the development of policies and implementation plans.

7.3 Introduction to New Urban Agenda

The New Urban Agenda was adopted in October 2016 at the United Nations Conference on Housing and Sustainable Urban Development (Habitat III) in Quito, Ecuador. Heads of states, local government leaders, government ministers, the private sector, civil society, experts and community members gathered and agreed to the policy framework document that set the global agenda for sustainable and inclusive urbanization for the next 20 years. The New Urban Agenda document incorporates both the *Quito Declaration on Sustainable Cities and Human Settlements for All* and the *Quito Implementation Plan for the New Urban Agenda*. The aims of the agenda rest on five pillars of implementation: national urban policies, urban legislation and regulations, urban planning and design, local economy and municipal finance, and local implementation (United Nations General Assembly 2017).

7.4 New Urban Agenda and Health-EDRM

Health-EDRM is present throughout the New Urban Agenda. The word "health" and its varieties are mentioned 29 times in the 29-page document, while "disaster" 22 instances and "climate" 23 instances (United Nations General Assembly 2017).

The shared vision of the New Urban Agenda (NUA), as outlined in Article 11 of the Quito Declaration, is as follows:

> We share a vision of cities for all, referring to the equal use and enjoyment of cities and human settlements, seeking to promote inclusivity and ensure that all inhabitants, of present and future generations, without discrimination of any kind, are able to inhabit and produce just, safe, **healthy**, accessible, affordable, **resilient** and sustainable cities and human settlements to foster prosperity and quality of life for all

Box 7.4 expands further on the shared vision. Among the expanded details of the vision, Articles 13(a) and 13(g) are particularly relevant to the health and disaster aspects of Health-EDRM.

Box 7.4 Shared Vision of the New Urban Agenda

Article 13. We envisage cities and human settlements that:

(a) Fulfil their social function, including the social and ecological function of land, with a view to progressively achieving the full realization of the right to adequate housing as a component of the right to an adequate standard of living, without discrimination, universal access to safe and affordable drinking water and sanitation, as well as equal access for all to public goods and quality services in areas such as food security and nutrition, health, education, infrastructure, mobility and transportation, energy, air quality and livelihoods;
(b) Are participatory; promote civic engagement; engender a sense of belonging and ownership among all their inhabitants; prioritize safe, inclusive, accessible, green and quality public spaces friendly for families; enhance social and intergenerational interactions, cultural expressions and political participation, as appropriate; and foster social cohesion, inclusion and safety in peaceful and pluralistic societies, where the needs of all inhabitants are met, recognizing the specific needs of those in vulnerable situations;
(c) Achieve gender equality and empower all women and girls by ensuring women's full and effective participation and equal rights in all fields and in leadership at all levels of decision-making; by ensuring decent work and equal pay for equal work, or work of equal value, for all women; and by preventing and eliminating all forms of discrimination, violence and harassment against women and girls in private and public spaces;
(d) Meet the challenges and opportunities of present and future sustained, inclusive and sustainable economic growth, leveraging urbanization for structural transformation, high productivity, value-added activities, and resource efficiency, harnessing local economies, and taking note of the contribution of the informal economy while supporting a sustainable transition to the formal economy;
(e) Fulfil their territorial functions across administrative boundaries, and act as hubs and drivers for balanced, sustainable and integrated urban and territorial development at all levels;
(f) Promote age- and gender-responsive planning and investment for sustainable, safe and accessible urban mobility for all, and resource-efficient transport systems for passengers and freight, effectively linking people, places, goods, services, and economic opportunities;
(g) Adopt and implement disaster risk reduction and management, reduce vulnerability, build resilience, and responsiveness to natural and human-made hazards, and foster mitigation of and adaptation to climate change;

(h) Protect, conserve, restore and promote their ecosystems, water, natural habitats, and biodiversity, minimize their environmental impact, and change to sustainable consumption and production patterns.

Source United Nations General Assembly (2017, p. 4)

Similar to the Sustainable Development Goals, the shared vision of NUA is guided by three principles on social inclusion, economic opportunity, and environmental protection. The latter principle addresses disaster risks to "[e]nsure environmental sustainability by promoting clean energy and sustainable use of land and resources in urban development, by protecting ecosystems and biodiversity, including adopting healthy lifestyles in harmony with nature, by promoting sustainable consumption and production patterns, by building urban resilience, **by reducing disaster risks** and by mitigating and adapting to climate change" (Article 14c).

The Quito Implementation Plan for the New Urban Agenda comprises of **transformative commitments for sustainable urban development**, which go through an extensive list of commitments revolving around the three guiding principles. These include sections on sustainable urban development for social inclusion and ending poverty (Articles 25–42), sustainable and inclusive urban prosperity and opportunities for all (Articles 43–62), and environmentally sustainable and resilient urban development (Articles 63–80). The section on the last principle of environmentally sustainable and resilient urban development (Articles 63–80) lists commitments of various ways to maintain a resilient development in cities. Particularly, it begins by "recogniz[ing] that cities and human settlements face unprecedented threats from… natural and human-made disasters, and climate change and its related risks, undermining the efforts to end poverty in all its forms and dimensions and to achieve sustainable development" (Article 63). Then, it progresses to acknowledge the vulnerabilities in urban settings:

> Article 64. We also recognize that urban centres worldwide, especially in developing countries, often have characteristics that make them and their inhabitants especially **vulnerable to the adverse impacts of climate change and other natural and human-made hazards**, including earthquakes, extreme weather events, flooding, subsidence, storms, including dust and sand storms, heatwaves, water scarcity, droughts, water and air pollution, vector-borne diseases and sea level rise, which particularly affect coastal areas, delta regions and small island developing States, among others.

Then it follows an Health-EDRM-esque commitment acknowledging the integration of health and well-being with DRR:

> Article 65. We commit ourselves to facilitating the sustainable management of natural resources in cities and human settlements in a manner that protects and improves the urban ecosystem and environmental services, reduces greenhouse gas emissions and air pollution and **promotes disaster risk reduction and management, by supporting the development of disaster risk reduction strategies and periodical assessments of disaster risk**

caused by natural and human-made hazards, including standards for risk levels, while fostering sustainable economic development and **protecting the well-being and quality of life of all persons through environmentally sound urban and territorial planning, infrastructure and basic services**.

This is furthered in subsequent articles, which commit to solutions that can address both health and DRR, namely:

- "well-connected and well-distributed networks of open, multipurpose, safe, inclusive, accessible, green and quality public spaces" that "improv[e] the resilience of cities to disasters and climate change, including floods, drought risks and heat waves", and improve "physical and mental health" (Article 67) and
- adequate investments in protective, accessible and sustainable infrastructure and service provision systems for water, sanitation and hygiene, sewage, solid waste management, urban drainage, reduction of air pollution and storm water management, in order to improve safety in the event of water-related disasters, improve health, ensure universal and equitable access to safe and affordable drinking water for all, as well as access to adequate and equitable sanitation and hygiene for all…. (Article 119)

Aligned with the Sendai Framework, NUA commits to strengthening the resilience of cities and their infrastructure and to

- adopting and implementing integrated, age- and gender-responsive policies and plans and ecosystem-based approaches in line with the **Sendai Framework for Disaster Risk Reduction 2015–2030**, (Article 77)
- holistic and data-informed DRR and management at all levels to reduce vulnerabilities and risk, especially in risk-prone areas of formal and informal settlements, including slums, (Article 77)
- **more proactive risk-based, all-hazards and all-of-society approaches** such as raising public awareness of risks and promoting ex-ante investments to prevent risks and build resilience, (Article 78)
- ensuring timely and effective local responses to address the immediate needs of inhabitants affected by natural and human-made disasters and conflicts, (Article 78); and
- integration of the "build back better" principles into the post-disaster recovery process to integrate resilience-building, environmental and spatial measures and lessons from past disasters, as well as awareness of new risks, into future planning. (Article 78)

The New Urban Agenda also specifically addresses conflict settings and the health and wellbeing of those who may be displaced due to conflicts or disasters:

> Article 19. …Special attention should also be given to countries in situations of conflict, as well as countries and territories under foreign occupation, post-conflict countries and countries affected by natural and human-made disasters [for the implementation of the New Urban Agenda].

It commits in Articles 28–29 to "respect[ing] for the human rights of refugees, internally displaced persons and migrants" and strengthening the coordination and collaboration "in the **provision of social and basic services for all**, including **generating investments in communities that are most vulnerable to disasters and those affected by recurrent and protracted humanitarian crises,**… to **promoting adequate services, accommodation and opportunities for decent and productive work for crisis-affected persons in urban settings** and to working with local communities and local governments to identify opportunities for engaging and developing local, durable and dignified solutions while ensuring that aid also flows to affected persons and host communities to prevent regression of their development." Furthermore, it commits to "support[ing] resilient urban services during armed conflicts… [and] reaffirm[ing] full respect for international humanitarian law" in Article 30.

Furthermore, the New Urban Agenda commits to climate action and adaptation planning, integrating these with disaster risk reduction into "age- and gender-responsive urban and territorial development and planning processes" (Article 101). Recognizing the interconnectedness and multi-sectoral nature of disaster risks, Article 101 continues, "...We will promote cooperation and coordination across sectors and build the capacities of local authorities to develop and implement disaster risk reduction and response plans, such as risk assessments concerning the location of current and future public facilities, and to formulate adequate contingency and evacuation procedures".

As can be seen, the New Urban Agenda outlines extensive vision and commitments for urbanization in the upcoming years. Acknowledging the interconnectedness of urban development with disaster risks, the document envisions cities and urban populations that are inclusive, healthy and resilient in disasters.

References

Asian Disaster Preparedness Center (2008) Impact of disasters on the education sector in Cambodia. Bangkok

Baker JL (2011) Climate change, disaster risk, and the urban poor: cities building resilience for a changing world. Washington DC

Banwell N, Rutherford S, Mackey B, Street R, et al (2018a) Commonalities between disaster and climate change risks for health: a theoretical framework. Int J Environ Res Public Health 15(3). https://doi.org/10.3390/ijerph15030538

Banwell N, Rutherford S, Mackey B, Chu C (2018b) Towards improved linkage of disaster risk reduction and climate change adaptation in health: a review. Int J Environ Res Public Health 15 (4):8–11. https://doi.org/10.3390/ijerph15040793

Bradshaw S, Fordham M (2013) Women, girls and disasters: a review for DFID

Buchanan-Smith M, Maxwell S (1994) Linking relief and development: an introduction and overview. IDS Bull 25(4):2–16

Busza J, Castle S, Diarra A (2004) Trafficking and health. BMJ 328:1369–1371

CARE (2011) Poverty causes disasters and disasters cause poverty, global platform for disaster risk reduction in Geneva. Available at: https://reliefweb.int/report/world/poverty-causes-disasters-and-disasters-cause-poverty. Accessed: 19 July 2018

Chan EY (2018) Building bottom-up health and disaster risk reduction programmes. Oxford University Press, New York

Chan EY, Ho JY (2019) Urban water and health issues in Hong Kong. In: Ray B, Shaw R (eds) Urban drought—emerging water challenges in Asia. Springer, Singapore

Checchi F (2010) Estimating the number of civilian deaths from armed conflicts. Lancet 375 (9711):255–257. https://doi.org/10.1016/S0140-6736(09)62128-0

Degomme O, Guha-Sapir D (2010) Patterns of mortality rates in Darfur conflict. Lancet 375 (9711):294–300. https://doi.org/10.1016/S0140-6736(09)61967-X

DFID (2006) Reducing the risk of disasters—helping to achieve sustainable poverty reduction in a vulnerable world: a DFID policy paper. Glasgow. Available at: http://www.reliefweb.int/rw/lib.nsf/db900sid/FBUO-7N8JJH/$file/benfield_march2001.pdf?openelement

EC–FAO Food Security Programme (2008) An introduction to the basic concepts of food security. Available at: www.foodsec.org

Elliott I (2016) Poverty and mental health: a review to inform the Joseph Rowntree Foundation's Anti-Poverty Strategy. London

Ferris E, Petz D, Stark C (2013) Disaster risk management: a gender-sensitive approach is a smart approach, the year of recurring disasters: a review of natural disasters in 2012. Washington DC. Available at: https://www.brookings.edu/research/disaster-risk-management-a-gender-sensitive-approach-is-a-smart-approach-chapter-4-of-the-year-of-recurring-disasters-a-review-of-natural-disasters-in-2012/

Food and Agriculture Organization of the United Nations (2016) Asia and the pacific regional overview of food insecurity. FAO, Bangkok

Garschagen M et al (2015) World risk report 2015. Berlin. Available at: http://collections.unu.edu/eserv/UNU:3303/WRR_2015_engl_online.pdf

Goldmann E, Galea S (2014) Mental health consequences of disasters. Annu Rev Public Health 35 (1):169–183. https://doi.org/10.1146/annurev-publhealth-032013-182435

IASC Inter-Agency Standing Committee (2017) The gender handbook for humanitarian action. Available at: https://interagencystandingcommittee.org/system/files/2018-iasc_gender_handbook_for_humanitarian_action_eng_0.pdf

Independent Evaluation Department (2013) Making infrastructure disaster-resilient. Philippines, Manila

Johnson SA (2017) The cost of war on public health: an exploratory method for understanding the impact of conflict on public health in Sri Lanka. PloS One 12(1):e0166674. https://doi.org/10.1371/journal.pone.0166674

Kelman I (2017) Linking disaster risk reduction, climate change, and the sustainable development goals. Disaster Prev Manag Int J 26(3):254–258. https://doi.org/10.1108/DPM-02-2017-0043

Krishnan S, Twigg J (2016) Menstrual hygiene: a 'silent' need during disaster recovery. Waterlines 35(3):265–276

Marx M et al (2006) Diarrheal illness detected through syndromic surveillance after a massive power outage: New York City, August 2003. Am J Public Health 96(3):547–553. https://doi.org/10.2105/AJPH.2004.061358

Matyas D, Pelling M (2012) Disaster vulnerability and resilience: theory, modelling and prospective. Foresight, Government Office for Science. https://doi.org/10.13140/rg.2.1.4684.2409

Nicogossian A et al (2012) Poverty, disparities, disasters and global burden of disease. World Med Health Policy 4(1). https://doi.org/10.1515/1948-4682.1231

Quinn J et al (2017) Public health crisis: the need for primary prevention in failed and fragile states. Cent Eur J Public Health 25(3):171–176. https://doi.org/10.21101/cejph.a4671

Redlener I, Reilly MJ (2012) Lessons from Sandy—preparing health systems for future disasters. N Engl J Med 367(24):2269–2271. https://doi.org/10.1056/NEJMp1213486

Rentschler JE (2013) Why Resilience Matters The Poverty Impacts of Disasters. World Bank Policy Research Working Paper. https://doi.org/10.1596/1813-9450-6699

Rohwerder B (2017) Conflict and gender dynamics in Yemen

Shaw R et al (2011) Disaster education: an introduction, community, environment and disaster risk management. Emerald Group Publishing Ltd. https://doi.org/10.1108/s2040-7262(2011)0000007007

Spiegel PB et al (2010) Health-care needs of people affected by conflict: future trends and changing frameworks. Lancet 375(9711):341–345. https://doi.org/10.1016/S0140-6736(09)61873-0

Steele P, Oviedo G, McCauley DE (2007) In: Steele P, Oviedo G, McCauley D (eds) Poverty, health, and ecosystems: experience from Asia. Asian Development Bank and International Union for Conservation of Nature, Gland, Switzerland and Manila, Philippines

UNDESA (2018) World urbanization prospects: the 2018 revision. doi: ST/ESA/SER.A/366
UNISDR (2008) Linking disaster risk reduction and poverty reduction. Available at: http://www.preventionweb.net/files/3293_LinkingDisasterRiskReductionPovertyReduction.pdf
United Nations (2009) Making disaster risk reduction gender-sensitive: policy and practical guidelines, p 163
United Nations Economic and Social Council (2010) Draft technical guidance for setting up, implementing and assessing surveillance systems of water-related disease. Bucharest
United Nations General Assembly (2015) Transforming our world: the 2030 agenda for sustainable development, A/RES/70/1
United Nations General Assembly (2017) New Urban Agenda, A/RES/71/256*
U.S. Food and Drug Administration (2018) FDA offers tips about medical devices and hurricane disasters. Available at: https://www.fda.gov/medicaldevices/safety/emergencysituations/ucm055987.htm. Accessed: 19 July 2018
World Health Organization (2000) Conflict and health, preventing violent conflict—the search for political will. Strategies and Effective Tools, Krusenberg
World Health Organization (2011a) Guidance on water supply and sanitation in extreme weather events. Copenhagen. Available at: http://www.euro.who.int/en/health-topics/environment-and-health/water-and-sanitation/publications/2011/guidance-on-water-supply-and-sanitation-in-extreme-weather-events
World Health Organization (2011b) Guidelines for drinking-water quality, 4th edn
World Health Organization (2016) World health statistics 2016: monitoring health for the SDGs sustainable development goals. World Health Organization, Geneva. Available at: https://books.google.com.hk/books?id=-A4LDgAAQBAJ&printsec=frontcover#v=onepage&q&f=false
World Health Organization (2018) Drinking-water. Available at: http://www.who.int/news-room/fact-sheets/detail/drinking-water. Accessed: 20 Nov 2018
World Health Organization and Public Health England (2017a) Health emergency and disaster risk management: overview. Health emergency and disaster risk management fact sheets. Available at: http://www.who.int/hac/techguidance/preparedness/factsheets/en/
World Health Organization and Public Health England (2017b) Safe hospitals: prepared for emergencies and disasters, disaster risk management for health fact sheet. Available at: http://www.who.int/hac/techguidance/preparedness/risk-management-safe-hospitals-december2017.pdf
Zimmerman C et al (2003) The health risks and consequences of trafficking in women and adolescents trafficking in women and adolescents: findings from a European Study. London School of Hygiene & Tropical Medicine (LSHTM), London. Available at: http://www.oas.org/atip/globalreports/zimmermantiphealth.pdf

Chapter 8
Bangladesh Public Health Issues and Implications to Flood Risk Reduction

Akiko Matsuyama, Fahmida Afroz Khan and Md. Khalequzzaman

Abstract Bangladesh is a disaster-prone country and suffers from increasingly frequent and devastating natural calamities owing to its geographical location and environmental situation. Climate change exacerbates its vulnerability and the country ranks sixth worst among the world's top 10 countries most affected by extreme weather events in the last 20 years. Seventy percent of its population live in flood-prone regions and 26% are affected by cyclones. This section reviews major threats to public health related to flood in Bangladesh, including communicable diseases such as water-borne and vector-borne diseases such as diarrhea, acute respiratory infection, and dengue fever. Effects of climate change including flood also affect agriculture and increase malnutrition, particularly among young children. Additionally, some health issues affected by disasters which have drawn scant attention are mental health, violence against children and women, chronic diseases, salinity of drinking water, and urban health; these problems are also discussed in the context of Bangladesh, with special emphasis on urban flooding and its adverse effects on the urban poor population. Primary health care system development in the country has generally indicated successful achievements. It, along with poverty reduction efforts, has contributed, to a substantial extent, to improvement of disaster management. However, urgent action is required for more coherent program to integrate health systems into H-EDRM.

Keywords Public health · Communicable diseases · Mental health · Violence · Chronic diseases · Salinity · Urban health

A. Matsuyama (✉)
Tsuda University, 2-1-1 Tsuda-machi, Kodaira-shi, Tokyo 187-8577, Japan
e-mail: akikomat@tsuda.ac.jp

F. A. Khan · Md.Khalequzzaman
Department of Public Health and Informatics, Bangabandhu Sheikh Mujib Medical University, Dhaka, Bangladesh

E. Ying Yang Chan and R. Shaw (eds.), *Public Health and Disasters*, Disaster Risk Reduction, https://doi.org/10.1007/978-981-15-0924-7_8

8.1 Introduction

Bangladesh is vulnerable to natural hazards because of its geographical location, environment situation, and population density with rapid urbanization. Figures 8.1 shows different types of natural hazards that have occurred in the last 25 years. Its population also faces effects of global climate change. It ranked sixth among the world's top 10 countries most affected by extreme weather events including storms, floods, and heatwaves, in the last 20 years, according to the Global Climate Risk Index in 2017 (Kreft et al. 2016).

Among natural hazards that repeatedly affect Bangladesh, floods and cyclones are the most devastating in terms of frequency and their effects on people's health. Approximately 70% of the country is prone to floods and 26% of the population are affected by cyclones (Disaster Management Bureau 2010). Ninety-five percent of the natural hazard-related death toll was due to cyclones and 3.6% was caused by flood.

Reviewing the major flood and cyclone events in the past 40 years, however, it should be noted that the death toll has been declined dramatically (Table 8.1). The number of deaths reported to be associated with such natural disasters was 500,000 in 1970, 138,000 in 1991, and 3300 in 2007, and decreased to 190 or less after that.

Such a substantial decline in mortality caused by cyclones and floods has been internationally praised as a role model in managing natural disasters (Qayyum 2016). Improvements in disaster management in Bangladesh involve key factors including construction of cyclone shelters, establishing an early warning system, evacuation plans, and coastal embankments, reforestation schemes, and increased community awareness (Khan 2008; Cash et al. 2013; Haque et al. 2012).

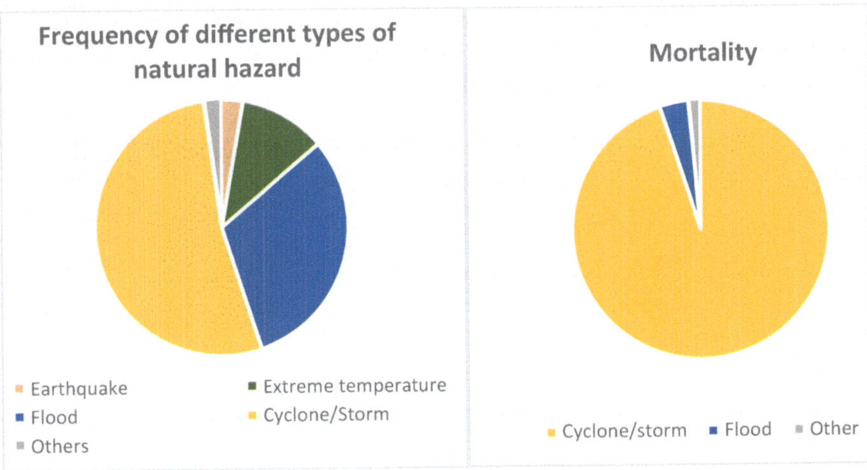

Fig. 8.1 Bangladesh's reported losses by different types of natural hazards between 1990 and 2014 (EM-DAT). Modified from figures of Bangladesh data 2014 in Prevention Web (2017); https://www.preventionweb.net/countries/bgd/data/ (Prevention Web 2017)

Table 8.1 List of major natural hazard (flood and cyclone)

	Type of hazard	Proportion (%) of total region or population affected (n)	Height of storm surge (m)	Deaths (n)
1970	Cyclone Bhola	–	6–10	225,000–500,000
1974	Floods	35.4%	NA	–
1985	Cyclone	167,500	3–4.6	11,000
1987	Flood	38.6%	NA	1657
1988	Flood	52.4%	NA	2379
1991	Cyclone Gorky	4.56 million	6–7.6	138,000
1998	Flood	67.4%	NA	918
2004	Flood	25%	NA	<800
2007	Cyclone Sidr	18.3 million	4.5–6.1	3300
2009	Cyclone Aila	3.9 million	2–3	190
2011	Flood and landslide	300,000	NA	10
2012	Flood and landslide	5 million	NA	139
2012	Storm	24,000 households	–	36
2014	Flood	2.8 million	NA	9
2015	Flood and landslide	200,000	NA	19
2015	Cyclone Komen	1.5 million	–	45
2016	Cyclone Roanu	500,000	–	21
2016	Flood and landslide	4.2 million	NA	106
2017	Cyclone Mora	3.3 million	–	6
2017	Flood and landslide	–	NA	135

Modified from Cash et al. and Center for Excellence in Disaster Management & Humanitarian Assistance (2017)

Furthermore, overall social and economic development in Bangladesh in the last 50 years is an essential contributor to the country's success story in disaster management. Diversification of local rice crops to both fall and winter harvest crops known as the green revolution and liberalization of trade in rice in 1994 have increased food security. NGOs' long-term development activities for poverty alleviation, particularly for the rural poor who were most vulnerable to natural disasters, are examples of many driving factors contributing to the community's resiliency (Moore et al. 2012).

From a public health perspective, overall gains in general health in the last decades are an inevitable contributor to decreased mortality during and after natural hazards. Some of the notable public health measures in Bangladesh include family planning, immunization program for children, oral rehydration therapy (ORT) for diarrheal diseases, maternal and child health program, early detection and treatment for tuberculosis including Direct Observed Treatment (DOT), and a vitamin A supplementation program for prevention of malnutrition in children and pregnant women (Chowdhury et al. 2013). These public health measures have achieved substantial health improvements in Bangladesh. As indicated in the progress of Millennium Development Goals (MDGs), an important health indicator, the mortality rate of children under five years old, declined from 146 (per 1000 live births) in 1990 to 45 in 2015 and the target was achieved (General Economic Division 2015). Again, overall health improvements should be discussed in the broader context of the nation's social and economic development. Nevertheless, risks posed by natural disasters to public health go beyond immediate health indicators such as mortality. The burden of health problems also disproportionally falls on more vulnerable groups such as the poor, ethnic minorities, people with disabilities, and old people (WHO 2015). In the next section, public health issues in relation to natural hazards, particularly floods and cyclones, will be discussed in more detail in the Bangladeshi social context, with special attention to the growing concern about rapid urbanization in the subsequent section.

8.2 Typology of Health Effects of Floods and Cyclones

Major public health problems related to floods and cyclones in Bangladesh can be categorized based on their characteristics shown in Table 8.2.

Referring to Table 8.2, some of the main public health problems in relation to floods in Bangladesh are described.

8.2.1 Communicable Diseases

Although child mortality has steadily declined in Bangladesh, diarrhea remains one of the major causes of death for children. During floods, increases in water temperature, precipitation frequency and severity, evaporation-transpiration rates, persistent humidity, and changes in coastal ecosystems promote water contamination with harmful pathogens and chemicals, result in increased human exposure and water-borne diseases. Previous studies found that the incidence of diarrhea peaked in the rainy season in tropical regions (Cairncross and Feachen 1993). The association between cholera and weather factors such as rainfall and temperature has been well established. As for non-cholera diarrhea caused by rotavirus, *Shigella*, *Salmonella*, *Campylobacter*, *Escherichia coli*, *Aeromonas*, and so on, such a

Table 8.2 Major public health problems related to flood and cyclone in Bangladesh

	Type of public health problem	Examples
Direct short-term	Mortality Morbidity –Injuries –Communicable diseases	Diarrhea Dengue fever Malaria Typhoid fever Respiratory infection Skin diseases
Indirect short-term	Disrupting health/medical infrastructure and interrupting continuous health care service	Adverse effects of pregnancy and childbirth outcomes
Indirect short- or mid-term	Water contamination Inadequate Sanitation	Decreased accessibility and availability of safe drinking water and adequate toilet affect communicable diseases
Indirect long-term	Child deaths due to drowning Undernutrition, malnutrition Toxic exposure Salinity Mental health Violence Non-communicable diseases (NCD) Disability and old people	Malnourished children Exposure to industrial and domestic waste, thus increase potential diseases Hypertension, hypertension disorder during pregnancy Depression Anxiety Child abuse Immediate partner's violence Deterioration of NCD Vulnerable in urban poor setting

Created by the authors

relationship is less clear. In Bangladesh, during the flooding in 1998, the number of cholera and non-cholera diarrhea cases was reported to be almost six and two times higher than non-flood times, respectively (Hashizume et al. 2008).

A study on typhoid fever, caused by a bacterium of the genus *Salmonella,* in Dhaka, also revealed distinct seasonality with the highest disease occurrences during monsoon months. The incidence increased with temperature, rainfall, and river water level (Dewan et al. 2013). It also indicated that it disproportionately infected males and very young children (ibid.). Typhoid fever, in general, is heavily affected by rainfall during flooding and water-logged periods particularly in low lying areas where people tend to rely on contaminated surface water and tube-wells for drinking and domestic chores.

Vector-borne diseases, such as dengue, dengue hemorrhagic fever, and malaria are other types of communicable diseases prevalent during and after flooding. Risks increase due to related expansions in vector numbers and ranges, shortening of pathogen incubation periods, and disruption and relocation of large human populations. There is strong evidence for an increase in dengue fever at high river levels (Hashizume et al. 2012; Sifat et al. 2015). Acute respiratory infection (pneumonia) is also common among people affected by floods (Milojevic et al. 2012). It can be worsened by crowding and exposure to indoor cooking using an open flame inside shelters, poor nutrition and lack of access to health facilities and adequate medicine for treatment.

8.2.2 Disruption to Health/Medical Infrastructure

Disruption to health infrastructure and interruptions to the continuity of health care services caused by natural hazards affect people in general but particularly affect the more vulnerable groups in the society (Veenema et al. 2017). In Bangladesh, utilization of maternal health services such as antenatal care, institutional delivery, and C-section delivery has generally increased and gaps in terms of both access and equity among socio-economic strata have also been improving (Quayyum et al. 2013; Anwar et al. 2015). However, the poorest, the most prone to be affected by floods, are still least likely to use adequate health care even during non-flood periods. Flooding is an additional factor responsible for their delay in seeking care. Moreover, lack of access to alternative health services such as pharmacies, traditional birth attendants (TBAs), homeopaths and traditional practitioners because of floods further worsens their health situation.

8.2.3 Water Contamination and Inadequate Sanitation

The impact of floods on water and sanitation is a significant public health issue for the flood-affected population. How contaminated water and water sources affected by flooding increase water-borne and vector-borne diseases have been previously discussed. Furthermore, people's coping behavior during a flood, such as using hanging latrines or boats and defecating directly into water bodies (Shimi et al. 2010) contaminates water sources for drinking and household chores. Fetching water is an important women's domestic activity. During floods and post-cyclone periods, they tend to spend more time and walk longer distances to access safer drinking water (Abedin et al. 2014) which may have adverse health effects as well as increase insecurity in more traditional rural environments ruled by *purdah*.

Additionally, high level of exposure to arsenic throughout the country seems to complicate the issue of drinking water during floods. In the arsenic affected area, people are directed to switch from tube-wells that are high (>50 μg/L) in arsenic to neighboring wells that are low in arsenic. Some studies indicate that shallow low arsenic wells are more likely to be contaminated with the fecal indicator *E. coli* than shallow high arsenic wells (Wu et al. 2011). During floods and water-logged periods, many wells are damaged or become unusable and people have little choice but to use water from whatever sources available, including wells high in arsenic or those low in arsenic but contaminated by more harmful pathogens and chemicals.

8.2.4 Drowning of Children

Drowning is the leading cause of death among children and over one-quarter of deaths among children aged 1–4 years in Bangladesh were due to drowning in 2003 and its proportion rose to 42% in 2011(Rahman et al. 2017). Up to 46 children drown daily and 16,500 drowning deaths occur each year. Most children are 2 years of age and most die within 20 meters of their homes (UNICEF 2017). While the percentage of drowning children was highest during the monsoon season (Rahman et al. 2017), water is present most of the time near their homes in low-lying areas and children tend to fall into surface water and die accidentally (Hanif et al. 2010; Ahmed et al. 1999).

8.2.5 Salinity

During the dry season, regions that are typically saturated with freshwater from rivers become susceptible to seawater moving upland with the occurrence of extensive flooding. While only 10% of land is affected during monsoon, the dry season can increase the affected area to 40% (Center for Excellence in Disaster Management & Humanitarian Assistance 2017). People are exposed to high levels of sodium in drinking water in coastal Bangladesh and it is an important public health problem. Salinity in drinking water is associated with increased risk of (pre) eclampsia and gestational hypertension among pregnant women (Khan et al. 2014). Hypertensive disorder during pregnancy is one of the major causes of maternal mortality in Bangladesh as elsewhere in the world (Firoz et al. 2011). It is also related to perinatal death. Additionally, excessive intake of sodium is linked to non-communicable diseases (NCDs), for example, hypertension, which is now widely recognized as a major burden of disease among the general population. Salinity also increases the risk of epidemics of diseases such as cholera. Cyclones or heavy rainfall mediated turbidity that changes jointly with salinity gradients can significantly influence the abundance and distribution of estuarine *Vibrio*. Extended salt intrusion and higher turbidities in tropical estuaries by stronger and more frequent storms and deforestation-delivered erosion could favor *Vibrio* growth (Lara et al. 2009).

8.2.6 Mental Health

Psychosocial consequences of disasters on mental health are a growing concern. Bangladesh has attained substantial achievements in responding to physical needs for health in disasters by providing food, shelter, and medical supplies over the decades. However, the mental health of disaster survivors has received scant attention, partially because of a lack of experts and trained workers in the area in Bangladesh. Moreover, it is difficult to attribute mental status after a disaster to a single cause because people's mental well-being before a disaster and socio-economic factors also intermingle with the experience of the event.

A study on mental health in survivors of a tornado-affected area in Bangladesh suggests important features of psychosocial aspects of disaster management. It was found that women and children were more psychologically traumatized than adult men (Choudhury et al. 2006). This article discussed a further series of other psychological problems for women in society, including failure to perform their role as a wife and/or mother in the family, and as a worker in the community, thus further depressing and isolating them.

8.2.7 Violence

Violence is a significant public health issue and has recently received more attention as a determinant of health. Effects of violence on children include medical and psychological problems such as depression, eating disorders, post-traumatic stress disorder (PTSD), chronic pain, chronic fatigue, and irritable bowel syndromes (Dalal 2008). While violence against children is a major public health problem in developing countries (WHO 2008), the Bangladesh Health and Injury Survey found that injuries and violence during floods in rural Bangladesh are the leading causes of death in post-infancy children (Director General of Health Services 2003). Another study on injuries and parental violence against children during floods in Bangladesh showed similar results. It found that 70% of mothers and 40% of fathers abused their children during a flood including emotional, physical, and sustenance (stopping daily food intake to the child) abuse. Risk factors for parental abuse included flood-caused economic hardship and intimate partner violence by the father against the mother. More than 86% of mothers were abused by their husbands during floods (emotional abuse 37%, physical abuse 48%, sustenance abuse 0.3%, and sexual abuse 0.5%) (Biswas et al. 2010).

8.3 Climate Change and Urban Floods

In the context of the diverse health issues caused by floods, urban flooding is responsible for the emerging public health threats faced by the vulnerable population. This is particularly true for Bangladesh with its rapid urbanization. This issue requires special attention as the nature and policy implication of floods in urban areas would be different from those in coastal areas.

8.3.1 Unplanned Rapid Urbanization

Increasing attention is now being paid to effects of climate change on the health of the urban population in Bangladesh. The country has been experiencing rapid urbanization with 34.3% of its population (World Bank 2015) dwelling in urban areas. This urban population is expected to rise to 58.4% by 2050 (DESA/Population Division 2018). Dhaka, the capital city, has 32% of the country's entire population (WHO 2015). The city, with its concentration of both domestic and foreign investment, attracts massive migration from rural areas for better working opportunities. Additionally, approximately 500,000 people who have been internally displaced by climate-related impacts including extreme weather, floods, and drought are flowing into urban slums annually (Mahbubur Rahman et al. 2017). There were about 3394 slum areas with approximately 176,000 households in Dhaka North and South (Bangladesh Bureau of Statistics 2014). The UNDP estimated that approximately 40% of the population in Dhaka reside in slums (UNDP 2014) and 60% of slum dwellers were vulnerable to recurrent floods (UN-Habitat 2009). Many people are living in already deteriorating environments with contaminated food and water supplies, air pollution, inadequate sanitation, and waste removal facilities, and poor housing conditions. The urban poor without basic amenities suffers from a vicious cycle of infections, malnutrition, and ill health (Afsana and Wahid 2013). It is also noted that the urban poor population, in comparison to its rural counterpart, has its own disadvantages. Unlike rural residents, the urban poor depend almost exclusively on the market for food and necessary items which make them vulnerable to price increases and other market shocks. Furthermore, their lifestyles include long commuting and working hours, in addition to traffic jams, which prevent them from easily accessing public health service including health care. They also have fewer shared assets and social networks (Rashid 2000) that can help during a crisis. Thus, it is conceivable that natural disasters such as floods would put the urban poor at further risk of disease and malnutrition.

8.3.2 Public Health Issues Affected by Urban Floods

An emerging public health problem in rapidly industrialized urban areas is the effect of toxic exposure on human health. Waste production by industrial activities and domestic waste from households affects the environment, soil, air, and water if improperly treated, stored or disposed of (Misra and Pandey 2005). Water pollution, particularly, is a serious problem in central Dhaka. The presence of a large population, unsanitary conditions, poorly regulated industrial discharges, and untreated domestic effluents have made many urbanized delta rivers highly polluted (Whitehead et al. 2018). The textile industry, which is a major economic booster in the country, is a heavy polluter in terms of waste gas, solids, water, and noise.

Wastewater is the most environmentally damaging, and effluent from textile plants is classified as the most polluting of all the industrial sectors (Chung 1983). Floods can spread polluted water and increase the risk of toxic exposure among the urban population.

Regarding industrial wastewater, it is highly alkaline because of its unusual chemical content, which causes skin irritations, rashes, and skin lesions (boils) upon contact (Ullah 2006; Halder and Islam 2015). Wastewater contains hydrogen sulfide (H2S), dioxins, formaldehyde, azo dyes, and heavy metals. H2S causes some allergic manifestations such as headaches, eye irritation, sore throats, cough, respiratory tract irritation, bronchial asthma, etc., and when inhaled in excess of tolerance limits, it has negative effects on pregnancy and cause increased numbers of spontaneous abortions and stillbirths following occupational exposure (Ullah 2006; Halder and Islam 2015; Chou 2003; Xu et al. 1998). Dioxin and Formaldehyde are known human carcinogens responsible for human reproductive carcinoma (Ca) and nasopharyngeal Ca, respectively (Kogevinas et al. 2001; Cogliano et al. 2005). Azo dyes are known human carcinogens that are responsible for bladder Ca, splenic Ca, and hepatosarcoma (OECD 2005; SCTEE Plenary Meeting 1999). On contact, it causes allergic manifestation to the skin as well as diarrhea, nausea, vomiting, hypertension, acute tubular necrosis, respiratory distress, etc. (OECD 2005). Azo dyes are also mutagenic to fish and decrease the growth and productivity of plants (Chung 1983; de Aragao Umbuzeiro et al. 2005). Heavy metals such as lead (Pb), cadmium (Cd), copper (Cu), zinc (Zn), mercury (Hg) are usually present in wastewater and are responsible for various hazardous health effects. Pb causes neurological problem and nervous system damage to children (Salem et al. 2000) as well as several kinds of organ toxicity (Galadima and Garba 2012); Cd is carcinogenic; Cu and Zn both cause liver and kidney damage (Salem et al. 2000; Nolan 1983; Krishnamurti and Vishwanathan 1991). Hg causes spontaneous abortions, congenital malformation, and neurological disorders (Duruibe et al. 2007; Leentech 2004).

Other public health issues in relation to floods include the deteriorating nutritional status of children and NCDs. Emerging evidence suggests links between a rising incidence of NCDs and climate change (Watts et al. 2017). NCDs are a growing concern in Bangladesh in general, with special concern for the urban population. The estimated total number of deaths due to NCDs in Bangladesh in 2012 was over 886,000 which was almost two-third of total deaths (WHO 2014). The majority of deaths from NCDs occurred in urban areas (Mathers and Loncar 2006). Natural hazards such as floods would worsen the health of people with NCDs and malnutrition by reducing access to health and medical care, medicine, and proper diets.

8.4 Conclusion

The historical review of public health issues related to floods in Bangladesh has revealed an overall improvement in physical health. However, immediate actions are necessary to deal with emerging and pressing public health concerns, including mental health, domestic violence, and urban flood caused by floods and have drawn scant attention till now. As the Sendai Framework for Disaster Risk Reduction suggests, an integrated and multi-hazard disaster risk reduction approach is required for improving disaster management systems and health should be at the center of future programs (UNISDR 2017). Community clinics in the country enforce such a direction and should be supported. Within flood-affected regions, there exist inequities in risk in terms of the level of exposure to floods, socio-economic situations, ownership of modern communication devices, and access to health care and education (Hanifi et al. 2010). In conclusion, not only integration of health components into every aspect of risk management but also identification of the most vulnerable groups in society to floods by gender, socioeconomic status, and specific health needs including disability are essential. The disaster management program, then, can render relevant means for each group with different needs to prevent, protect, and promote their health to be resilient against floods. Gender-sensitive approaches should also be an integral part of such efforts.

References

Abedin MA, Habiba U, Shaw R (2014) Community perception and adaptation to safe drinking water scarcity: salinity, arsenic, and drought risks in coastal Bangladesh. Int J Disaster Risk Sci 5:110–124

Afsana K, Wahid SS (2013) Health care for poor people in the urban slums in Bangladesh. Lancet 382(December 21/28)

Ahmed MK, Rahman M, Ginneken J (1999) Epidemiology of child deaths due to drowning in Matlab, Bangladesh. Int Epidemiol Assoc 28:306–311

Anwar I, Nababan HY, Mostari S, Rahman A, Khan JA (2015). Trends and inequities in use of maternal health care services in Bangladesh, 1991–2011. PLoS ONE 10(3):e0120309

Bangladesh Bureau of Statistics (2014) Census on Slum Dwellers. Ministry of Planning, Government of Bangladesh, Dhaka

Biswas A et al (2010) Unintentional injuries and parental violence against children during flood: a study in rural Bangladesh. Rural and Remote Health 10(1199)

Cairncross S, Feachen RG (1993) Environmental health engineering in the tropics: an introductory text. Wiley, Chichester

Cash RA et al (2013) Reducing the health effect of natural hazard in Bangladesh. Lancet 382:2094–2103

Center for Excellence in Disaster Management & Humanitarian Assistance (2017) People's Republic of Bangladesh

Chou C (2003) Hydrogen sulfide: human health aspects. Concise International Chemical Assessment Document 53. WHO, Geneva

Choudhury WA et al (2006) Mental health and psychosocial aspects of disaster preparedness in Bangladesh. Int Rev Psychiat 18(6):529–535

Chowdhury AM, Bhuiya A, Chowdhury ME, Rasheed S, Hussain Z, Chen L (2013) The Bangladesh paradox: exceptional health achievement despite economic poverty. Lancet 382:1743–1745

Chung K-T (1983) The significance of azo-reduction in the mutagenesis and carcinogenesis of azo dyes. Mutat Res/Rev Genet Toxicol 114(3):269–281

Cogliano VJ et al (2005) Meeting report: summary of IARC monographs on formaldehyde, 2-butoxyethoanol, and 1-tert-butoxy-2-propanol. Environ Health Perspect 113(9):1205–1208

Dalal K (2008) Causes and consequences of violence against child labour and women in developing countries. Karolinska Institution, Stockholm

de Aragao Umbuzeiro G et al (2005) The contribution of azo dyes to the mutagenic activity of the Christais River. Chemosphere 60(1):55–64

DESA/Population Division, U. N. (2018) World urbanization Prospects 2018

Dewan AM, Corner R, Hashizume M, Ongee ET (2013). Typhoid fever and its association with environmental factors in the Dhaka Metropolitan area of Bangladesh: a spatial and time-series approach. PLOS ONE Negl Trop Dis 7(1):e1998

Director General of Health Services, Institute of Child and Mother Health, UNICEF, The Alliance for Safe Children (2003) Bangladesh Health and Injury Survey: Report on Children. Dhaka, Director General of Health Services, Institute of Child and Mother Health, UNICEF, The Alliance for Safe Children

Disaster Management Bureau (DMB), Department of Disaster Management, Ministry of Disaster Management and Relief, Government of Bangladesh (2010) Bangladesh: national plan for disaster management 2010–2015. Bangladesh, DMB, Government of Bangladesh, Dhaka

Duruibe J et al (2007) Heavy metal pollution and human biotoxic effects. Int J Phys Sci 2(5):112–118

Firoz T et al (2011) Preeclampsia in low- and middle-income countries. Best Pract Res Clin Obstet Gynaecol 25:537–548

Galadima A, Garba Z (2012) Heavy metals pollution in Nigeria: causes and consequences. Elixir Pollut 45(7917–22)

General Economic Division, Government of Bangladesh (2015) Millennium development goals Bangladesh progress report 2015. General Economic Division, Government of People's Republic of Bangladesh, Dhaka

Halder JN, Islam MN (2015) Water pollution and its impact on the human health. J Environ Hum 1(2):36–46

Hanifi SMA et al (2010) High concentration of childhood deaths in the low-lying areas of Chakaria HDSS, Bangladesh: findings from a spatial analysis. Glob Health Act Suppl 1

Haque U et al (2012) Reduced death rates from cyclones in Bangladesh: what more needs to be done? Bull World Health Organ 90:150–156

Hashizume M, Wagatsuma Y, Farque ASG, Hayashi T, Hunter PR, Armstrong B, Sack DA (2008) Factors determining vulnerability to diarrhea during and after severe floods in Bangladesh. J Water Health 6(3):323–332

Hashizume M, Dewan A, Sunahara T, Rahman MZ, Yamamoto T (2012) Hydroclimatological variability and dengue transmission in Dhaka, Bangladesh: a time-series study. BMC Infect Dis 12(98)

Khan SA (2008) Disaster preparedness for sustainable development in Bangladesh. Disaster Prev Manage Int J 17(5):662–671

Khan AE et al (2014) Salinity in drinking water and the risk of (pre)eclampsia and gestational hypertension in coastal Bangladesh: a case-control study. PLoS ONE 9(9):e108715

Kogevinas M et al (2001) Human health effects of dioxins: cancer, reproductive and endocrine system effects. Hum Reprod Update 7(3):331–339

Kreft S, Eckstein D, Melchior I (2016) Global climate risk index 2017: who suffers most from extreme weather events? Weather-related loss events in 2015 and 1996 to 2015. GERMANWATCH

Krishnamurti C, Vishwanathan P (1991) Copper in the Indian environment and its human health implications. Tata McGraw Hill Pub Camp Ltd

Lara RJ et al (2009) Influence of catastrophic climatic events and human waste on vibrio distribution in the Karnaphuli Estuary, Bangladesh. EcoHealth 6:279–286
Leentech R (2004) Water treatment and air purification. Rotterdamseweg, Netherland, Lenntech
Mahbubur Rahman M et al (2017) Exploring the impact of rural-urban migration on urban land use and land cover: a case of Dhaka city, Bangladesh. Migr Dev
Mathers CD, Loncar D (2006) Projections of global mortality and burden of diseases from 2002–2030. Plos Med 3(11):e442
Milojevic AB, Amstrong B, Hashizume M, McAllister K, Faruque A, Yunis Md, Kim SP, Moji K, Wilkinson P (2012) Health effects of flooding in rural Bangladesh. Epidemiology 23(1):107–115
Misra V, Pandey S (2005) Hazardous waste, impact on health and environment for development of better waste management strategies in future in India. Environ Int 31(3):417–431
Moore M, Chandra A, Feeny KC (2012) Building community resilience: what can the United Stated learn from experiences in other countries. Disaster Med Public Health Prep 7(3):292–301
Nolan KR (1983) Copper toxicity syndrome. J Orthomol Psychiatr 12(4):270–282
OECD (2005) Environmental requirements and market access. OECD
Prevention Web, U. (2017) Bangladesh disaster & risk profile. Retrieved 31 May 2018, from https://www.preventionweb.et/countries/bgd/data
Qayyum MA (2016) Disaster management in Bangladesh: Reducing vulnerabilities. The Daily Star, Dhaka
Quayyum Z et al (2013) Can community level interventions have an impact on equity and utilization of maternal health care-Evidence from rural Bangladesh. Int J Equity Health 12(22)
Rahman A, Alonge O, Bhuiyan AA, Agrawal P, Salam S, Talab A, Sadeq-ur Rahman Q, Hyder AA (2017) Epidemiology of drowning in Bangladesh: an update. Int J Environ Res Public Health 14(488)
Rashid SF (2000) The urban poor in Dhaka City: their struggle and coping strategies during the floods of 1998. Disaster 24(3):240–253
Salem H et al (2000) Heavy metals in drinking water and their environmental impact on human health. ICEHM 2000(September):542–556
SCTEE Plenary Meeting (1999) The risk of cancer caused by textile and leather goods coloured with azo dyes. CSTEE Plenary Meeting, Brussels
Sifat S, Viennet VE, Kathryn Glass, David Harley (2015) The emergence of dengue in Bangladesh: epidemiology, challenges and future disease risk. Trans R Soc Trop Med Hyg 109:619–627
Shimi AC, Ara Parvin G, Biswas C, Shaw R (2010) Impact and adaptation to flood: a focus on water supply, sanitation and health problems of rural community in Bangladesh. Disaster Prev Manage 19(3):298–313
UN-Habitat (2009) State of the World Cities: Harmonious Cities 2008/2009. UN-Habitat, London
UNICEF B, Media Centre (2017) Shielding children from severe injuries
UNISDR Prevention Web (2017) Bangladesh disaster & risk profile. Accessed May 31, 2018.
Ullah ANZ et al (2006) Human health and industrial pollution in Bangladesh. Dhaka
UNDP (2014) The human development report: urbanization challenges and opportunities. UNDP
Veenema TG et al (2017) Climate change-related water disaster's impact on population health. J Nurs Scholarsh 49(6):625–634
Watts N et al (2017) The Lancet Countdown: tracking progress on health and climate change. Lancet 389(10074):1151–1164
Whitehead P, Bussy G, Hossain MA, Dolk M, Das P, Comber S, Peters R, Charles KJ, Hope R, Sarwar Hossain Md (2018) Restoring water quality in the polluted Turag-Tongi-Balu river system, Dhaka: Modelling nutrient and total coliform intervention strategies. Sci Total Environ 631–632:223–232
WHO (2008) World report on child injury prevention. WHO, Geneva
WHO (2014) Noncommunicable diseases country profile. WHO, Geneva

WHO (2015) Disaster risk management in post-2015 development goals. WHO, Geneva
World Bank (2015) World Development Indicators. World Bank, Washington, DC
Wu J et al (2011) Increase in diarrheal disease associated with arsenic mitigation in Bangladesh. PLoS ONE 6(12):e29593
Xu X et al (1998) Association of petrochemical exposure with spontaneous abortion. Occup Environ Med 55(1):31–36

Chapter 9
Smart Water Solutions to Address Salinity, Drinking Water and Health Issues in Coastal Bangladesh

Md. Anwarul Abedin, Bhaswati Ray, Mohammad Golam Kibria and Rajib Shaw

Abstract World water resources are facing dramatic changes as a result of global climate change, high water demands, population growth, industrialization and urbanization, and safe drinking water is one of the most pressing commodities in the developing world. Among others, Bangladesh provides an excellent setting for investigation; with two-thirds of its land area less than 5 m above sea level, it is one of the countries most threatened by sea level rise and saltwater intrusion. Consumption of saline drinking water in coastal regions of Bangladesh is a major threat considering the health issues, viz. physical and mental aspect of human health including maternal and child health and about 30 million people are at risk. With this contrast, this book chapter highlights extent and consequence of drinking water salinity on human health, how smart water solution including GIS tool, i.e., geographic data of safe water collection source, will assist to cope with the adverse impact of health scenario in the coastal community. Finally, this chapter will emphasize community participation on co-design and co-develop of this smart water solution and customization with local context to diminish health risk by providing safe drinking water.

Keywords Drinking water salinity · Health consequence · Smart water solution · Coastal community

Md.A. Abedin (✉) · M. G. Kibria
Department of Soil Science, Bangladesh Agricultural University, Mymensingh 2202, Bangladesh
e-mail: m.a.abedin@bau.edu.bd

B. Ray
SivanathSastri College, University of Calcutta, Kolkata, India

M. G. Kibria
School of Agriculture and Environment, University of Western Australia, Perth, Australia

R. Shaw
Graduate School of Media and Governance, Keio University, Fujisawa, Japan

E. Ying Yang Chan and R. Shaw (eds.), *Public Health and Disasters*, Disaster Risk Reduction, https://doi.org/10.1007/978-981-15-0924-7_9

9.1 Introduction

Water is a valuable natural resource facing a serious threat at present. Water crisis is a severe problem across the globe, but this problem is much acute in developing countries such as Bangladesh (Panos 1998). The water resource and its quality are facing significant changes with the changing environment from local level to global aspects (Easterling et al. 2000). Water supply and availability are being severely influenced by water crisis problem because of many uncertain effects of climate change (Baker 2013). Among the developing countries, Bangladesh is facing acute water crisis problem due to saline water intrusion in coastal areas (Rahman and Bhattacharya 2006). Coastal areas of Bangladesh are already established as a climate hot spot area because of fresh drinking water crisis (Khan et al. 2011). Reduced river flow, long spell of dry season, low annual rainfall and sea level rise aggravate the problem of water salinity in coastal belt of Bangladesh. People all over the country are facing the problem of contaminated water as a consequence of climate change (Kabir et al. 2016), but coastal belts are becoming the worst hit compared to rest other parts of the country due to water crisis as well as frequent exposure of different natural disasters. Poor community people in coastal belt are more vulnerable to water crisis because they do not have any other sources of fresh drinking water for regular consumption (Vineis et al. 2011). Around 97% people of coastal belt of Bangladesh are directly dependent on groundwater sources for drinking purposes (Shamsudduha 2013), and around 20 million people are affected due to regular consumption of saline drinking water (Khan et al. 2011). Arsenic-contaminated groundwater is another severe problem of saline water in coastal areas of Bangladesh. There are 12 districts in Bangladesh with high arsenic contaminated groundwater; among those, seven districts in coastal areas are included with high arsenic contamination. Regular consumption of saline water for drinking and irrigation purposes aggravates the problem of arsenic contamination and causes serious health problem due to continuous exposure of this contaminated water.

Fresh drinking water scarcity in coastal belt of Bangladesh negatively influenced human health (Islam 2013), and at present, it is great challenge for the government to supply freshwater sources for the community people to ensure healthy community (Khan et al. 2011). Researchers reported that regular consumption of saline water leads to severe health problem (McMichael 2003), which is not clearly understood yet (Nahian et al. 2018). Increased blood pressure and hypertension lead to different cardiovascular diseases which are associated with the consumption of saline drinking water (Dennison-Himmelfarb et al. 2013). Regular consumption of high concentration of sodium and arsenic with saline water aggravates the problem of hypertension (DGHS 2013). According to WHO, 200 mg/L sodium and 250 mg/L chloride is considered as saline water, but in Bangladesh chloride level in drinking water can be accepted up to 1000 mg/L (Ahmed and Rahman 2000). This threshold level of sodium and chloride was fixed considering the taste of water, not based on health consideration (WHO 2012). Each year saltwater affected areas are increasing due to sea level rise, which leads to saline water intrusion in coastal belt and makes the groundwater saline as well as unfit to use. The present trend of saline water intrusion with freshwater sources aggravates the water crisis problem, and the coastal people are becoming highly vulnerable to the problem of hypertension and high blood pressure associated with the consumption of saline drinking water with high arsenic contamination.

Smart water solution can be the best option to supply fresh drinking water for the coastal people in Bangladesh. It combines the community people participation with the local government and web-based interface to facilitate the process of water management system. After installing the web-based interface by the scientists and technocrats, it will be finally owned by the community people and will also be governed by the community people with active collaboration between local government and non-government organizations. In this smart water solution system, location of fresh and safe drinking water source will be indicated with the GPS location, so that community people can collect water from their nearest sources. Besides, it will be updated by the community people with the help of local government if any freshwater source is identified. Thus, this continuous updating database will help the local people to be aware about the safe drinking water sources in their nearest ease.

The above information regarding the fresh drinking water scarcity in the coastal areas of Bangladesh clearly indicates the importance of finding a solution to combat this problem. Due to the important role of water for achieving sustainable development, it is mandatory to ensure drinking water and waste or problem water efficiently in different era of climate variability and population growth. The water management system must be sustainable and resilient to long term and consistent increase in water consumption levels and this can be achieved by smart water solution. Therefore, this chapter will highlight the scope of smart water solution to combat water crisis problem and health issues of the people living along the coast of Bangladesh.

9.2 Coastal Areas of Bangladesh

Bangladesh is a developing country, where a major portion of people are living below the poverty level, and it is highly vulnerable climate change because of low-lying country located in Ganges–Brahmaputra Delta (Allison et al. 2003). The coastal region of Bangladesh covers around 20% of total land area and approximately 30% of total arable land in the country (Minar et al. 2013). Bangladesh is a low-lying riverine country with a coastline of 580 km (360 mi) on the northern seaside of the Bay of Bengal. The coastal belt of Bangladesh is one of the climate hotspot areas of Bangladesh, and the community people are facing lots of challenges due to the adverse effects of climate change.

Bangladesh's coastal area is not uniform or static; it is dynamic like the people of Bangladesh (Brammer 2014). The intrusion of saline waters in coastal areas of Bangladesh contaminates the sources of fresh drinking water. Thus, the people living in coastal areas are facing a severe crisis of fresh drinking water, and agricultural production is drastically decreasing due to saline soil and lack of freshwater irrigation. Due to the change in the climatic condition, sea level is rising in coastal areas of Bangladesh and it is continuously increasing. The areas that are exposed to saline water in coastal areas are increasing over time. The problem of salinity in coastal areas becomes higher in dry season, and the severity of the problem of salinity is less in rainy season (Carter 1975). The figure of coastal belt of Bangladesh shows the exposed coast to saline water, which is the study area for the saline drinking water.

9.3 Causes of Freshwater Crisis in Coastal Areas of Bangladesh

The main reason for saline drinking water problem in coastal areas of Bangladesh is the intrusion of saline water in the coast, and each year the area exposed to saline water is increasing due to sea level rise. Groundwater generally occurs under confined conditions at a depth ranging from 20 to 400 m. The upper group of aquifers contains saline water at a depth of 20–168 m, with the salinity level above the permissible limit of 600 ppm. Salinity in these aquifers is due to sodium chloride (NaCl) concentration derived from seawater ingress and saltwater entrapped in the sediments under marine conditions. The freshwater-bearing aquifers occur within a depth range of 160–490 m below ground level separated by a blanket of clay whose thickness varies between 40 and 150 m (Gayen and Zaman 2013). In coastal belt, saline water intrusion is quite common and is accentuated by global warming and sea level rise, unplanned dam and embankment construction and spread of shrimp farming. Wetland shrimp cultivation increases the salinity of adjacent freshwater ponds and shallow aquifers through seepage of saltwater. Thus, all the freshwater sources are getting contaminated with the saline water, and these water sources are becoming unsuitable for drinking, because excessive intake of saline water can cause serious health diseases.

9.4 Salinity Status of Drinking Water in Coastal Areas

People living in coastal belt of Bangladesh are facing serious problem of water salinity in both surface and groundwater sources (Sakamoto 2017; Abedin et al. 2014), causing severe crisis of fresh drinking water. Therefore, women in coastal and haor areas need to go miles to collect water for their regular drinking purposes. Most of the community people in coastal areas of Bangladesh are poor, and piped water is not available in this region. Therefore, the people in coastal belt of Bangladesh are dependent on tube wells or ponds for drinking water sources, and those are highly susceptible to saline water intrusion. The salinity level depends on the depth of tube well and location (distance from the coast). Shallow tube wells are highly susceptible to saline water intrusion, and the tube wells those are installed deeper than 200 m are likely to be salt free (Sakamoto 2017).

Drinking water from natural sources in coastal Bangladesh has become contaminated by varying degrees of salinity due to saltwater intrusion from rising sea levels, cyclone and storm surges, and upstream withdrawal of freshwater (Khan et al. 2011). The level of water salinity in the western part of the southwest region is higher than the eastern part of coastal belt of Bangladesh. Average salinity is higher in the dry season than the monsoon because of inadequate supply of freshwater from the upstream. The salinity problem generally builds up from October to May and remains higher during the dry season, and salinity drops after May due to upstream slow and rainfall (IWM 2014).

9.5 Problems Associated with Saline Water Consumption

Comprising crops, livestock, forestry, aquaculture, fisheries, and other side products are facing serious problem due to the problem of water salinity in the coastal areas of Bangladesh. Agricultural and economic sectors are severely influenced due to freshwater crisis in coastal areas of Bangladesh.

- *Agricultural sectors*: Saline water in the coastal areas causes significant decrease in agricultural production. It affects agricultural production by interfering with nitrogen uptake, reducing plant growth and pastures reproduction. Plant growth stopped due to excessive accumulation of sodium and chloride ions those are highly toxic to plants. Plants take up nutrients from soil water and water moves into plant roots by osmosis, which is controlled by the level of salts in the soil water and in the water within the plant. If salt concentration in the soil water is too high, water may flow from the plant roots back into the soil. This results in the dehydration of the plant, causing yield decline or even death of the plant. Dryland salinity is also closely linked to other soil degradation issues, including soil erosion. Salinity is often associated with prolonged wetness and lack of surface cover and therefore increases the vulnerability of soils to erosion.
- *Economic loss*: Infrastructures are significantly damaged due to continuous exposure of saline water, and it increases repair and maintenance costs for a range of services provided for public use such as roads and buildings. Road and bridge damages are caused by shallow and saline groundwater involves huge cost to repair and maintain. Other than these, saline water also causes significant damage to footpaths, parks, sewage pipes, housing and industry. It also includes additional costs incurred by the community people to minimize salinity and rising water table problems such as installing rainwater harvesting tanks, development of subsurface drainage system, use high specification materials during road and building construction.

All the community people and stakeholders are negatively influenced by freshwater scarcity in coastal belt of Bangladesh. Table 9.1 summarizes the general impacts of regular saline water consumption on different aspects. Physical infrastructures, agriculture, human health, economic status, etc., are significantly influenced by water salinity.

Table 9.1 Impact of saline water in coastal areas of Bangladesh

Sectors	Consequences
Infrastructures	Continuous exposure of different physical infrastructures to saline water causes permanent damage to houses, roads, playing fields, etc. The lifespan of road pavements decreases significantly when they remain submerged for couple of days under saline water due to sea level rise. Saline water corrodes the properties of bitumen, concrete and brick structures, which decreases their values and stability
Agriculture	Plants normally take up water from soil through their root systems by osmotic process which is controlled by the level of salt concentration in soil solution and water contained in plants. High salt concentration in soil solution makes water to flow from the plant roots back into the soil, leading to dehydration causing significant yield loss
Economic stability	Due to significant loss in agricultural production, people in coastal areas become affected in terms of their economic stability, which ultimately affects national economy. Reconstruction of damaged infrastructures (roads, buildings, etc.) also involved huge cost each year
Human health	Regular consumption of saline water causes different health problems among the community people such as hypertension, high blood pressure and cardiovascular diseases. Due to high arsenic concentration in groundwater of the coastal belts, the problems are getting more aggravated within the poor community people in that area
Biodiversity	Most of the plant species cannot withstand excess salt concentration in their root zone and getting destroyed, which is responsible for biodiversity loss in coastal belt of Bangladesh. Destruction of natural habitat and fragmentation of wildlife corridors cause significant change to terrestrial biodiversity and landscape pattern in coastal belts
Soil erosion	Soil erosion is very common due to dryland salinity, which is responsible for different issues related to soil degradation. Saline soil is associated with prolonged wetness and lack of surface cover which makes the soil more vulnerable to become eroded
Water quality	Deterioration of water quality is the most severe impact of saline water intrusion in coastal areas, which causes salinization of previously freshwater sources. High level of salt in drinking water affects its taste and hard to consume. High concentration of sodium chloride and magnesium sulfate level in drinking water may cause laxative effects and have serious economic, social and environmental consequences

9.6 Health Impact Due to Saline Drinking Water

Regular consumption of saline water negatively influences different health aspects of community people in coastal areas of Bangladesh. Researchers reported that around 75% of the community people are suffering from gastrointestinal problems and diarrhea and around 75% also complained of suffering from black spots on skin while nearly 40% respondents complained of hyperkeratosis (Islam 2013). About hypertension, they were unsure whether it is caused by any of the contaminants or they were genetically prone to such conditions. Apart from the health problems, the respondents also complained of social stigma faced during marriage negotiations. Human body can tolerate the salt concentration up to 5ppt, but salt concentration in saline water has now reached 20ppt (Islam 2013). Table 9.2 summarizes the common health problems found among the community people in coastal areas of Bangladesh found from different researches.

Table 9.2 Health impact of saline water consumption on human health

Causes	Health problems
High salt concentration ($NaCl$, $MgSO_4$)	**Hypertension**, the most acute problem is due to saline water consumption, and this problem is very prominent among the female of high age (>35 years) group and pregnant women
	Cardiovascular diseases, high concentration of salt intake (around 2000 mg/L Na) increases the risk of cardiovascular diseases by 17 and 23% greater risk of stroke
	High blood pressure is a major health problem found among the community people in coastal belt of Bangladesh. Regular consumption of high salt concentration (5 g per day) increases the systolic blood pressure by approximately 9 mmHg. This high BP is responsible for approximately 50% of deaths from coronary heart disease and over 60% of those from stroke
	Kidney stone, bone metabolism and hypercalciuria become prominent among the people in coastal areas of Bangladesh due to the regular consumption of saline water. As a consequence of high blood pressure, associated abnormalities of calcium metabolism lead to these health problems
	Stomach cancer and asthma problems are becoming prominent among the community people due to regular consumption of saline water
Arsenic contamination	**Melanosis and keratosis** is an early stage of arsenic problem due to intake of arsenic contamination of drinking water which is acute among the people in the coastal area of Bangladesh
	Black foot disease (Gangrene) are becoming a very acute problem in some parts of the coastal belt of Bangladesh due to high arsenic concentration in groundwater sources. High arsenic accumulation leads to infection in blood vessels and causes this serious health problem

Different research reported the potential risk of health impact due to saline water consumption, and those are summarized here. High blood pressure (prehypertension and hypertension) is one of the major health problem associated with saline water consumption. Regular consumption of slightly saline and moderately saline drinking water increases the chances of being hypertensive by 17% and 42%, respectively, compared to those who consume fresh drinking water. Women had 31% higher chance of being hypertensive than men, and older people are more likely to be hypertensive compared to younger people. Seasonal variation in drinking water salinity is negligible in coastal areas of Bangladesh, but the propensity of high blood pressure is maximum among the coastal community during dry season. Water salinity and associated occurrence of hypertension were found higher for deep aquifer (21.6%) compared to shallow aquifer (20.8%) (Abedin et al. 2014). The health problem due to regular consumption of saline can be managed by adopting the smart water solution technique. It is very new approach for modern water management system to ensure the supply of safe water for the people living in coastal belt of Bangladesh.

9.7 Concept of Smart Water Solution

Smart water solution combines information and communications technology (ICT) and real-time data and responses as an integral part of the solution for water management challenges (Fig. 9.1). This technique uses citizen science with the active collaboration between local government and community people. It includes water treatment techniques such as ICT-based smart water solution technologies that ensure accurate and available real-time data on freshwater sources to combat

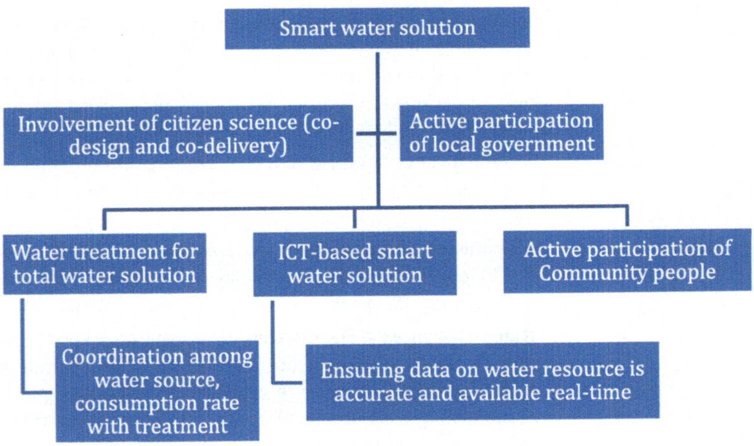

Fig. 9.1 Smart water solution with the involvement of community people (*Source* Authors)

the problem of water crisis in coastal belts of Bangladesh. Smart water solution has the potential to mitigate the freshwater crisis in coastal areas, which includes water quality, efficient irrigation, water flow, and so on to supply healthy environment for the community people. Smart water management can positively influence water quality, agricultural production by providing fresh irrigation source and environment.

Smart water management is mainly designed for providing the information regarding the freshwater resources to mitigate the freshwater crisis of the community people in coastal areas. Smart water solution can maximize the social and economic benefit by integrating ICT for continuous survey and monitoring of fresh and saline water bodies, which can be used for optimizing the water distribution network. Smart water solution involves data collection, recording, integration of data using sensor networks, distribution of data among the community people, processing as well as storage of data and finally modeling, analysis and visualization of decisions support web-based tools (Shahanas and Sivakumar 2016). Smart water solution is an immerging approach for management of water resource that requires more tuning to be well adopted with the system of saline water management system in coastal areas of Bangladesh.

9.8 Application of Smart Water Solution Through Citizen Science

Citizen science approaches provide opportunities to support smart water management system to supply safe drinking water for the community people living in coastal areas of Bangladesh. Even though the term citizen science is relatively new, the practice of participatory research and volunteered science has a long tradition in a wide range of disciplines such as environmental monitoring and life science. (Haklay 2013). Citizen science bridges the co-design and co-delivery system to supply fresh drinking water in coastal belt of Bangladesh (Fig. 9.2). The co-design process requires the development of map highlighting the total water demand by the community people with the smart water mobile application for easiest management system. It also includes the participation of community people for safe drinking water management system. On the other hand, co-delivery covers the identification of freshwater source with the GPS location to assist in the smart water mobile application system with all the nearest drinking water sources, and considering these two points, the authority need to decide on the appropriate water management policy.

In the application of smart water solutions, citizen science can address the problem and provide the solution through data analysis and locate freshwater sources with their updated details through mapping (Fig. 9.3). Citizen science aims to strengthen the linkage between science and society by engaging citizens in community works (Haklay 2013). Citizens can participate in choosing or defining

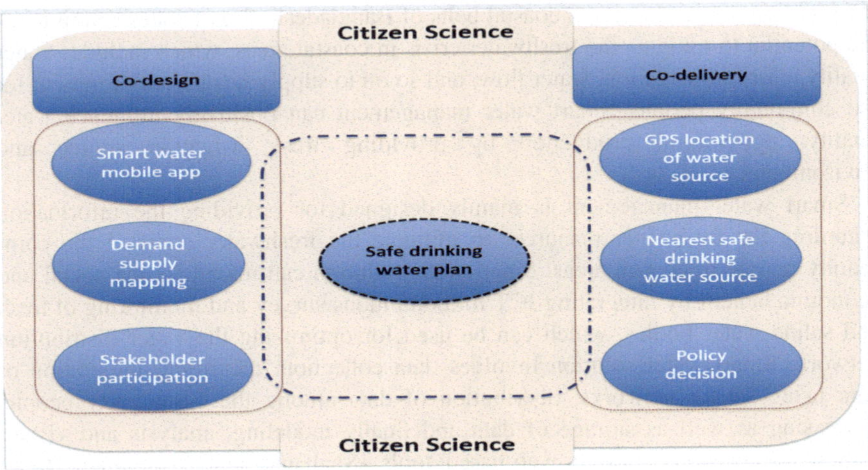

Fig. 9.2 Smart water solution framework through citizen science (*Source* Authors)

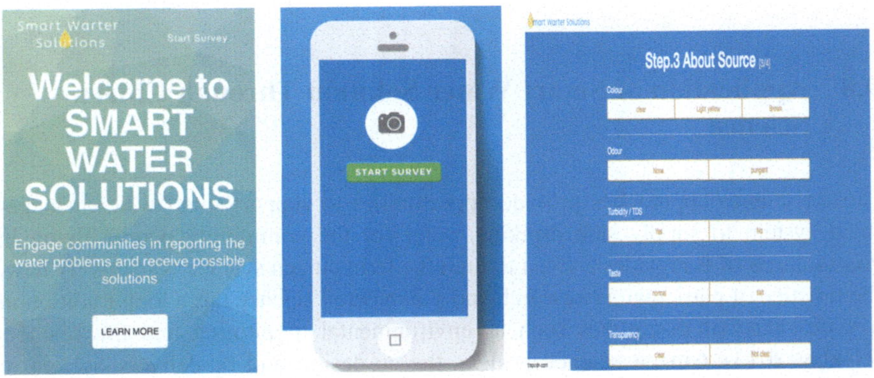

Fig. 9.3 Smart water solution: Web-based interface and mobile application (*Source* Authors)

the problem of saline water in coastal areas, gather information and resources for getting freshwater, develop explanations, design methods to treat the water, collect samples and record data, interpret data and draw conclusions, and disseminate results for the betterment of the community people. The Internet helps projects to reach broad audiences by increasing visibility and allowing interested participants to find topics or projects. Further, the development of social media, mobile devices, online networks and computational facilities multiplies the capacity for data collection, storage, integration, analysis and dissemination (Schroter et al. 2017).

9.9 Combat Water Scarcity and Health Problem Using Smart Water Solution in Coastal Bangladesh

Regular consumption of saline water by the local people in coastal areas creates serious health problem. Therefore, safe drinking water is imperative to mitigate health problems. Smart water solution provides web-based interface having map of available nearest safe water sources for the local community and easy access to the source points. This also gives the information of the alternate options of safe water sources where the water scarcity is acute. Hence, community has the choice of interest regarding freshwater sources in his/her locality. If the information regarding available safe water sources is visible, it will help community to find nearby safe water points, which will ultimately diminish the health problem of the water community associated with water scarcity.

The supply of safe drinking water in the coastal areas of Bangladesh is possible by using smart water solution techniques. This smart water management will be introduced first by experts, and then, it will be handed to the community people, those are the main responsible person to operate it with the current and updated databases of any new location of freshwater sources in the map of web-based interface with GPS location. Thus, community people can ensure the supply of fresh drinking water by helping themselves. By this system, people will be able to identify which source is contaminated with saline water and which one is suitable for drinking purposes. This easiest system will ensure safe water consumption by each and every community people and thus will help to reduce the potential risk of different health issues due to saline water consumption.

9.10 Strategies to Address the Problem of Safe Drinking Water Crisis

9.10.1 Rainwater Harvesting

Rainwater harvesting is one of the feasible options to supply fresh drinking water sources in the coastal areas of Bangladesh, and government is working to promote and install rainwater harvesting systems for coastal communities. The problem of safe drinking water scarcity become more prominent during and after disaster; people tries to solve this problem using advance technologies. The roof top rainwater harvesting system seems suitable for developing both household- and institutional-based RWH systems to get safe drinking water. As rainfall is high from March to October, the harvested rainwater can satisfy the water demand of the household easily. If there is no appropriate storage facility of excess rainwater, people will have to suffer during dry season (November to February). Rainwater harvesting units are installed in some locations in both individual level and

Fig. 9.4 Rainwater harvesting units for safe drinking water (*Source* Authors)

community level (Habiba et al. 2013) at small scale and large scale, respectively (Fig. 9.4). However, rainwater harvesting and pond sand filters are becoming popular in Bangladesh.

Rainwater harvesting is a very sound and economically feasible ways to get safe drinking water, which depends on rainfall, roof size, roofing material, etc. Rainwater conservation system can be used as an effective tool to mitigate freshwater crisis in coastal areas of Bangladesh. Both public and private organizations should be engaged together with different projects to promote, educate and involve communities for rainwater harvesting practice as a very first step of implementation. Partnership with different organization will help to maintain, monitor and ensure the aimed benefit of rainwater harvesting in an effective way. The best option for saline-prone areas is to preserve rainwater in artificial ponds and distribute it to communities. Installation of filtration and desalination plants involves huge cost, which is very difficult to get accepted among the community people. So, installing sand filter systems could be a possible solution, in which hand pumps are used to suck water from artificial ponds through a filter that makes the water potable.

9.10.2 Community People Participation with GOs and NGOs

An integrated approach of community people with local government and non-government organizations is essential for getting safe drinking water. Communities' adaptation methods should be supported by local governments and NGOs to make them both more effective and environmentally friendly. Although a number of organizations are working together at field level, but the system is not getting effective due to the lack of coordination with community people and the lack of integrated approach of the community, government and development organizations. The government should help different organization to coordinate with community people in order to insure proper implementation at the grassroots level.

The smart water solution technique comprises of these strategies to supply safe drinking water for the community people in coastal areas. The sources of safe drinking water such as rainwater harvesting plants, safe ponds and sand filters with overhead tanks or others will be updated in the web-based interface and in mobile applications with the GPS location. So that people can easily get the information of safe drinking water location in a single click and can have drinking water from their nearest accessible locations. This will ultimately reduce the efforts to collect safe drinking water and thereby mitigate the potential risk of different health problems associated with the consumption of saline water.

9.11 Conclusions and Recommendations

Groundwater is the major source of drinking water for coastal people of Bangladesh and substantially contributes to overall sodium intake among coastal communities because of drinking saline water. Poor socioeconomic conditions along with water salinity are the main reason for increasing drinking water scarcity in the coastal areas of Bangladesh. Coordination is essential among individual, community and institutional level to supply fresh drinking water for the community people in coastal areas of Bangladesh. To combat the water crisis in the coastal areas of Bangladesh, smart water management is imperative and can be recommended as a new approach for water management using modern technologies. Smart water solution includes web-based interface map and mobile apps to locate both safe and unsafe drinking water sources. Therefore, people can know which water source is safe for drinking purpose and which one is not safe as a drinking water source. Using the information, the community people in coastal belt can collect safe drinking water from their nearest sources, which will ultimately help to reduce the risk of water-related health problems of the community people in the coastal areas of Bangladesh. Based on the literature, few recommendations are listed below to provide fresh and safe drinking water for the community people:

- Installation of rainwater harvesting systems at individual family level by active coordination between local government and non-government organizations;
- Introducing solar-powered deep tube wells with overhead tanks to supply piped water;
- Proper monitoring and maintenance of community-based fresh pond sand filters to assure the supply of safe drinking water.

Acknowledgements Part of the research was supported by Keio University research project "*Smart Water Solutions in Arsenic and Salinity Prone areas in Coastal Bengal*".

References

Abedin MA, Habiba U, Shaw R (2014) Community perception and adaptation to safe drinking water scarcity: salinity, arsenic, and drought risks in coastal Bangladesh. Int J Disaster Risk Sci 5:110–124

Ahmed MF, Rahman M (2000) Water supply and sanitation: rural and low income urban communities. ITN-Bangladesh. Center for water supply and waste management, BUET, Dhaka, Bangladesh

Allison MA, Khan SR, Good Bred SL Jr., Kuehl SA (2003) Stratigraphic evolution of the late Holocene Ganges Brahmaputra lower delta plain. Sediment Geol 155:317–342

Baker C (2013) Water and international security. Talk for CSCAP 2nd meeting for study group on water. Siem Reap, Cambodia

Brammer H (2014) Bangladesh's dynamic coastal regions and sea-level rise. Climate Risk Manage 1(2014):51–62

Carter DL (1975) Problems of salinity in agriculture. In: Poljakoff-Mayber A, Gale J (eds) Plants in saline environments. Ecological studies (analysis and synthesis), vol 15. Springer, Berlin

Dennison-Himmelfarb C, Handler J, Lackland DT, Lefevre ML, Mackenzie TD, Ogedegbe O, Smith SC Jr, Svetkey LP, Taler SJ, Townsend RR (2013) Evidence-based guideline for the management of high blood pressure in adults report: from the panel members appointed to the Eighth Joint National Committee (JNC 8)

DGHS (2013) National guidelines for management of hypertension in Bangladesh. Director General of health Services, Ministry of Health and Family Affairs, Dhaka, Bangladesh

Easterling DR, Meehl GA, Parmesan C, Changnon SA, Karl TR, Mearns LO (2000) Climate extremes: observations, modeling, and impacts. Science 289(5487):2068–2074

Gayen A, Zaman A (2013) Groundwater development and sustainable management in the delta front area of Sundarban Islands. Int J Emerg Technol Adv Eng 3(11):606–611

Habiba U, Abedin MA, Shaw R (2013) Salinity induced livelihood stress in southwestern part of Bangladesh. In: Abedin MA, Habiba U, Shaw R (eds) Water insecurity: a social dilemma, 1st edn. Emerald Publishers, Bingley, pp 139–165

Haklay M (2013) Citizen science and volunteered geographic information—overview and typology of participation in Sui. In: Sui DZ, Elwood S, Goodchild MF (eds) Crowdsourcing geographic knowledge: volunteered geographic information (VGI) in theory and practice. Springer, Berlin, pp 105–122

Islam S (2013) Safe drinking water disappearing fast in Bangladesh. Retrieved from https://www.theguardian.com/global-development/2013/may/07/safe-drinking-water-disappearing-bangladesh

IWM (2014) Salinity in the South West Region of Bangladesh. Institute of Water Modelling, Bangladesh

Kabir R, Khan HTA, Ball E, Caldwell K (2016) Climate change impact: the experience of the coastal areas of bangladesh affected by cyclones Sidr and Aila. J Environ Public Health 9. Article ID 9654753

Khan AE, Ireson A, Kovats S, Mojumder SK, Khusru A, Rahman A, Vineis P (2011) Drinking water salinity and maternal health in coastal bangladesh: implications of climate change. Environ Health Perspect 119(9):1328–1332

Mcmichael AJ (2003) Climate change and human health: risks and responses. World Health Organization

Minar MH, Belal Hossain M, Shamsuddin MD (2013) Climate change and coastal zone of Bangladesh: vulnerability, resilience and adaptability. Middle-East J Sci Res 13(1):114–120

Nahian MA, Ahmed A, Lázár AN, Hutton CW, Salehin M, Streatfield PK (2018) Drinking water salinity associated health crisis in coastal Bangladesh. Elem Sci Anth 6(1):2

Panos (1998) Liquid assets: is water privatization the answer to access? Panos Media Briefing No. 29, p 1. Retrieved from http://www.oneworld.org/panos/briefing/water.htm

Rahman M, Bhattacharya A (2006) Salinity intrusion and its management aspects in Bangladesh. J Environ Hydrol 14:1–8

Sakamoto M (2017) Saline drinking water and salt in diet: an approximate picture of the situation in a coastal area of southeastern Bangladesh. Int J Disaster Risk Sci 8:109

Schroter M, Kraemer R, Mantel M, Kabisch N, Hecker S, Richter A, Neumeier V, Bonn A (2017) Citizen science for assessing ecosystem services: status, challenges and opportunities. Ecosyst Serv 28:80–94

Shahanas KM, Sivakumar PB (2016) Framework for a smart water management system in the context of smart city initiatives in India. Proc Comput Sci 92(2016):142–147

Shamsudduha M (2013) Groundwater-fed irrigation and drinking water supply in Bangladesh: challenges and opportunities. In: Zahid A, Hassan MQ, Islam R, Samad QA (eds) Adaptation to the impact of climate change on socio-economic conditions of Bangladesh. Alumni Association of German Universities in Bangladesh, German Academic Exchange Service (DAAD), Dhaka

Vineis P, Chan Q, Khan A (2011) Climate change impacts on water salinity and health. J Epidemiol Global Health 1:5–10

WHO (2012) Guideline: Sodium intake for adults and children. World Health Organization, Geneva, Switzerland

Chapter 10
Health Issues and Disaster Risk Reduction Perspectives in China

Emily Ying Yang Chan and Sida Liu

Abstract With environmental degradation and rapid urbanisation, China is one of the most severely affected countries in terms of the number of disasters, human causalities and economic losses. As one of the most important indicators of disaster risk reduction, disaster-related death has been reduced in recent years due to the improvement of capacity on disaster management. As key components of primary prevention under the Health Emergency and Disaster Risk Management (Health-EDRM) framework, the emergency medical and public health response and disease surveillance system have been largely strengthened. However, the health sector in China was often only involved in the later stage of disaster response. As highlighted in the Sendai Framework for Disaster Risk Reduction 2015–2030, the international community of disaster risk reduction is calling for a widened emphasis for the health sector to employ a more proactive approach in future disaster risk reduction-related affairs. Although the newly established Ministry of Disaster Management of China has integrated the duties previously scattered in many government departments, the health sector remains not directly included in its structure. Preventive measures including primary care resilience, health infrastructure safety code and hospital emergency plan are still not in place in many areas of the country. There is an urgent need to build a framework and mechanism to ensure a better involvement of health into new China disaster management system, as highlighted in the Health-EDRM framework.

Keywords China · Disaster management system · Health sector · Ministry of Disaster Management of China · Preventive measures · Surveillance system

E. Y. Y. Chan (✉) · S. Liu
Collaborating Centre for Oxford University and CUHK for Disaster and Medical Humanitarian Response, JC School of Public Health and Primary Care, The Chinese University of Hong Kong, Shatin, Hong Kong
e-mail: emily.chan@cuhk.edu.hk

E. Ying Yang Chan and R. Shaw (eds.), *Public Health and Disasters*, Disaster Risk Reduction, https://doi.org/10.1007/978-981-15-0924-7_10

10.1 Introduction

As a country with vast territory and the world's largest population, China is prone to almost all kinds of natural disasters. According to United Nations Office for Disaster Risk Reduction (UNISDR), China is the country with the highest number of natural disasters during 2004–2015 and incurred the second-highest disaster-related economic loss in the world (UNISDR 2015a). Due to the frequent and intensive effect of disasters, China government has been attaching much importance to disaster management since its establishment in 1949. China has made enormous progress in disaster response and risk reduction. However, along with the gradually intensified effect of climate change and the rapid urbanisation during the last 40 years, China is confronting with increased disaster health risks.

Disaster risk reduction (DDR) is a complex task and requires a multidisciplinary approach. Traditionally, DDR was mostly dominated by experts from geophysical, hydrological and metrological science, who attempted to use mathematical models to estimate and predict the time and scale of disaster. Despite disasters' significant impacts on people's health, medical and health professionals were only involved in the later stage of disaster response, including the provision of emergency medicine and infectious disease control. As recognised by UNISDR in the Sendai Framework for Disaster Risk Reduction 2015–2030, health sector should play a more important role in future DRR and should be actively integrated into all stages and processes of decision-making in DRR (UNISDR 2015b). In this chapter, we attempt to discuss the current health impact of disaster and how the government's new policy could build the DRR capacity as guided by the primary prevention concept under the framework of Health Emergency Disaster Risk Management (Health-EDRM).

10.1.1 Health Impact of Disasters in China

China has the largest population and is one of the biggest countries in the world, covering a land area of approximately 9.6 million km^2, a sea area of more than 4.7 million km^2. China also has the longest range of latitude in the world (5500 km), spans across six climate zones, including cold temperate, temperate, warm temperate, subtropical, and tropical zones, as well as plateau zone in the Qinghai-Tibetan area. Moreover, China has a coastline of more nearly 15,000 km along the Yellow Sea, the East China Sea and the South China Sea. Under these unique landscapes, China faces the threats of all types of natural disasters, ranging from extreme low and high temperatures in the northern and southern regions of China to tropical cyclones and typhoons in the southeast coastline, as well as drought and sandstorm in the northwest. China also lies in the world's major seismic belts (e.g. Circum-Pacific Seismic belt) and is prone to earthquakes.

In 2017, Asia remained the continent with the highest number of disasters occurred, within which China is the most disaster-affected country in the world

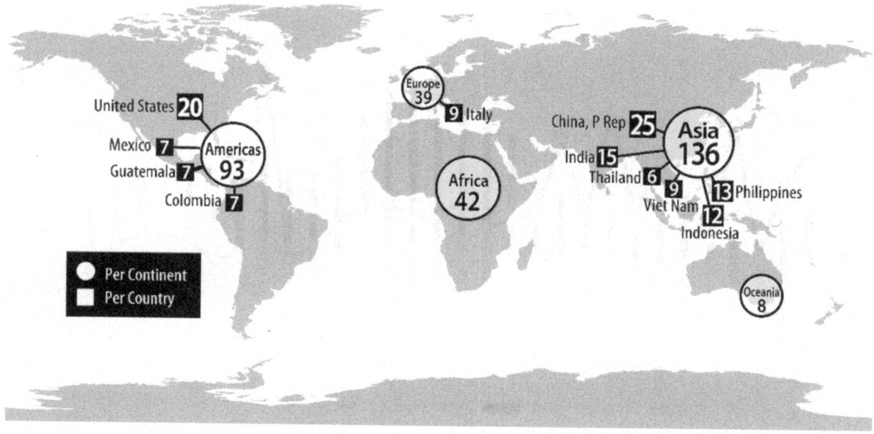

Fig. 10.1 Number of disasters by continent and top 10 countries in 2017. *Source* Below and Wallemacq (2018)

Table 10.1 Total number of mortalities, injured, affected during 2003–2017

	No. death		No. injured		No. affected (millions)	
1	Haiti	237137	Peru	1827734	**China**	**13574.82**
2	Indonesia	180891	**China**	**652398**	India	5468.80
3	Myanmar	139556	Haiti	595024	Philippines	1194.05
4	**China**	**109280**	Sri Lanka	179300	USA	1070.68
5	Pakistan	83588	Philippines	172951	Bangladesh	984.60
6	Russian	56518	Indonesia	157097	Thailand	642.43
7	India	50711	Pakistan	146457	Pakistan	480.55
8	Sri Lanka	37855	Japan	142982	Ethiopia	458.91
9	Iran	28761	Bangladesh	94195	Brazil	403.76
10	France	24388	Viet Nam	84063	Viet Nam	271.97

Data source EM-DAT, The International Disaster Database (2018)

(Fig. 10.1). The health impact of those disasters has also been substantial. Given the relatively large population, China suffered a high number of death tolls and injuries during the last 15 years 3 (EM-DAT International Disaster Database 2018). China also had the highest number of people affected by natural disasters in the world, more than the remaining top 10 countries combined (Table 10.1).

During the last 30 years, China's revolutionary economic reform had led to one of the most rapid economic development in human history. However, this was at the expense of environmental degradation. As a result, both natural and technological disasters became more frequent (Fig. 10.2). To enhance emergency management, the Chinese government issued The State Plan for Rapid Response to Public Emergencies in early 2006. It is a national-level guideline to streamline overall disaster migration and response. Later in April 2006, the Emergency

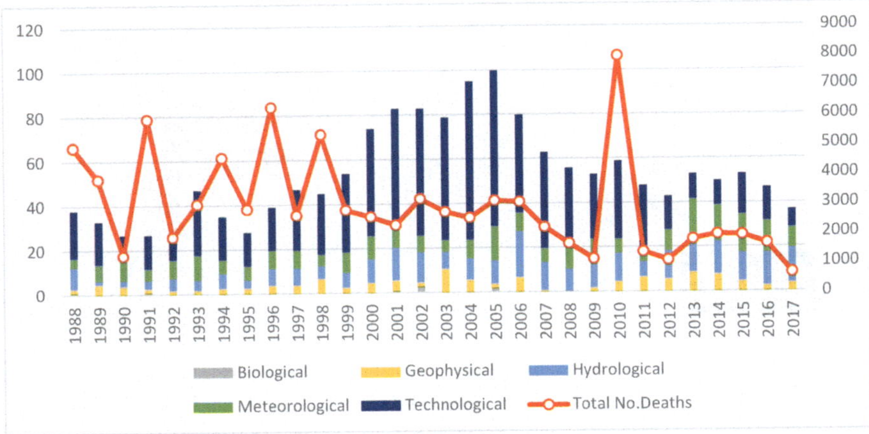

Fig. 10.2 The disaster occurrence trends in China by type of disasters, 1988–2017 (Stacked bar = Disaster occurrence, Red line = Number of disaster-related deaths). *Data source* EM-DAT, The International Disaster Database (2018). *Note* The deaths caused by the Wenchuan earthquake in Sichuan Province in 2008 were excluded, which caused a death toll of more than 80,000. The soaring of death count in 2010 was mainly due to two major disasters, namely Yushu earthquake in Qinghai Province and Zhouqu debris flow in Gansu Province, with more than 5000 killed or missing

Management Office of State Council was also established to ensure implantation of the plan. Since then, more than 100 disaster-related, disaster type-specific laws and decrees were issued (Wang 2008). During the latter half of the last 30 years, the number of disasters was decreased dramatically, mainly due to the reduction of technological/man-made disasters.

In terms of health impact, a significant reduction trend was also observed in disaster-related death (Fig. 10.3). In 2017, the figure dropped to 662, the lowest point in the recent 30 years. Hydrological and technological disasters have been the deathliest cause, followed by geophysical hazards. There is a decreasing trend observed in the recent 10 years, particularly for technical disasters (e.g. mining explosion). This may be due to a series of laws and more restricted governance applied to the mining industry. Many small and non-governmentally managed mines were forced to close after a number of severe incidents occurred during 2004–2005. An exception is 2005 when three major incidents occurred in the same year (442 deaths in the sinking of Dong Fang Zhi Xing cruise, 173 deaths in the port of Tianjin explosion and 73 deaths in the Shenzhen construction waste landslide). Although mortality is traditionally the highest concern of the government and the public, the health impact of disaster on society could be far more beyond it.

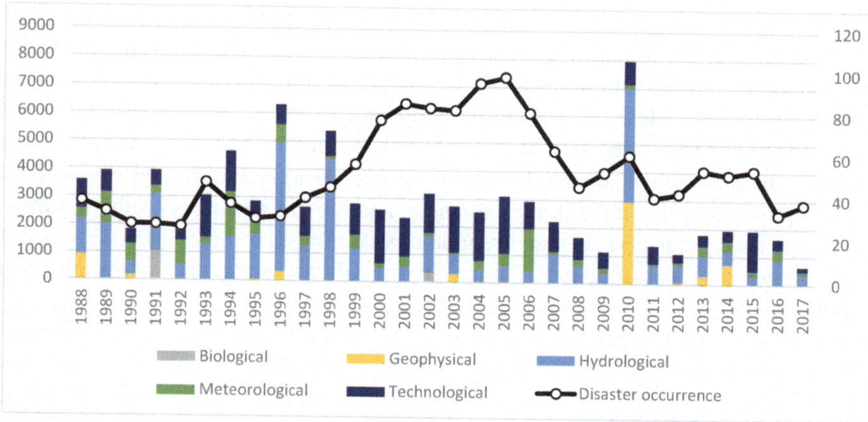

Fig. 10.3 The number of deaths caused by different types of disasters in China, 1988–2017 (Stacked bar = number of deaths, black line = total disaster occurrence). *Data source* EM-DAT, The International Disaster Database (2018). *Note* The deaths caused by the Wenchuan earthquake in 2008 were excluded, which caused a death toll of more than 80,000. The soaring of death count in 2010 was mainly due to two major disasters, namely Yushu earthquake in Qinghai Province and Zhouqu debris flow in Gansu Province, with more than 5000 killed or missing

10.2 Disaster Management and Its Relationship with Health in China

10.2.1 Disaster Management System in China

In China, the term "disaster" refers to an emergency incident. According to the State Plan for Rapid Response to Public Emergencies, emergency incidents are classified into four categories, including natural disasters, accidental disasters, public health incidents and social security incidents. Due to the complex nature of disaster management, 34 ministries and departments, as well as military organisations and social groups were used to involve in this task under the Emergency Management Office of the State Council (Asian Disaster Reduction Centre 2018). However, this office is only active in the emergency phase, especially when a severe catastrophe occurs. Other important aspects such as disaster risk assessment, monitoring, forecasting and public education before the onset of disaster were left to certain ministries or departments to manage and coordinate. For instance, disaster preparedness for industrial incidents was mainly led by State Administration of Production Safety Supervision and Management while Ministry of Public Security was responsible for the coordination of social security incidents. Although this mechanism was designed to promote a multidisciplinary approach, it was hard to perform with effective coordination in a real situation. Under this decentralised governance system, different government agencies held different responsibilities and mandates, which often resulted in ineffective information sharing, unclear accountability and delayed response. More importantly, with those scattered administrative power, the aim of disaster risk reduction had been an enormous challenge to accomplish since individual ministries were lack of interest to promote disaster prevention given the limited financial and human resources. Although there have been more than 100 laws and regulations coming into force since 2006, most of them were agency-specific, making it difficult to implement when multiple departments were involved (Yi et al. 2012).

In recognition of those issues and in order to unify the leadership and management of disaster, the new Ministry of Emergency Management was formed in May 2018, which marked the new era of reform in China's disaster risk reduction system. This infrastructure building as a key strategy of primary prevention under the Health-EDRM framework is exemplary and can become a valuable lesson for other disaster-prone developing countries. This new ministry amassed and integrated the authorities and resources, including functional units, professionals and materials, from more than 13 ministerial departments (Fig. 10.4). It is now a sole government agency directly under State Council (Ministerial level) that responsible for almost all types of disaster/emergency at all phases of the disaster cycle. Previously separated duty for mitigation and response of earthquake, flooding, drought and firefighting (including forest fire) has been integrated under the same leadership with more effective coordination and a clearer boundary of responsibility.

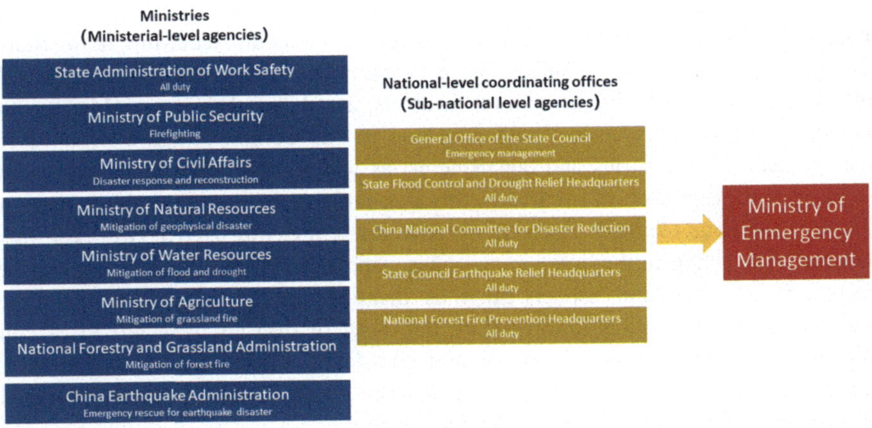

Fig. 10.4 The responsibilities and functions assigned to the Ministry of Emergency Management

10.2.2 Health Sector in Disaster Management and Risk Reduction in China

With the ongoing global disaster and health challenges in the previous decades, the global understanding of the complexity of disaster risk in the twenty-first century has been evolving. The international community of disaster risk reduction is calling for a widened emphasis for the health sector to employ a more proactive approach in future disaster risk reduction-related affairs. All sectors including health should actively engage and collaborate in the decision-making process of preventing, preparing for, responding to and recovering from disaster. This has been reflected in the new global commitment of Sendai Framework in 2015 (UNISDR 2015b).

Health impacts such as mortality and injury are traditional and most notable indicators of disaster management and reduction. Similar to the situation in other parts of the world, the health sector in China is mainly engaged during the later phase of the disaster cycle, usually after the disaster onset. As one of the four categories of emergency incident in China, the public health incidents were mainly related to prevention and response to infectious diseases such as SARS in 2003 and H7N9 in 2007. The management of those issues was mainly led by National Health Commission of PRC (former Minister of Health before 2013) and Chinese Center of Disease Control and Prevention (China CDC).

After the SARS outbreak in 2003, the former Ministry of Health conducted a full review and reform of China's disease surveillance system. The China CDC was established with a specific duty of monitoring, preventing and controlling infectious disease outbreaks in China. Since then, the capacity of responding to public health emergencies in China has been significantly improved. In 2004, a web-based, real-time disease reporting system - National Disease Reporting System was established, which can monitor and provide early warning for 39 notifiable infectious diseases (Yang et al. 2011). Geographically, this system comprises of 605 surveillance points selected by stratified cluster random sampling, covering all 31 mainland provinces, autonomous regions, and municipalities. During the Wenchuan earthquake in 2008, mobile phone-based emergency reporting system was first implemented and provided valuable risk assessment data for disease control in disaster-affected areas (Yang et al. 2009). Meanwhile, China has also actively incorporated the International Health Regulation (IHR) 2005 into its own health emergency capacity building process. According to the WHO progress monitoring report in 2013, China's compliance rate of IHR core public health emergency response indicators has reached 98%, significantly higher than the world's average (Liu et al. 2015). With the effort in recent years, the incidence of most infectious diseases, particularly vector-borne, gastrointestinal and bacterial infections were significantly reduced (Zhang and Wilson 2012).

In addition to infectious disease control and public health, the capacity for providing timely clinical healthcare service under emergency was also improved. First, a number of regional and provincial emergency medical teams (EMT) were established under the leadership of the National Health Commission. Those teams are vehicle-based, self-sufficient and able to carry out emergency medical services such as triage, outpatient care and simple surgery in the field. In recent years, China has also been actively involved in international emergency reliefs, including the Nepal earthquake in 2014 (Lin et al. 2017). As of June 2018, China has built three international level EMT teams (from Shanghai, Guangdong and Sichuan Provinces) that were recognised by WHO. The latest Sichuan team is the first Chinese medical team to be accredited as Type 3, the highest level of WHO EMT Initiative, capable of providing medical care for 200 outpatients and carrying out 15 major surgical operations (WHO 2018).

Despite significant improvement in China's emergency health response infrastructure, many issues and challenges in Health-EDRM primary prevention remain to be resolved:

- Although the new Ministry of Emergency Management centralises the functions and duties of many disaster-related departments, it does not include health units in its organisation structure. It may be because health involves highly specialised disciplines and may better be handled by professionals. However, preventive efforts such as safety code of healthcare infrastructure, local primary care capacity, public first aid skills and general/disaster health literacy are well recognised as important building blocks of DRR. Health sector shall be involved to contribute more in the future.
- An effective disaster response can only be achieved by adequate preparedness prior to disaster occurrence. However, the involvement of the health sector in the disaster risk reduction of the three categories of incidents other than public health has been relatively limited. The health sector in China is still playing a reactive role rather than a proactive one. Previous research indicates that only 65.9% of the tertiary hospitals in 2012 were capable of surging hospital beds in 24 h, with a capacity of adding 12.52% of regular beds (Zhong et al. 2014a).
- A challenge remains on the lack of Health-EDRM-related education in China. In recent years, many universities and institutions have developed courses or departments specifically for disaster medicine training. However, those courses are mainly designed for providing clinical care during disasters. There is a clear shortage of experts capable of conducting preventive public health investigation and service in disaster settings. The field training course with the concept of experiential learning shall be developed as a supplementary component in the future curriculum of faculty of medicine. This will provide the opportunity for future medical professionals to have field/disaster exposure during their university training.
- During the Wenchuan earthquake in 2008, the health infrastructure in the affected area was severely demeaned (Redmond and Li 2010), resulting in a significant medical service gap in the early stage after the disaster occurred.
- The allocation of funding recourses is imbalanced. The funding and healthcare service gap between urban and rural areas, western and eastern regions, emergency response and daily basis services widely exist. The local primary care

significantly lags behind in economically deprived areas due to lack of budget and expertise.

- There is also a lack of incentive for doctors and healthcare institutions to promote disaster preparedness and response. After the healthcare reform in the 1980s, the national finance support to public health sector was significantly reduced. Healthcare providers such as hospitals and clinics have become more self-sufficient and profit-driven. The hospitals and doctors are paid much more for their daily service than disaster-related work (Zhong et al. 2014b).
- A previous study of the Wenchuan earthquake shows that 77% of evacuated patients who required care had underlying chronic medical conditions (Chan 2008). Lack of supply in chronic disease medication and emphasis on emergency/trauma care were part of the reasons. As of 2018, there is no official technical guideline for control and management of chronic diseases in disaster settings available.

10.3 Conclusion

Health is a basic right of human beings, which can be severely threatened by disasters. In order to protect health of the population during and after a disaster, wider determinants of health should be addressed, especially through adequate preparedness toward health-related challenges in the form of primary prevention and DRR infrastructure building. The latest international development and DRR frameworks epitomised in the concept and framework of Health-EDRM has recognised health as an essential component for overall national DRR strategy. Effective health emergency and disaster risk management require health sector to perform a more proactive role in the future. Over the previous decades, China has made a great progress on its capacity on health emergency management, and the establishment of Ministry of Emergency Management marks the new era of China's DRR, with a new focus on primary prevention of disaster health risk. With this unprecedented reform underway, there is, however, an urgent need to ensure the building of a Health-EDRM supporting policy framework to facilitate effective implementation mechanisms to enhance health risk management and to protect well-being of Chinese citizens.

References

Asian Disaster Reduction Centre, Information on Disaster Risk Reduction of the Member Countries. Available at: http://www.adrc.asia/nationinformation.php?NationCode=156&Lang=en&Mode=country. Accessed on 5 Oct 2018

Below R, Wallemacq P (2018) Annual disaster statistical review 2017. Centre for Research on the Epidemiology of Disasters, Brussel, Belgium. https://cred.be/sites/default/files/adsr_2017.pdf

Chan EY (2008) The untold stories of the Sichuan earthquake. Lancet 372(9636):359–362

EM-DAT International Disaster Database (2018). Centre for Research on the Epidemiology of Disasters. Data retrieved from https://www.emdat.be/database

Lin X, Liu K, Zhang Y, Dan Y, Xing D, Chen L, Du D (2017) China medical team: medical rescue for "4.25" Nepal earthquake. Chin J Traumatol 20(4):235–239

Liu B, Sun Y, Dong Q, Zhang Z, Zhang L (2015) Strengthening core public health capacity based on the implementation of the International Health Regulations (IHR) (2005): Chinese lessons. Int J Health Policy Manag

Redmond A, Li J (2010) The UK medical response to the Sichuan earthquake. Emerg Med J 28 (6):516–520

UNISDR (2015a) The economic and human impact of disaster in the last 10 years, available at: https://www.flickr.com/photos/isdr/16111599814/. Accessed on 15 Sep 2018

UNISDR (2015b) Sendai framework for disaster risk reduction 2015–2030. UNISDR, Geneva. Available at: https://www.unisdr.org/files/43291_sendaiframeworkfordrren.pdf. Accessed on 3 Oct 2018

Wang XJ (2008) The comparative research of natural disaster management system in China and other countries. Master dissertation. Lanzhou Lanzhou University Press, vol 5, pp 31–33

WHO (2018) China International Emergency Medical Team (Sichuan). Available at: https://extranet.who.int/emt/content/china-international-emergency-medical-team-sichuan. Accessed on 3 Oct 2018

Yang CH, Yang J, Luo XS, Gong P (2009) Use of mobile phones in an emergency reporting system for infectious disease surveillance after the Sichuan earthquake in China. Bull World Health Organ 87:619–623

Yang W, Li Z, Lan Y, Wang J, Jin L, Sun Q et al (2011) A nationwide web-based automated system for outbreak early detection and rapid response in China. Western Pac Surveill Response J 2(1):10–15

Yi L, Ge L, Dong Z, Zhou J, Gao Z (2012) An analysis on disasters management system in China. Nat Hazards 60:295–309

Zhang L, Wilson DP (2012) Trends in notifiable infectious diseases in China: implications for surveillance and population health policy. PLoS ONE 7(2):e31076

Zhong S, Hou X-Y, Clark M, Zang Y-L, Wang L, Xu L-Z et al (2014a) Disaster resilience in tertiary hospitals: a cross-sectional survey in Shandong Province, China. BMC Health Serv Res 2014(14):135

Zhong S, Clark M, Hou XY, Zang Y, FitzGerald G (2014b) Progress and challenges of disaster health management in China: a scoping review. Glob Health Action 2014:7

Chapter 11
Health and Disaster Risk Management in India

Supriya Krishnan and Ila Patnaik

Abstract India has been rapidly urbanizing. Its state of health, well-being, and infrastructure capacity are in a period of transformation. Through the perspective of a rapidly urbanizing nation, the chapter presents an overview of India's health capacity in managing disaster risks. It looks at demographic, epidemiological, and developmental transitions in India and how that is impacting decision-making for the health sector. It reflects upon relevant experiences and the current status of healthcare provisioning to identify issues aiding and ailing the achievement of health outcomes in times of disasters and otherwise.

Keywords Emergency response · Systematic reviews · Public health · Disaster risk reduction · Hospital safety · Developmental risk

11.1 Introduction

Health is an essential indicator of a country's development. The level of development partly determines the impact of any disaster on health. Robust health infrastructure is crucial for effective response and long-term recovery benefits. This is especially important for a disaster-prone, resource-stretched, and high-density country like India which sees recurrent small and medium disasters. These impact both health infrastructure and health outcomes. Firstly, India has suffered expensive economic consequences on health infrastructure due to disasters. Health infrastructure damage and reconstruction costs post the Indian Ocean Tsunami (2004) amounted to USD 30 million (Carballo et al. 2005). The 2001 Bhuj earthquake collapsed a 281-bed civil hospital leading to 172 deaths (Hengesh et al. 2002). Large-scale failure of health systems is also known to result into a negative impact

S. Krishnan (✉) · I. Patnaik (✉)
National Institute of Public Finance and Policy, New Delhi 110067, India
e-mail: supriya.kr09@gmail.com

I. Patnaik
e-mail: ilapatnaik@gmail.com

E. Ying Yang Chan and R. Shaw (eds.), *Public Health and Disasters*, Disaster Risk Reduction, https://doi.org/10.1007/978-981-15-0924-7_11

on a region's economy (ADB 2005). The government bears much of these costs of damaged health infrastructure and poor health consequences. Hence, it is imperative that health infrastructure is planned to respond well to disaster cycles. Secondly, before the discussion on health infrastructure, it is important to take stock of developmental risk factors that impact general health outcomes and well-being.

This century has seen a remarkable transformation in the development landscape of India. Over the last 35 years, India has achieved an average Gross Domestic Product (GDP) growth of 6.3% per year; this is doubling every 11 years. While, on the one hand, the economic prosperity of the country improves income, education, and the ability to purchase health care, development resulting from rapid growth has the potential to increase health risks from disasters. For instance, massive rural to urban migration is expected to add 300 million new urban residents in India by 2050. Accommodating this population influx to will accumulate risk in the form of unregulated growth, lack of mindful planning, environmental degradation, inefficient services such as sewage and waste disposal systems, high rate of infectious diseases and pollution. These risks ail the general well-being of the environment which indirectly hampers health outcomes. There is a need for mindful development that minimizes risk and improves the overall response capacity of the region.

11.2 India's Burden of Disease

11.2.1 Traditional Burden

India is home to one-sixth of the world's population but shoulders a substantial 21% of the world's burden of disease (WHO 2012). As a traditionally low income, developing country, India's health risks were a function of its poverty, illiteracy, and poor public health services. Conditions included water-borne and vector-borne infections such as tuberculosis, cholera, leprosy, dengue, and *chikungunya* in addition to acquired immunodeficiency syndrome and sexually transmitted diseases. Child health, malnutrition, infant mortality, maternal mortality, nutrition, and gender equity issues continue to be persistent concerns. As a result, health priorities have focused on the provision of curative health care and institutional delivery. Since the early 1990s, substantial measures to address communicable diseases and nutrition have produced better outcomes. However, there is a long road ahead before India catches up with global baseline indicators for average performance in health-related outcomes.

11.2.2 *Emerging Burden*

India's emergence of *first-order* health problems is an outcome of its burgeoning anthropogenic and climatological risks. As a fast-growing economy, India is seeing a spike in chronic diseases. This ranges from cardiovascular diseases, respiratory diseases, metabolic diseases, cancer, and mental illnesses. Non-communicable diseases (NCDs) accounted for 60% of total deaths in 2015 (WHO 2015) indicating a protracted epidemiological transition. The following issues highlight the impacts on health caused by the process of this rapid development.

11.2.2.1 Air Pollution

In 2015, air pollution became the most significant risk factor for health in India causing more death and disability than dietary risks or child and maternal malnutrition put together. Fourteen Indian cities featured in World Health Organization's (WHO) list of the 20 most polluted cities in the world by particulate matter concentration. This was also the year when India overtook China as the leader of deaths due to air pollution.[1]

11.2.2.2 Sanitation

In India, around 40 million people are affected by water-borne diseases annually leading to a loss of 73 million workdays (Bush et al. 2011). Open defecation and poor sanitation practices are a prevalent problem. The "*Swachh Bharat Abhiyan*" (Clean India Mission) is a national program launched in 2014 in support of Sustainable Development Goal no. 06 to "*Ensure availability and sustainable management of water and sanitation for all.*"[2] The campaign is taking actions to promote environmental cleanliness and end open defecation through the construction of toilets by 2020. In 2016, the Union Budget allocated USD 1.3 billion toward the mission (GoI 2016). However, sanitation and cleanliness are more significant problems in behavioral change in government parlance. Media is playing an essential role in improving citizen ownership and sense of engagement.

11.2.2.3 Vector-borne Diseases

As a predominantly tropical landscape with heavy rainfall and inefficient water management systems, the occurrences of rodent and vector-borne diseases see a

[1]Data summarized from WHO Global Ambient Air Quality Database (updated 2018): http://www.who.int/airpollution/data/cities/en/.

[2]Clean India Mission: http://www.swachhbharatmission.

seasonal aggravation. Outbreaks of malaria, Japanese encephalitis (JE), dengue, leptospirosis, and diarrheal diseases significantly increase patient load in healthcare facilities (Ahern et al. 2005). These conditions are also common in the aftermath of hydrological disasters. WHO estimated the disability-adjusted life years lost because of vector-borne diseases in India to be 4.2 million, and malaria is believed to be responsible for nearly half of the cases (Kumar et al. 2007).

11.2.2.4 Urbanization

The *Smart Cities Mission* is a flagship urban renewal and retrofit program of the Government of India. It proposes to develop 100 cities across India as models for guiding future urbanization.[3] The objective of this mission includes provisioning of core infrastructure and improving quality of life through accessible services including health. A spatial analysis of 90 proposed smart cities illustrates that 76% of these cities lie in zones that are at medium to high risk of floods, earthquakes, and winds or all three (see Annexure 11.3, Fig. 11.2). Any new development in these cities is an opportunity to mainstream disaster risk resilience to minimize the creation of new risks. However, as a scheme that is meant to be "*open to the interpretation of city authorities,*" disaster resilience does not figure as a prominent development priority.

11.2.2.5 Infrastructure Development

In India, responsibility for provisioning infrastructure services for health, water, sanitation, sewage, and power is assigned to different levels of development authorities. The authorities are at the national, state (provincial), district, and local levels (municipal, *panchayat*[4]). At the lowest level, municipal authorities have the power to make service provisions within their "*legally notified boundaries.*" Massive rural to urban population flow in India is creating regions known as "*census towns.*"[5] These are places satisfy the legal criteria of a town but are not notified as statutory towns. This means that these are neither regulated by municipal bylaws nor can the local authority make development decisions. A paper on "*Regulating Infrastructure Development for India*" brings forth this lack of a comprehensive power devolution as reasons for poor planning and regulation of

[3]Smart cities India: http://smartcities.gov.in/content/.

[4]Village council.

[5]Places with a minimum population of 5000 and a density of at least 400 km^{-2}, and at least 75% of the male working population engaged in non-agricultural pursuits.

services (Sapatnekar et al. 2018). This has direct consequences on the quality of the environment, health, and coping capacities for disasters. As of 2011, census towns constituted as much as 49% of emerging urban towns in India.[6]

11.3 India's Health and Hazard Landscape

More than half of India's landmass is vulnerable to more than one hazard of high intensity. In the period between 1970 and 2015, India witnessed 614 recorded disasters such as floods, earthquakes, extreme temperature, drought, landslides, and storms leading to more than 198,000 deaths; affecting more than 2 billion people; and causing losses of USD 93 billion (Louvain (UCL)-CRED 2015) (see Table 11.1).

While India has consistently ranked high on the list of fatalities due to natural disasters, the economic losses to infrastructure are seeing a steep rise. The last decade has highlighted the frailty of health systems in India. Floods in the state of Kerala (2018) damaged a 125-year-old hospital that served 350,000 people. This was similar to the Chennai floods of 2015 where 18 patients died due to a power failure or the high risk of infections during annual urban flooding in Mumbai (Barnagarwala 2017). A notably tragic example of disaster impact to a health facility was the collapse of the 281 bed civil hospital during the Gujarat earthquake (2001) that killed 172 people. This incident led to the adoption of the base isolation technology for earthquake resistance of buildings. This was later included in the revision of the Indian Seismic Code IS 1893: 2002 for Criteria for Earthquake Resistant Design of Structures (Indian Standards 2002). An overview of seismic vulnerability of 1.6 lakh public health facilities across India indicates that more than 54% fall in moderate to very high risk zones (see Annexure 11.1, Table 11.7).

11.3.1 Existing Literature

Public health emergencies and disasters are sparsely researched subject in India. The majority of publicly available reports look at individual health conditions, impacts from specific disaster events, or on evaluating existing policies. Evolution of this research is in relation to India's evolving policy discourse on disasters which developed further with the occurrence of major disasters (see Annexure 11.2, see Table 11.8) The Odisha Super Cyclone (1999), the Gujarat earthquake (2004), and the Indian Ocean Tsunami (2004) are the most researched subjects for their impact on health. Before the Super Cyclone, there are studies on impacts of vector-borne diseases such as leptospirosis (WHO 2000), cholera outbreaks (Sur et al. 2000) and

[6]Census India 2011: http://censusindia.gov.in/2011-Common/CensusData2011.html.

Table 11.1 Major disaster events in India

Year	Type	Name	Deaths	Affected	Damages (USD '000)
1971	Storm	E: Odisha, Andhra Pradesh	9658	6900000	30000
1972	Drought	N, W: Rajasthan, Himachal and Uttar Pradesh	n.a.	200000000	100000
1977	Tropical cyclone	S: Andhra Pradesh cyclone	14204	14469800	498535
1978	Flood	E: West Bengal floods	3800	32000000	165000
1987	Drought	North, West, and Eastern India	300	300000000	Not available
1993	Earthquake	W: Latur earthquake	9748	30000	280000
1999	Tropical cyclone	E: Super Cyclone	9843	12628312	2500000
2001	Earthquake	W: Gujarat earthquake	20005	6321812	2623000
2002	Drought	All India	n.a.	300000000	910722
2004	Earthquake	S: Indian Ocean Tsunami	16389	654512	1022800
2013	Riverine flood	North India floods	6054	504473	1100000
2013	Tropical cyclone	E: Cyclone Phailin	47	13230000	633471

Source EM-DAT Disaster Database
N North, *S* South, *W* West, *E* East

a study on psychiatric disorders in survivors of the 1993 Latur earthquake that killed over 10,000 people (*graded VIII on the Mercalli intensity scale*) (Sharan et al. 1996).

After 2000, India's ongoing epidemiological transition from communicable to non-communicable diseases saw studies emerge on the threat of chronic diseases, cardiovascular diseases (Shah and Mathur 2010; Rastogi et al. 2004), diabetes, cancers, hypertension (Reddy et al. 2005; Ghaffar et al. 2004; Patel et al. 2011; Misra et al. 2011), and an overview of mental health priorities, also reflecting on impacts from disasters (Khandelwal et al. 2004). A relevant study on health systems was on the epidemic preparedness in public health that looked at environmental forecasting and disease surveillance methods to promote epidemic prevention control (Myers et al. 2000). The Odisha Super Cyclone, which impacted more than 10,000 people, triggered several studies on mental health vulnerability (Kar et al. 2004) and post-traumatic stress disorder (Sharan et al. 1996; Kar et al. 2007).

Studies on outbreaks of vector and water-borne diseases saw sustained interest with a focus on single conditions that typically result from flooding such as leptospirosis (Rao et al. 2003; Vijayachari et al. 2008; Sehgal et al. 2002), diarrhea (Mondal et al. 2001), chikungunya (Mavalankar et al. 2007), and general challenges of managing infectious diseases (John et al. 2011). (Jha et al. 2016) studied evolution of disaster policies in India that brought about the massive reduction of mortality during a disaster through better early warning and evacuation systems.[7] The 2001 Gujarat earthquake that killed around 18,000 and injured over 150,000 led to focused research on health sector development for disaster preparedness (Bremer 2003) and the importance of community medical response to reduce casualties (Roy et al. 2002).

11.3.2 The State of Health System During a Disaster: The Case of the Gujarat Earthquake 2001

The vulnerability of the population and deficiencies in the existing health system aggravates the pressure on healthcare facilities during a disaster. An on-ground situation study was conducted after the Gujarat earthquake (2001). It found that the military medical unit had to step into respond to health emergencies. Initially equipped to treat 750 casualties, the units handled thousands of patients out of four tents. 11 surgical teams came in from across the country to perform 2002 major and 7524 minor operations. There was a shortage of critical surgical equipment, linen with the total absence of essential services such as water, fuel and power and bio-waste disposal (Bremer 2003).

11.3.3 Climate Change and Health

Climate change and its impact on health took a central position on the international agenda (Griggs and Noguer 2002). Research emerged on the impact of temperature (Akhtar et al. 2007), rainfall (Ahern et al. 2005), tsunamis (Carballo et al. 2005), changing pattern of vector-borne diseases (Bhattacharya et al. 2006; Kumar et al. 2007) and extreme events on human health (Bush et al. 2011; Dhara et al. 2013). Recent climate emergencies in India included a heat wave in Orissa 2004, a cold wave in Uttaranchal and Uttar Pradesh 2004, a tsunami affecting Tamil Nadu, Andhra, Kerala, and the Andaman Nicobar Islands 2004, floods in Madhya Pradesh and Gujarat 2005, rains and floods in Maharashtra (2005), and a cyclone in Andhra Pradesh (2005) (WHO 2005). The Indian Ocean Tsunami (2004) that killed 10,000 people in India and rendered several homeless brought forth further research on the economic impact on public health infrastructure due to disasters (ADB 2005). The magnitude of human loss in both the tsunami and the earthquake brought further studies on mental health and psychosocial care for adults (Becker 2007; Sharan et al. 1996) and women (Becker 2009) survivors of the disaster.

---

[7]Sustained efforts of the government in Odisha brought down the rate of mortality to 0.5% in Cyclone Phailin 2013 as compared to Super Cyclone (1999).

Studies on strengthening emergency medical care and trauma care see limited research as they also do not have a central organization system (Razzak and Kellermann 2002; Joshipura 2008; Joshipura et al. 2003; Kobusingye et al. 2005; Pandian et al. 2006; Ramanujam and Aschkenasy 2007). The importance of education on emergency medicine is elaborated in the INDO-US joint working paper on academic medicine (Das et al. 2008).

11.4 Role of Government

11.4.1 Public Health System

As per the Constitution of India (GoI 1956), public health, sanitation, hospitals, and dispensaries fall under the purview of the state. Larger welfare items such as medical education, family welfare, and quality control fall under the purview of both the center and the state. As a federal system, India allocates decision-making power to the center and administrative units under it, i.e., the States (provinces). The Ministry of Health and Family Welfare (MoHFW) is responsible for running programs and establishing policies for health. The system of health service delivery is through a three-tier system based on the level of care and population it serves (see Fig. 11.1 and Table 11.2):

1. **Primary healthcare**: Sub-Centers (SC) and Primary Health Centers (PHC) for villages and tribal areas;
2. **Secondary healthcare**: District hospitals (DH) and Community Health Centers (CHC) at the block level; and
3. **Tertiary healthcare**: Advanced medical research institutions.

Management of health during disasters falls under the larger umbrella of public health. Since 1950 (post-independence), India has amalgamated delivery of public health and medical services. A *Model Public Health Act* was developed (in 1955 and 1987) to promote preventive public healthcare ecosystem (monitoring, inspecting, regulating). However, it did not gain leverage for adoption by states. There were single focus health schemes for prevalent problems such as leprosy, tuberculosis, malaria and high maternal and child mortality (Lakshminarayanan 2011). The focus on curative medical services reduced the emphasis on preventive public health measures.

Health has several social, economic, and environmental determinants. However, there was separation of governance of key sectors such as public health, environment, water resources, and sanitation. International donor agencies also focused on more measurable programs such as vaccination, mortality, and disease surveillance. This has increased the incentive to design narrow focus programs whose outcomes can be more easily measured as opposed to broader programs for mitigation and prevention that need longer time horizons (Das Gupta et al. 2009).

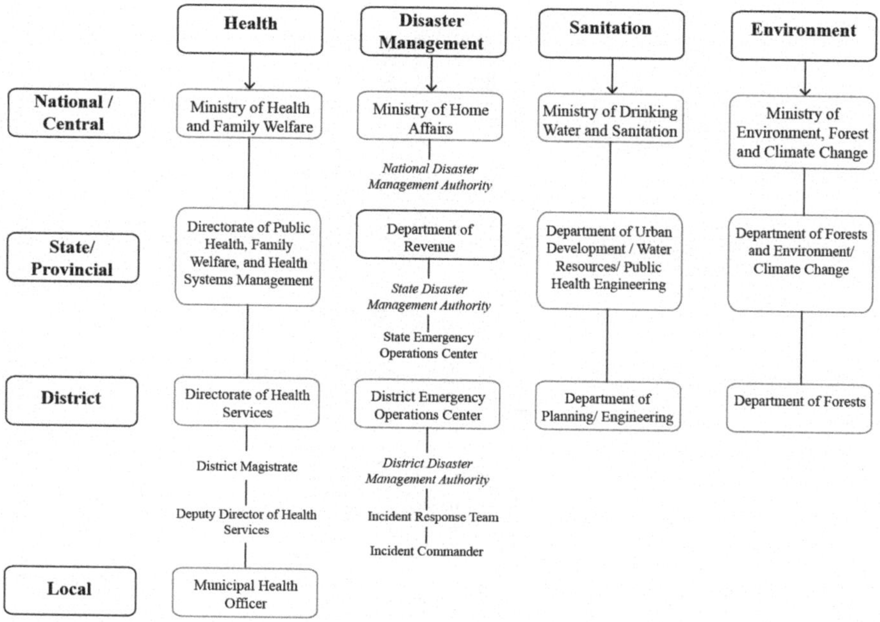

Fig. 11.1 Public health and disaster management in India (drawn by authors with data from Planning Commission and allied documents)

To take a comprehensive overview of health, the National Health Mission (NHM) was launched in 2013 to strengthen health systems and improve public health delivery across India.[8] NHM takes a macro view of areas affecting health such as water, sanitation, education, nutrition, social, and gender equality and develops separate funding pools for each. NHM publishes the Indian Public Health Standards (IPHS) enumerating compliance requirements for public health facilities to build consistency in healthcare infrastructure planning and up gradation.[9] The guidelines require all facilities to be built to be resilient to flooding, earthquakes, and emergencies aligned with the guidelines set out in the National Building Code.

In order to improve service delivery, capacity building, assessment of expenditure framework and accountability, support is sought from multilateral development agencies. The Asian Development Bank supports the National Health Mission (Bank 2018) whereas the World Bank has provided support to a range of health system strengthening projects in several states of India (Uttar Pradesh, Odisha, Uttarakhand, Karnataka, Nagaland).

[8]National Health Mission: http://www.nhm.gov.in/.

[9]Indian Public Health Standards: http://nhm.gov.in/nhm/nrhm/guidelines/indian-public-health-standards.html.

Table 11.2 Hierarchy of public healthcare facilities in India and the population they serve

Health institution	Avg. population to be served	Beds	Avg. population being served				
			Odisha	Assam	UP	Bihar	Punjab
District Hospital	35,000–30,000,000	101–500	1,446,461	1,246,771	1,513,730	2,731,701	1,458,118
Sub-divisional Hospital	5,00,000–6,00,000	31–100	1,613,360	2,833,570	19,981,234	3,460,155	791,550
Community Health Center	80,000–1,20,000	30	116,520	145,651	217,187	1,526,539	184,695
Primary Health Center	20,000–30,000	6	74,112	83,118	190,116	156,332	64,881
Sub-centers	3000–5000	2	28,096	36,370	76,999	56,786	9,388

Source Indian Public Health Standards

11.4.1.1 Public Health in the State of Tamil Nadu

The state of Tamil Nadu (TN) has set an example of efficient management of public health systems. It has consistently produced better health outcomes with the same average per capita expenditure on health as India. Unlike other states, TN does not merge the delivery of public health and medical services, gaining from respective economies of scope (Gupta et al. 2010). Since 1922, it has legislated for an independent Directorate of Public Health with an administrative authority board and budget (GoI 1939). The Madras Public Health Act enables the state to adopt a macro view on public health and has mandated to address any situation that poses a "*credible health threat.*" This includes maintenance of drains, solid waste management, poorly designed infrastructure, animals in unhealthy conditions, etc. The Directorate has a dedicated cadre of professionals who are trained in activities related to public health such as managerial executives, grassroots health workers, entomologists, laboratory staff and laborers for sanitation activities. The Directorate works with larger budgets than medical services and medical education.

This method of governance allows the state to conduct annual "*anticipatory planning*" to prepare for recurring disasters such as floods, endemic diseases, and public health emergencies. Tamil Nadu's highly organized response to the Indian Ocean tsunami was a result of this well-oiled public health machinery.[10] Despite the magnitude of the disaster that killed around 18,000 people in India, no significant disease outbreaks were reported in India (World Health Organization 2013).

11.4.1.2 National Health Policy (NHP)

The National Health Policy (NHP) was an outcome of the *Alma-Ata Declaration* that guides approach for health in national plans. It aims at achieving universal health coverage and delivering affordable preventive, promotive, and rehabilitative health care. NHP 2002 (clause 5.2) recognized the need for a robust disaster management plan to cope with calamities. However, NHP 2017 elaborated creation of a unified emergency response system and maintenance of excess capacities in health infrastructure to better respond to both manmade and natural disasters. It proposes that from the level of the Community Health Centers, all institutions must develop and maintain mass casualty management protocol. To better ensure effective emergency management, it upholds capacity building of local governments and the community as first responders. It envisages provision of a trauma management and life support ambulances for one per *3 million* people in urban areas and one per *1 million* people in rural areas (GoI 2017b).

[10]WHO South Asia tsunami situation reports.

11.4.2 Disaster Management

The governance response to manage natural disasters in India is the Disaster Management Act 2005 (DM Act). Until the early 2000s, the focus of India's disaster management policy was on relief and rescue operations (Jha et al. 2016). The role of a healthcare institution in mitigating disaster impact did not receive a clear mandate. The National Disaster Management Authority (NDMA) was formed in 2005 under the Ministry of Home Affairs (MHA) to manage economic, environmental, and developmental issues relating to disasters. As a signatory to the Sendai Framework for Disaster Risk Reduction, India was one of the first countries to devise a National Disaster Management Plan (2005). It was drafted with an aim to achieve a substantial reduction of disaster risk and reduction of disruptions to critical services (GoI 2005).

The years after 2005, the newly formed National Disaster Management Authority (NDMA) published several standards on specific hazards. This included guidelines for minimum standards of relief (NDMA, n.d.), management of earthquakes (NDMA 2007b), tsunamis (NDMA 2010b), floods (NDMA 2008b), urban floods (NDMA 2010c), cyclones (NDMA 2008a), drought (NDMA 2010a), landslides (NDMA 2009a), nuclear emergency (NDMA 2009b), and chemical disasters (NDMA 2007a). These guidelines included a section on medical preparedness and mass casualty management in the case of each of these disasters. NDMA also published comprehensive guidelines on psychosocial health (NDMA 2009c). In 2016, NDMA published guidelines for hospital safety (India 2016).

11.4.2.1 Guidelines for Hospital Safety

The DM Act requires state governments to formulate State Disaster Management Plans (SDMPs) which detail how to prepare, mitigate, respond, and recover from disasters. A component of these plans is medical preparedness and mass casualty management. It recommends that State may provide health care and services by following the standards laid down by the National Authority *(Part II, point 24(d))*. The Act prescribes every hospital to have an emergency plan that must be updated regularly with periodic checks and mock drills.

From 2007, NDMA published hazard-specific management guidelines. The hazards covered are earthquakes (NDMA 2007b), tsunamis (NDMA 2010b), floods (NDMA 2008b), urban floods (NDMA 2010c), cyclones (NDMA 2008a), drought (NDMA 2010a), landslides (NDMA 2009a), nuclear emergency (NDMA 2009b), and chemical disasters (NDMA 2007a). Hazard-specific guidelines provide details on appropriate medical preparedness and healthcare delivery system to deal with injuries, conditions, and diseases associated with that hazard. It further assigns responsibilities to departments in the state, district, state health departments, private hospitals, and urban local bodies to enable multi-agency collaboration.

The NDMA guideline for earthquakes recommends selective seismic strengthening and retrofitting of lifeline structures in earthquake-prone areas (hospitals and health facilities, tertiary care centers, and all hospitals designated as major hospitals). It requires them to be updated on the India Disaster Resource Network (IDRN) database (NDMA 2007b). The guidelines for urban flooding recommend involvement of the corporate sector to improve the delivery of relief measures (NDMA 2010c). NDMA published its "*minimum standards of relief*" for persons affected by disasters including rehabilitation, shelter, food, health, water, sanitation, and vulnerable groups.

In 2016, NDMA laid down guidelines for *hospital safety* to mainstream disaster prevention, mitigation, preparedness, and response activities into the health sector (India 2016). The document brings together from a spectrum of guidelines from national (National Building Code, Bureau of Indian Standards, Clinical Establishment Act, Indian Public Health Standards) and international (Pan American Health Organization, World Health Organization) sources. It empaneled domain experts to further improve upon these guidelines and provides frameworks for implementation. It elaborates in detail on the some of the following parameters for hospitals and disasters:

1. Awareness generation for hospital safety
2. Hospital disaster preparedness
3. Design and safety of hospital buildings
4. Fire safety
5. Licensing and accreditation.

Based on the learnings from the above components, the document lays down a "*National Action Framework for Hospital Safety.*" It highlights five priority areas for action along with highlighting gaps, recommending interventions, estimating a timeline and assigning work to requisite agencies (see Table 11.3). However, in practice, it has been found that there is a gap in compliance with building codes, lack of planning and preparedness and variation in quality of medical facilities. There is no statutory provision to regulate and standardize disaster response plans for hospitals. Hence, hospitals do not maintain disaster management plans. In a survey, it was found that only 26% of trauma care health facilities had a well-documented disaster management plan (Joshipura 2008; Mehta et al. 2006).

Table 11.3 Priority areas of action

Priority Area	Action
Priority Area I	Strengthening institutional mechanisms
Priority Area II	Advocacy, awareness generation, and education
Priority Area III	Capacity building
Priority Area IV	Preparedness, response, and recovery
Priority Area V	Risk reduction and structural mitigation

Source National framework for hospital safety

In an onsite survey of primary health facilities in a flood-prone district, it was found that basic response utilities such a power backup, a line of standard operating procedures and equipment were missing. Monitoring these aspects of hospital management is the responsibility of States and administrative units under them (Phalkey et al. 2012).

11.4.2.2 Health in State Disaster Management Plans (SDMP)

Under the DM Act, each State is required to prepare a State Disaster Management Plan (SDMP) to carry forward national goals for disaster management. The SDMP is expected to cover the state's vulnerabilities, measures for the prevention and mitigation of disasters, capacity building and assignment of responsibilities to relevant departments in the event of a calamity (see Table 11.4). Hospitals are identified as critical/lifeline facilities. The SDMP document addresses all phases of a disaster (preparedness, mitigation, response). The level of detail of a State Disaster Management Plan did not seem proportional to the disaster proneness of the state or the number of health facilities at risk (see Annexure 11.1). Plans were also out of date. Even though the law requires states to update their plans annually, only eight states had updated their plans till 2016. Plans dedicate the majority of their sections to respond to a disaster, rather than prepare for one in their standard operating procedures (SOPs).

While plans alone will not determine the quality of response to a disaster, lack of a well-drafted plan will result in poorly implemented practices when both time and resources are limited. A SDMPs lean heavily to the "response" phase of a disaster. Standard operating procedures (SOPs) for concerned departments also elaborates

Table 11.4 Principal departments responsible for aspects related to health and disaster risk at the State level

Department	Responsibility
Health and Family Welfare	Healthcare infrastructure, response, contingency plans, preparedness and mitigation, vulnerability assessments, mass casualty plan, ambulance network, hospital networking, first aid, training, and capacity building
Public Health Engineering Department	Drinking water supply and sanitation, extreme events, groundwater resources
Revenue and Disaster Management	Buildings and lifeline infrastructure operations and construction compliance with Indian construction codes, training, and capacity building
Urban Development and Urban Local Bodies	Development of built stock and provision of services such as drainage, sewage, and drinking water

Source Various State Disaster Management plans in India

more steps for response and rehabilitation. In comparison with global frameworks indicators for the non-structural and functional resilience of health are poorly covered. A review of State Disaster Management Plans of 22 states in India presents the following findings:

Structural indicators are clearly communicated and find mention in 75% of the documents. For example, the state of Himachal Pradesh mentions that 48% of its medical institutions are located in highly vulnerable districts and must comply with codes of the Bureau of Indian Standards (BIS). Punjab recommends assigning a quality auditor agency to monitor construction in seismic zones 3, 4, and 5 (medium to very high earthquake risk). The collapse of the civil hospital during the Bhuj earthquake triggered the last revision of the Indian Seismic Code for Earthquake Resistant Design of Structures (IS 1893: 2002). This has also improved the inclusion of structural indicators in most SDMPs as there are both legal mandates and evolved guidelines for facilities to follow.

Non-structural indicators are the least addressed in all documents. Less than 50% refer to even one of the indicators from the WHO Safe Hospitals indicators. Points on the safety of medical equipment, furniture, and backup supplies are mentioned as part of larger checklists for response but do not provide action points. Independent agencies in India lay down plans focusing of strengthening non-structural utilities (GeoHazards 2009). The NDMA guidelines on hospital safety elaborate on this. As a relatively recent document, it has not seen adoption in state plans yet. However, the SDMPs do not refer to any other universal guidelines that hospitals may follow for safety of non-structural aspects.

Functional indicators find a mention in 75% of the documents. All states recommend the preparation of a medical preparedness plan, mass casualty management plan, and checklists to train health workers for emergencies. A fundamental requirement to enable functional continuity of hospitals during emergencies is a list of all available health facilities and supporting services (such as power station, police station, ambulances). A mere 50% of documents provide any information on health facilities in the state. Assam and Odisha provide details on the population being served. Odisha also highlights provisioning of a dedicated high tension power line to the district headquarter hospitals for uninterrupted communication with the health control room.

Functional indicators for post-disaster psychosocial support and mental health are missing across documents, except Meghalaya. At least one-third of the survivors of the super cyclone in the state of Odisha suffered disabling psychiatric symptoms (Kar et al. 2007). NDMA has recognized this issue as "*a continuum of the interventions in disaster situations*" and laid down guidelines on Psychosocial Support and Mental Health Services (PSSMHS) in Disasters (Becker 2009).

Other indicators such as mobile hospitals, media management, district level data, and standard operating procedures are well addressed. 16 of the 22 states mention utilizing the India Disaster Resource Network (IDRN). It is an online portal that includes data of health professionals and medical equipment to accelerate decision-making during a disaster. Assam and Gujarat have established a functional State Disaster Resource Network (SDRN).

11.5 Elements of Health and Disaster Risk Management in India

11.5.1 Resource Management

Health response during and after a disaster may be hampered by the shortage of or unavailability of human resources and equipment. To maintain a centralized resource inventory for disaster response, Government of India in collaboration with the United Nations Development Program (Disaster Risk Management) launched the India Disaster Resource Network (IDRN).[11]

IDRN is a web-based inventory that captures and collates data systematically regarding resources available at the level of the line departments in a district (local administrative unit). The Web site hosts a country-wide query system where one can look for availability of medical equipment, skilled health workers and critical health-related supplies at the level of the district. The portal is monitored by the National Institute of Disaster Management (NIDM) and offers authorized access to officers of the government including emergency officers, relief managers, district collectors/magistrates, and other officers to upload and access data (see Table 11.5).

Currently, the IDRN offers information for more than 75% of districts in India. To enable contribution of equipment from the private sector, the portal hosts a module where the Confederation of Indian Industry (CII)[12] and Builders Association of India (BAI)[13] can upload resource details to the database. Selected states are pioneering the use of this online inventory format with improved databases of resource inventories that are compatible with the national interface. The state of Gujarat has taken the lead with a State Disaster Resource Network which has information for resources in 97% of its villages (GSDMA 2003). However, the quality and comprehensiveness of data are in a process of continual improvement.[14]

---

[11]India Disaster Resource Network: www.idrn.gov.in.

[12]Confederation of Indian Industry: https://www.cii.in/.

[13]Builders Association of India: https://www.baionline.in/.

[14]Gujarat SDRN: http://117.239.205.164/sdrnguj/.

Table 11.5 India Disaster Resource Network (IDRN)

Resource type	Items
Skilled human resources	Anesthetist, general physician, gynecologist, lab technicians, medical first responders, OT assistants, firefighting team, paramedics, radiologists, surgeons, and trauma specialists
Equipment type	Air compressor, aluminum ladder, aspects blanket, breathing apparatus, CT scan, defibrillator, detection kit: for poison in water, first aid kits, rubber gloves, light ambulance van, medium ambulance van, mechanical ventilators, mobile blood bank, mobile laboratory service, mobile medical van, mobile OT unit, MRI, face mask, portable alpha monitor, portable ECG, portable decontamination apparatus, portable oxygen cylinders, portable suction unit, portable ultra-sound, portable ventilators, and portable x-rays
Critical supplies	Anti-snake venom, bronchodilators, chlorine tablets, decontamination solution, vaccines, and Iodate tablets

Source IDRN website

11.5.2 Disease and Epidemic Surveillance

Lack of robust surveillance systems, low immunization coverage, and poor sanitation systems has led to an inefficient disease control system in India. Seasonally, high incidences of vector-borne diseases, water-borne diseases, and endemic diseases are frequent. After the Bhuj earthquake (2001), WHO set up an onsite surveillance system with 620 reporting sites (Bremer 2003). To detect and monitor disease outbreaks systematically, the Ministry of Health and Family Welfare with the support of the World Bank set up the Integrated Disease Surveillance Program (IDSP).[15] In 2004. The IDSP was preceded by the National Surveillance Programme for Communicable Diseases (NSPCD) which was set up in 1997 (see Annexure 11.2).

IDSP monitors trend in occurrences of communicable and non-communicable diseases across the country. Its plan includes integrating existing surveillance systems, strengthening of public health laboratories, training and deploying of Rapid Response Teams (RRT) for timely action during possible epidemic situations (see Table 11.6). For example, after the Indian Ocean Tsunami (2004), a one-page surveillance instrument was developed to monitor 10 priority health conditions among the displaced population (Nsubuga et al. 2006). A decentralized model is adopted where data collection, capture, and analysis are conducted at the local administrative units which then flow upwards to the national level. This is done in weekly intervals in three formats (*suspected, presumptive, laboratory-confirmed*). In order to facilitate faster reporting and put out early warning signals for outbreaks, a satellite-based communication and learning network: Education Satellite (EduSat) was established by the National Centre for Disease Control (NCDC) and Indian

[15] Integrated Disease Surveillance Program: http://www.idsp.nic.in.

Table 11.6 Integrated Disease Surveillance Program (IDSP): List of diseases under regular surveillance

Disease type	Disease name
Vector-borne	Malaria, dengue, Japanese encephalitis (JE), chikungunya, etc.
Water-borne	Acute diarrheal disease (Cholera), typhoid, gastroenteritis, hepatitis
Respiratory	Tuberculosis
Vaccine preventable	Measles
Under eradication	Polio
International commitments	Plague, yellow fever
Unusual clinical syndromes	Meningo-encephalitis, respiratory distress, hemorrhagic fevers, other undiagnosed conditions
Sexually transmitted diseases (under sentinel surveillance)	Blood-borne, HIV/HBV, HCV
Non-communicable diseases	Cardiovascular, diabetes, strokes, fluorosis, etc.
Others	Water quality, outdoor air quality, road traffic accidents

Source IDSP website

Space Research Organisation (ISRO). It connects communication terminals at national, state headquarters, district surveillance units, and government medical colleges (Raut and Bhola 2014).

IDSP is vital to evaluate the performance of public health interventions and resource allocations. Since the program depends on local level data collection, it is only as effective as the quality of local capacity to report and maintain records. The government needs to further invest in rigorous technical capacity building to improve the quality of surveillance. Selected states are pioneering the use of technology to track specific health conditions for better response. Karnataka has developed a digital database of spatial and temporal distribution of mortality and morbidity of vector-borne diseases.[16] Punjab has created a portal for online reporting of climate-related cases of illness[17] (Dasgupta et al. 2016).

11.5.3 Mobile Medical Units and Sanitation

A critical health component of disaster management is containment of epidemics. This requires a system that provides clean water, sanitation (in situ, ex situ) and promotes good hygiene. NDMA presents "*Guidelines on Minimum Standards of Relief*" covering fundamental service requirements at relief and rehabilitation camps such as temporary shelter, food, water, sanitation, medical cover and needs of

[16]State of Karnataka: http://nvbdcp.gov.in/index4.php?lang=1&level=0&linkid=506&lid=3783.

[17]State of Punjab: https://www.punjabnvbdcp.in/denguelogin.php.

vulnerable groups (NDMA, n.d.). The Ministry of Health and Family Welfare recently released Operational Guidelines for Mobile Medical Units that outlines the nature of service, quality, monitoring, and financing mechanism essential for such a fleet (GoI 2018). In 2011, the Ministry of Rural Development released the standard operating procedure (SOP) for drinking water supply and sanitation services during natural hazards (GoI 2011). The Bureau of Indian Standards offers guidelines for construction of some commonly use sanitation apparatus during disasters such as Leaching Pits for Rural Communities (IS: 12314) and Installation of Septic Tanks (IS: 2470) (Indian Standards 1993; Bureau of Indian Standards 2007).

11.5.4 Trauma Care

Post-disaster, a system must be able to provide emergency care for physical trauma (falls, injuries, burns) and maintain referral networks of facilities to cater to various emergencies. In India, a trauma-related death occurs every 1.9 min, mainly due to traffic accidents (Madan 2006) which are estimated to cause economic losses of up to 3% of GDP (Joshipura et al. 2003). Hence, work on trauma care leans heavily toward road traffic accidents. Trauma care is provided as part of casualty and emergency departments of hospitals. However, existing trauma care facilities or their professionals have no standard process of accreditation and training leading to disparity in levels of treatment. The National Health Policy (2002) envisaged the establishment of a "hub-spoke" trauma care network in large urban areas to reduce mortality (GoI 2002). This was followed by a national program on *Capacity building for developing Trauma Care Facilities in Government Hospitals on National Highways*.

To offer skilled human resources for disaster medicine, the Medical Council of India offers 37 seats in various colleges for a postgraduate degree in Accident and Emergency Medicine (as of July 2014).[18] A "National Trauma Management Course" accredited by the International Association of Trauma Surgery and Intensive Care (IATSIC) is available in larger cities (Joshipura 2008).

11.5.4.1 Training on Trauma Life Support

In 2012 NDMA, in collaboration with the Apex Trauma Center at the All India Institute of Medical Sciences (AIIMS) conducted the Pilot Project on Capacity Building for Advanced Trauma Life Support in India (NDMA 2013). This was a capacity building project for selected doctors, nurses, and paramedics on various

[18]Emergency Medicine India: http://www.emergencymedicine.in/current/articles.php?article_id=56.

protocols and guidelines of trauma care followed internationally. Assam, Bihar, and Andhra Pradesh were the chosen states. The training included:

1. Advanced Trauma Life Support (ATLS)
2. Advanced Trauma Care for Nurses (ATCN)
3. Pre-Hospital Trauma Life Support (PHTLS)
4. Rural Trauma Team Development Course (RTTDC).

The trained professionals are required to disseminate this knowledge in their respective states. Around 129 doctors and 50 nurses completed the various courses. Training included procedures for various kinds of trauma (head, musculoskeletal, spinal cord, burns, cold) and various age groups (pediatrics, elderly, women).

11.5.4.2 Emergency Medical Service (EMS)

Improving health service delivery during disasters requires a responsive and time-sensitive Emergency Medical Service (EMS). This includes transporting patients, offering Basic Life Support (BLS) and Advanced Life Support (ALS) before actual care may begin. India does not have a centralized body to manage operations of Emergency Medical Services.

Currently, EMS systems are a consolidation of initiatives by different states with different models of delivery with wide variability in dispatch and transport capabilities. It was reported that in 2006 that only 12% of stroke patients used ambulances to reach a hospital in an urban city (Pandian et al. 2006). The existing national scheme (*toll-free number: 108*) is monitored by the Emergency Management and Research Institute (EMRI).[19] EMRI works on a public–private partnership model and operates more than 2600 ambulances. Individual states have modified versions of this system to offer transport, pre-hospital stabilization services, referral facilities for healthcare professionals, or all of the above (Sharma and Brandler 2014).

Emergency medicine was recognized as a specialized training by the Medical Council of India in 2009. Formal training in emergency medicine is provided in collaboration with international organizations such as RCGP UK (Royal College of General Practitioners, UK), CEM UK (College of Emergency Medicine, UK), and GWU USA (George Washington University, USA) (Subhan and Jain 2010).

11.5.4.3 Mental Health and Psychosocial Support

Good mental health is an essential aspect of the rebuilding process for a population that has experienced events beyond its coping capacity (Becker 2007). Health facilities are responsible for managing both short-term and long-term emotional

[19]Emergency Management and Research Institute: http://www.emri.in/ideology/.

consequences of disasters. At least one-third of the survivors of the super cyclone in the state of Odisha suffered disabling psychiatric symptoms (Murthy et al. 2003).

India's mental health response during disasters has evolved from identifying and treating individual psychiatric cases to strengthening the coping abilities of survivors in a community (Kishore Kumar et al. 2000). The National Mental Health Programme (NMHP) includes strategies for Psychosocial Care and Mental Health Services in disasters (PSMHS). NDMA has laid down guidelines to improve coping capacities of disaster-struck communities by offering appropriate support to rebuild their lives. The service networks consist of psychiatric units of tertiary healthcare facilities, and educational institutions, clinical psychologists, social workers, NGOs, paramedical professionals, community level workers, and volunteers. It encompasses an "*all-hazard*" health plan for response, relief and rehabilitation aspects of different kinds of disasters (NDMA 2009c).

11.5.4.4 Heat Stress

Temperature trends have been rising steadily in the last 15 years, 2016 was India's hottest year on record (Bush et al. 2011). To better enforce healthcare systems during heat waves, the Ahmedabad Heat Action Plan was formulated in 2017. As a first of its kind in the South Asian region, one of its chief pillars is dedicated to capacity building among health professionals and training of medical staff to manage heat-related illness (GoI 2017a, b). There are plans to scale this up to other Indian cities (Knowlton et al. 2014).

11.5.4.5 Social Resilience

Grass roots and community level initiatives actively support health provisions during disasters. In the flood-prone state of West Bengal, flood micro-plans have been made at the district level with information on alternate health service points in case of flooding. Urban governing bodies have designed response mechanisms to reduce the risk of outbreaks after flooding. Post the Mumbai floods of 2005, the municipal corporation provided comprehensive healthcare services through 130 specially constituted medical teams. Over 300000 patients were treated virtually at their doorstep through outreach camps (Dasgupta et al. 2016; Gupta 2007).

11.6 Way Forward

Despite its large population, India has demonstrated its potential for efficient delivery and campaigning for public health services through various schemes. However, maintaining service quality over extended periods of time has been a continuing issue. The diversified institutional landscape has proven to be both an opportunity for innovation at the local level and a challenge for implementing frameworks at the central levels. Evaluation of the effectiveness of central health schemes and their impact on health infrastructure, health outcomes and emergency services needs sustained investigation. This would set the space for broader environmental and societal goals that improve systemic coping capacity in the event of a disaster. The above overview of the health and disaster risk management of India throws up some critical challenges that need to be addressed immediately:

- **Information infrastructure**: Good decision-making may be correlated with better data and analysis. Better and organized registries for infectious diseases, trauma care and geospatial database on hospitals need to be comprehensively updated and maintained.
- **Capacity building**: The shortage in capacities of healthcare facilities and expertise in specialized areas such as trauma care, mental health, immunization, and emergency medicine needs to be better integrated.
- **Risk assessment**: Better resource management during an emergency (such as the golden hour) requires systematic documentation and assessment of the situation. Combining data assessment, surveillance and monitoring across all states is essential for evidence-based decisions and reducing biases in the allocation of resources.
- **Climate variability**: Moving from a post-disaster reactive to an integrated approach will involve adopting an "*all-hazard*" approach that applies to health facilities in a specific region. Changes in weather variables are known to affect incidents of seasonal, climate-sensitive vector-borne and water-borne diseases, such as malaria and diarrheal illness (Dasgupta et al. 2016). Hence, ensuring resilience will require systematic mapping of variability in climate to analyze its impact on health.

Building resilience of health for disasters is a subset of the general well-being of the health system. This is an especially enormous challenge when a country is still working its way to navigate deficiencies in the system and achieve baseline indicators for health. Concentrated and sustained efforts and investments in the medical community, health infrastructure and the society at large will be critical to building robust systems responsive to disasters.

Annexure 11.1: Seismic Vulnerability of India's Public Health Facilities

A study was conducted to analyze seismic vulnerability of public health facilities in India. Data on geo-locations of 160,000 public health facilities including primary, secondary, and tertiary care facilities was analyzed for which seismic zone they fall in. 54.25% of facilities fall in medium to very high-risk earthquake zones. A state wise scheme of number of health facilities in each zone is presented in Table 11.7.

Table 11.7 Seismic vulnerability of public health facilities in India

State	Predominant seismic zone	Zone 3	Zone 4	Zone 5	Total
A&N Islands	5	0	0	31	31
Andhra Pradesh	3	775	0	0	775
Andhra Pradesh Old	3	579	0	0	579
Arunachal Pradesh	5	0	0	160	160
Assam	5	0	0	884	884
Chandigarh	4	0	46	0	46
Chhattisgarh	3	159	46	0	159
Dadra & Nagar Haveli	3	13	0	0	13
Daman & Diu	3	7	0	0	7
Delhi	4	0	601	0	601
Goa	3	39	0	0	39
Gujarat	3/4	1697	141	78	1916
Haryana	3/4	223	360	0	583
Himachal Pradesh	4/5	0	317	314	631
Jammu & Kashmir	4/5	2	506	291	799
Jharkhand	3	315	34	0	349
Karnataka	3	366	0	0	366
Kerala	3	1265	0	0	1265
Lakshadweep	3	9	0	0	9
Madhya Pradesh	3	608	0	0	608
Maharashtra	3	1893	169	0	2062
Manipur	5	0	0	114	114
Meghalaya	5	0	0	176	176
Mizoram	5	0	0	83	83
Nagaland	5	0	0	153	153
Odisha	3	456	0	0	456
Puducherry	3	4	0	0	4
Punjab	3/4	279	429	5	713
Rajasthan	3	439	167	1	607
Sikkim	5	0	0	32	32
Tamil Nadu	3	887	0	0	887
Telangana	3	175	0	0	175
Tripura	5	0	0	70	70
Uttarakhand	4	1	273	51	325
West Bengal	3	914	295	27	1236
Total	–	11105	3338	2470	16913
Percentage of total number of hospitals	–	35.62%	10.71%	7.92%	–

Health data from www.data.gov.in and data on seismic zones from National Disaster Management Authority

Annexure 11.2: Relevant Programs and Policies in Health and Disaster Risk Management in India

See Table 11.8.

Table 11.8 Chronology of relevant programs and policies in health and disaster risk management in India

Year	Institution/Policy/Programme
1911	Indian Council of Medical Research [1949 (renamed to ICMR)]
1933	Medical Council of India
1966	Indian Standard Code for Criteria for Earthquake Resistant Design of Structures; 1st revision
1976	Ministry of Health and Family Welfare
1983	National Health Policy
1987	Modern Public Health Act
1995	National Institute of Disaster Management (training and capacity development programs)
1997	Integrated Disease Surveillance Program (pilot)
1998	National Surveillance Programme for Communicable Diseases
2002	Indian Standard Code for Criteria for Earthquake Resistant Design of Structures; 5th revision
2003	National Vector-Borne Disease Control Programme
2004	Integrated Disease Surveillance Program
2005	Disaster Management Act
2005	National Disaster Management Authority (NDMA) established
2005	National Rural Health Mission
2007	Indian Public Health Standards (IPHS) for public healthcare infrastructure planning and upgradation
2007	Department of Health Research under the Ministry of Health & Family Welfare
2007	NDMA guidelines on Medical Preparedness and Mass Casualty Management; Preparation of State Disaster Management Plans; Earthquakes
2008	NDMA guidelines for Cyclones and Floods
2009	National Health Bill
2009	National Disaster Management Policy
2009	NDMA guidelines for Floods; Landslides and Snow Avalanches; Nuclear and Radio-logical Emergencies; Psychosocial Support and Mental Health Services
2010	NDMA guidelines on Management of urban flooding; Tsunamis; Drought; Management of dead in the aftermath of disaster; Incident Response System
2012	NIDM/UNDP Report on Mainstreaming Disaster Risk Reduction into Health
2013	Mental Healthcare Bill
2013	Pilot Project on Capacity Building for Advanced Trauma Life Support in India
2015	National Health Mission
2016	NDMA guidelines for Hospital Safety
2017	National Health Policy
2017	Draft Public Health Bill (prevention, control and management of epidemics, bio-terrorism and disasters)

Source Various official government web sites

Annexure 11.3: Proposed Smart Cities at Risk of Earthquakes, Floods and Cyclonic Winds

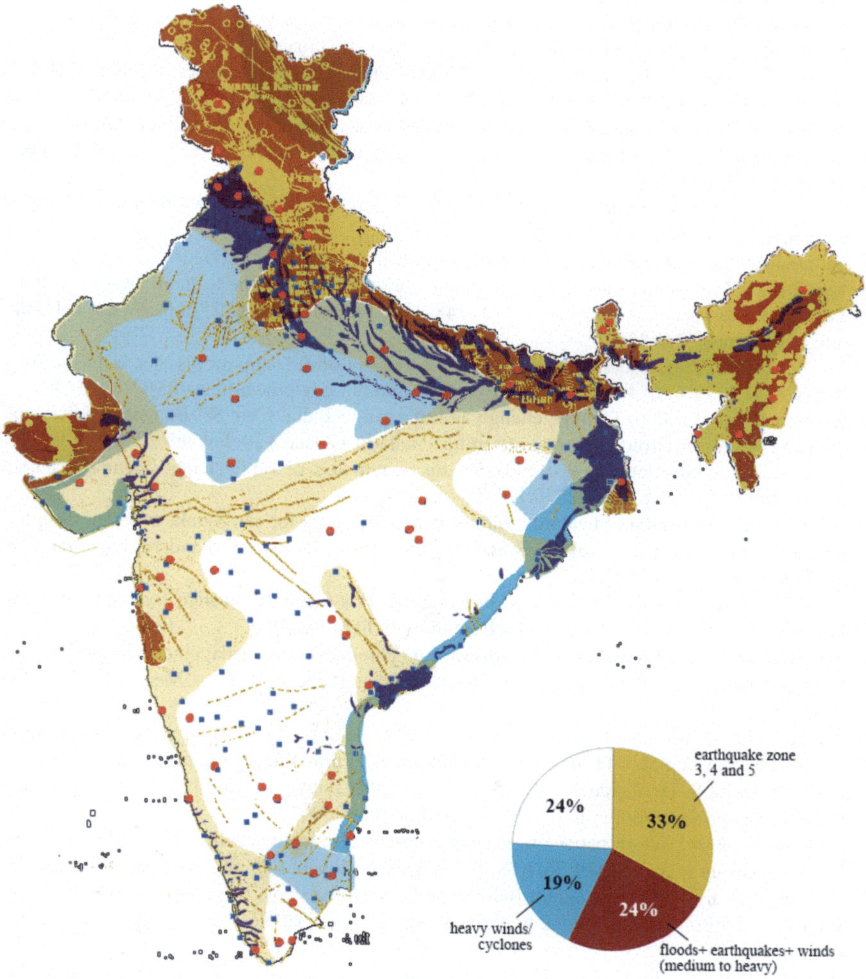

Fig. 11.2 Multi-hazard vulnerability map of India indicating locations of proposed Smart Cities (drawn by authors with data from NDMA, www.data.gov.in and www.mapsofindia.com)

References

ADB (2005) India Post Tsunami Recovery Program: preliminary damage and needs assessment. Report. Asian Development Bank, United Nations and World Bank. https://www.tnrd.gov.in/externallyaidedprojects/Tsunami_rehabilitation/india-assessment-full-report%5B1%5D.pdf

ADB, Asian Development Bank (2018). India: supporting national urban health mission. Technical report. https://www.adb.org/projects/47354–003/main%

Ahern M et al (2005) Global health impacts of floods: epidemiologic evidence. Epidemiol Rev 27 (1):36–46. ISSN: 1478–6729. https://academic.oup.com/epirev/article/27/1/36/520815

Akhtar R et al (2007) Climate change and health and heat wave mortality in India. Glob Environ Res-Engl Ed 11(1):51. http://www.airies.or.jp/attach.php/6a6f75726e616c5f31312d31656e67/save/0/0/11_1–08.pdf%

Barnagarwala T (2017). Mumbai rains: Govt hospitals flooded, patients face infection risk. https://indianexpress.com/article/cities/mumbai/mumbai-rains-govt-hospitals-flooded-patients-face-infection-risk-4820020/%

Becker SM (2007) Psychosocial care for adult and child survivors of the tsunami disaster in India. J Child Adolescent Psychiatr Nurs 20(3):148–155. ISSN: 1073–6077. https://www.ncbi.nlm.nih.gov/pubmed/17688552

Becker SM (2009) Psychosocial care for women survivors of the tsunami disaster in India. Am J Public Health 99(4):654–658. https://www.ncbi.nlm.nih.gov/pmc/articles/PMC2661489/%

Bhattacharya S et al (2006) Climate change and malaria in India. Curr Sci-Bangalore 90(3):369. http://www.indiaenvironmentportal.org.in/files/malaria%20and%20climate%20change.pdf%

BIS (1993) Indian Standard code of practice for installation of septic tanks. Standard. https://archive.org/details/gov.law.is.2470.1.1985

BIS (2002) Bureau of Indian Standards: Indian standard criteria for earthquake resistant design of structures, Part 1: general provisions and buildings (Fifth Revision). Standard. https://archive.org/details/gov.law.is.1893.1.2002

BIS (2007) Bureau of Indian Standards, Indian Standards Code of Practice for Sanitation with Leaching Pits for Rural Communities Bureau of Indian Standards

Bremer R (2003) Policy development in disaster preparedness and management: lessons learned from the January 2001 earthquake in Gujarat, India. Prehosp Disaster Med 18(4):372–384. ISSN: 1945–1938

Bush F et al (2011) Impacts of climate change on public health in India: future research directions. Environ Health Perspect 119(6):765. https://www.ncbi.nlm.nih.gov/pubmed/21273162

Carballo M, Daita S, Hernandez M (2005) Impact of the Tsunami on healthcare systems. http://journals.sagepub.com/doi/pdf/10.1177/014107680509800902%

Das AK et al (2008) White paper on academic emergency medicine in India. INDO-US Joint Working Group (JWG). JAPI 56:789–797. https://www.ncbi.nlm.nih.gov/pubmed/19263706

Das Gupta M et al (2009) How might India's public health systems be strengthened? The World Bank. https://openknowledge.worldbank.org/bitstream/handle/10986/4332/WPS5140.pdf?sequence=1%

Dasgupta P, Ebi K, Sachdeva I (2016) Health sector preparedness for adaptation planning in India. Clim Change 138(3–4):551–566. https://link.springer.com/article/10.1007/s10584-016-1745-7%

Dhara VR, Schramm PJ, Luber G (2013) Climate change infectious diseases in India: Implications for healthcare providers. Indian J Med Res 138(6):847. https://www.ncbi.nlmnih.gov/pmc/articles/PMC3978971/

GeoHazards Intl. (2009) Reducing earthquake risk in hospitals from equipment, contents, architectural elements and building utility systems. http://docs.wixstatic.com/ugd/08dab1_0125e46ba28c4f479e403c53822f2665.pdf

Ghaffar A, Srinath Reddy K, Singhi M (2004) Burden of non-communicable diseases in South Asia. BMJ: British Med J 328(7443):807. https://www.ncbi.nlm.nih.gov/pmc/articles/PMC383378/%

GoI (1939) Items and norms of assistance from the State Disaster Response Fund (SDRF) and the National Disaster Response Fund (NDRF) for the period 2015–2020. http://www.sanchitha.ikm.in/sites/default/files/MadrasPublicHealth_%20Act1939.pdf%
GoI (1956) Constitution of India, Seventh Schedule, Article 246. https://www.mea.gov.in/Images/pdf1/S7.pdf%
GoI (2002) National health policy. https://www.nhp.gov.in/sites/default/files/pdf/NationaL_Health_Pollicy.pdf%
GoI (2005) Disaster management act. Legal rule or regulation. https://ndma.gov.in/images/ndma-pdf/DM_act2005.pdf
GoI (2011) Standard operating procedures for responding to natural disasters. Technical report. Ministry of drinking water and sanitation. https://mdws.gov.in/sites/default/files/Standard_Operating_Procedures_for_responding_to_natural_disasters.pdf
GoI (2016) Union budget speech 2016–2017. https://www.indiabudget.gov.in/budget2016-2017/ub2016-17/bs/bs.pdf%
GoI (2017) Ahmedabad Heat Action Plan. Report. Government of India. https://www.nrdc.org/sites/default/files/ahmedabad-heat-action-plan-2017.pdf
GoI (2017) www.cdsco.nic.in/writereaddata/National-Health-Policy.pdf%
GoI (2018) Operational guidelines for mobile medical units. Report. Ministry of Health and Family Welfare, Government of India. http://nhm.gov.in/images/pdf/programmes/mmu/Mobile_Medical_Units.pdf
Griggs DJ, Noguer M (2002) Climate Change 2001: the scientific basis. Contribution of working group I to the third assessment report of the intergovernmental panel on climate change. https://www.ipcc.ch/ipccreports/tar/wg1/pdf/WGI_TAR_full_report.pdf%
GSDMA (2003) State disaster resource network: comprehensive repository. Pamphlet. http://10.24.36.189/sdrn
Gupta K (2007) Urban flood resilience planning and management and lessons for the future: a case study of Mumbai, India. Urb Water J 4(3):183–194. https://www.tandfonline.com/doi/abs/10.1080/15730620701464141%
Gupta MD et al (2010) How might India's public health systems be strengthened? Lessons from Tamil Nadu. Economic and Political Weekly, pp 46–60. https://www.jstor.org/stable/pdf/25664195.pdf%
Hengesh JV et al (2002) Bhuj, India earthquake of January 26, 2001-reconnaissance report. http://www.iitk.ac.in/nicee/skj/Research_Papers/Bhuj_26Jan2001.pdf
Jha A, Basu R, Basu A (2016) Studying policy changes in disaster management in India: a tale of two cyclones. Disaster Med Public Health Preparedness 10(1):42–46. ISSN: 1935–7893. https://doi.org/10.1017/dmp.2015.116
John TJ et al (2011) Continuing challenge of infectious diseases in India. The Lancet 377 (9761):252–269. http://www.emergencymedicinemims.com/downloads/Infectious%20Disease%20in%20India_Lancet_2011.pdf%
Joshipura MK et al (2003) Trauma care systems in India. Injury 34(9):686–692. https://doi.org/10.1016/S0020-1383(03)00163-3%
Joshipura MK (2008) Trauma care in India: current scenario. World J Surg 32(8):1613. ISSN: 0364–2313. https://link.springer.com/article/, https://doi.org/10.1007/s00268-008-9634-5
Kar N, Jagadisha PSVN et al (2004) Mental health consequences of the trauma of super-cyclone 1999 in Orissa. Indian J Psychiatr 46(3):228. https://www.ncbi.nlm.nih.gov/pmc/articles/PMC2951648/pdf/IJPsy-46-228.pdf%
Kar N, Mohapatra PK et al (2007) Post-traumatic stress dis-order in children and adolescents one year after a super-cyclone in Orissa, India: exploring cross-cultural validity and vulnerability factors. BMC Psychiatry 7(1):8. https://bmcpsychiatry.biomedcentral.com/articles/, https://doi.org/10.1186/1471-244x-7-8%
Khandelwal SK et al (2004) India mental health country profile. Int Rev Psychiatry 16(1-2):126–141. https://www.ncbi.nlm.nih.gov/pubmed/15276945%
Kishore Kumar K et al (2000) Psychosocial care for community level helpers: super cyclone manual. Bangalore, India: Bangalore Books for Change

Knowlton K et al (2014) Development and implementation of South Asia's first heat-health action plan in Ahmedabad (Gujarat, India). Int J Environ Res Public Health 11(4):3473–3492. www.ncbi.nlm.nih.gov/pmc/articles/PMC4024996/%

Kobusingye OC et al (2005) Emergency medical systems in low-and middle-income countries: recommendations for action. Bull World Health Org 83:626–631. https://www.scielosp.org/article/bwho/2005.v83n8/626-631/%

Kumar A et al (2007) Burden of malaria in India: retrospective and prospective view. Am J Trop Med Hygiene 77(6 Suppl):69–78. ISSN: 0002-9637. https://www.ncbi.nlm.nih.gov/books/NBK1720/?report=reader

Lakshminarayanan S (2011) Role of government in public health: current scenario in India and future scope. J Family Commun Med 18(1):26. https://www.ncbi.nlm.nih.gov/pmc/articles/PMC3114612/%

Louvain (UCL)-CRED, Universite catholique de (2015) EM-DAT: the emergency events database. https://www.emdat.be%

Madan VS (2006) Road traffic accidents: emerging epidemic. Indian J Neurotrauma 3(01):1–3. http://medind.nic.in/icf/t06/i1/icft06i1p1.pdf%

Mavalankar D, Shastri P, Raman P (2007) Chikungunya epidemic in India: a major public-health disaster. Lancet Infect Dis 7(5):306–307. https://www.thelancet.com/journals/laninf/article/PIIS1473-3099(07)70091-9/abstract%

Mehta S et al (2006) Disaster and mass casualty management in a hospital: How well are we prepared? J Postgrad Med 52(2):89. http://www.jpgmonline.com/article.asp?issn = 0022-3859%3Byear%3D2006%3Bvolume%3D52%3Bissue%3D2%3Bspage%3D89%3Bepage%3D90%3Baulast%3DMehta%

Misra A et al (2011) Nutrition transition in India: secular trends in dietary intake and their relationship to diet-related non-communicable diseases. J Diabet 3(4):278–292. https://www.ncbi.nlm.nih.gov/pubmed/21649865%

Mondal NC, Biswas R, Manna A (2001) Risk factors of diarrhea among flood victims: a controlled epidemiological study. Indian J Public Health 45(4):122–127. https://www.ncbi.nlm.nih.gov/pubmed/11917333%

Murthy RS et al (2003) Evaluation report on psychosocial care of survivors of supercyclone in Orissa. In: Bhubaneswar: Sneha Aviyan, Action Aid.

Myers MF et al (2000) Forecasting disease risk for increased epidemic preparedness in public health. Adv Parasitol 47:309. https://www.ncbi.nlm.nih.gov/pmc/articles/PMC3196833/

NDMA (n.d.) National disaster management guidelines on minimum standards of relief. Technical report. https://ndma.gov.in/images/guidelines/guideline-on-minimum-standard-of-relief.pdf%

NDMA (2007a) National disaster management guidelines: management of chemical disasters. Technical report. http://nidm.gov.in/pdf/guidelines/new/chemicaldisaster.pdf%

NDMA (2007b) National disaster management guidelines: management of earthquakes. Technical report. http://nidm.gov.in/pdf/guidelines/new/earthquakes.pdf%

NDMA (2008a) National disaster management guidelines: management of cyclones. Technical report. http://nidm.gov.in/pdf/guidelines/new/cyclones.pdf%

NDMA (2008b) National disaster management guidelines: management of floods. Technical report. http://nidm.gov.in/pdf/guidelines/new/flood.pdf%

NDMA (2009a) National disaster management guidelines: management of landslides and snow avalanches. Technical report. http://nidm.gov.in/pdf/guidelines/new/landslidessnowavalanches.pdf%

NDMA (2009b) National disaster management guidelines: management of nuclear and radiological emergencies. Technical report. http://nidm.gov.in/pdf/guidelines/new/managementofnuclearradiologicalemergencies.pdf%

NDMA (2009c) Psycho-social support and mental health services in disasters. Report. National disaster management authority. https://ndma.gov.in/images/guidelines/pssmhsguidlines.pdf

NDMA (2010a) National disaster management guidelines: management of floods. Technical report. http://nidm.gov.in/pdf/guidelines/new/droughtguidelines.pdf%

NDMA (2010b) National disaster management guidelines: management of Tsunamis. Technical report. http://nidm.gov.in/pdf/guidelines/new/ndmaguidelinesmanagementofsunamis.pdf%

NDMA (2010c) National disaster management guidelines: management of urban flooding. Technical report. http://nidm.gov.in/pdf/guidelines/new/management_urban_flooding.pdf%

NDMA (2013) Pilot Project on capacity building for advanced trauma life support in India. Technical report. https://ndma.gov.in/images/cbt/ndma-ignou/Project%20Completion%20 Report.pdf%

NDMA, National Disaster Management Authority (2016) National disaster management guidelines hospital safety. Technical report. http://www.ndma.gov.in/images/guidelines/ Guidelines-Hospital-Safety.pdf

Nsubuga P et al (2006) Public health surveillance: a tool for targeting and monitoring interventions. Dis Control Prior Dev Countries 2:997–1018. https://www.ncbi.nlm.nih.gov/ books/NBK11770/%

Pandian JD et al (2006) Factors delaying admission to a hospital-based stroke unit in India. J Stroke Cerebrovasc Dis 15(3):81–87. https://www.ncbi.nlm.nih.gov/pubmed/17904057%

Patel V et al. (2011) Chronic diseases and injuries in India. Lancet 377(9763):413–428. http:// www.who.int/choice/publications/Chronic_diseaseIndia.pdf%

Phalkey R et al (2012) Prepared to react? Assessing the functional capacity of the primary healthcare system in rural Orissa, India to respond to the devastating flood of September 2008. Global Health Action 5(1):10964. https://www.tandfonline.com/doi/full/10.3402/gha.v5i0. 10964%

Ramanujam P, Aschkenasy M (2007) Identifying the need for pre-hospital and emergency care in the developing world: a case study in Chennai, India. JAPI 55. https://www.researchgate.net/ profile/Miriam_Aschkenasy/publication/5936532_Identifying_the_need_for_pre-hospital_ and_emergency_care_in_the_developing_world_A_case_study_in_Chennai_India/links/ 55ddd1a508ae45e825d39080.pdf%

Rao RS et al (2003) Leptospirosis in India and the rest of the world. Braz J Infect Dis 7(3):178–193. http://www.scielo.br/scielo.php?pid=S1413-86702003000300003&script = sci_arttext%

Rastogi T et al (2004) Physical activity and risk of coronary heart disease in India. In: Int J Epidemiol 33(4):759–767. https://academic.oup.com/ije/article/33/4/759/665434%

Raut DK, Bhola A (2014) Integrated disease surveillance in India: way forward. Glob J Med Public Health 3(4). http://www.gjmedph.com/uploads/R2-Vo3No4.pdf

Razzak JA, Kellermann AL (2002) Emergency medical care in developing countries: is it worthwhile? Bull World Health Organ 80:900–905. https://www.scielosp.org/scielo.php?pid= S0042-96862002001100011&script=sci_arttext%

Reddy KS et al (2005) Responding to the threat of chronic diseases in India. Lancet 366 (9498):1744–1749. https://pdfs.semanticscholar.org/047c/3682624c83c75c308217645f1a40e1 f086a8.pdf%

Roy N et al (2002) The Gujarat earthquake (2001) experience in a seismically unprepared area: community hospital medical response. Prehosp Disaster Med 17(4):186–195. https://www. researchgate.Net/profile/Ralph_Coughlin/publication/10604429_The_Gujarat_Earthquake_ 2001_Experience_in_a_Seismically_Unprepared_Area_Community_Hospital_Medical_ Response/links/0c96052b9d86ab9356000000/The-Gujarat-Earthquake-2001-Experience-in-a-Seismically-Unprepared-Area-Community-Hospital-Medical-Response.pdf%

Sapatnekar S, Patnaik I, Kishore K (2018) Regulating infrastructure development in India. Report. http://www.nipfp.org.in/media/medialibrary/2018/05/WP_230.pdf

Sehgal SC, Sugunan AP, Vijayachari P (2002) Outbreak of leptospirosis after the cyclone in Orissa. Natl Med J India 15(1):22–23. https://europepmc.org/abstract/med/11855587%

Shah B, Mathur P (2010) Surveillance of cardiovascular disease risk factors in India: the need & scope. Indian J Med Res 132(5):634. https://www.ncbi.nlm.nih.gov/pmc/articles/PMC3028 945/%

Sharan P et al (1996) Preliminary report of psychiatric disorders in survivors of a severe earthquake. Am J Psychiatry 153(4):556. https://www.ncbi.nlm.nih.gov/pubmed/8599406%

Sharma M, Brandler ES (2014) Emergency medical services in India: the present and future. Prehospital Disaster Med 29(3):307–310
Subhan I, Jain A (2010) Emergency care in India: the building blocks. Int J Emerg Med 3(4):207–211. https://link.springer.com/article/10.1007/s12245-010-0223-7%
Sur D et al (2000) Severe cholera outbreak following floods in a northern district of West Bengal. Indian J Med Res 112:178. https://www.ncbi.nlm.nih.gov/pubmed/12452126%
Vijayachari P, Sugunan AP, Shriram AN (2008) Leptospirosis: an emerging global public health problem. J Biosci 33(4):557–569. https://www.ias.ac.in/article/fulltext/jbsc/033/04/0557-0569%
WHO (2000) Leptospirosis, India: report of the investigation of a post-cyclone outbreak in Orissa, November 1999. Weekly Epidemiol Record 75(27):217–223. https://apps.who.int/iris/bitstream/handle/10665/231209/WER7527_217-223.PDF?sequence=1%
WHO, World Health Organisation (2005) Human health impacts from climate variability and climate change in the hindu kush-himalaya region: report of an inter- regional workshop. http://apps.who.int/iris/handle/10665/205297%
WHO (2012) WHO country cooperation strategy summary: India 2012–2017. Report. World Health Organization. Regional Office for South-East Asia
World Health Organization (2013) Tsunami 2004: a comprehensive analysis. Technical report. http://apps.searo.who.int/PDS_DOCS/B4998.pdf%
WHO (2015) India: WHO statistical profile. Report. World Health Organization. www.who.int/gho/countries/ind.pdf

Chapter 12
Safe Drinking Water Solutions in Parts of West Bengal, India: Combating Health Issues Through Participatory Water Management

Bhaswati Ray, Md. Anwarul Abedin and Rajib Shaw

Abstract The world is facing a global water crisis with the capacity of the hydrological cycle to supply water challenged by excessive demand, pollution of water resources and poor water management. Many districts in West Bengal are affected by diminishing water reserves and exposure to contaminants in the groundwater system with serious health impacts. Exposure to polluted water can cause diarrhoea, skin irritation and respiratory problems, depending on the pollutant present in the water body. Given the vital role of water in sustainable development, it is important that the community takes up the responsibility of providing safe drinking water. The emphasis would be on the voluntary participation of the users, the providers and the science community in finding solutions to health issues that result from contamination of drinking water. As no universal model may be suited to combat health problems, community-specific sustainable options need to be adopted. Disseminating scientific and correct information among the population coupled with increased community participation and education are recognized as necessary adjuncts for an intervention to be successful and sustainable. The objectives of this research include creating awareness about health impacts of poor quality drinking water, to assess the socio-cultural factors that determine the risk of exposure to contaminated drinking water, provision of an interactive platform for information sharing, co-designing solutions involving the local communities and the local governance and to suggest policy interventions to resolve water issues.

B. Ray (✉)
Sivanath Sastri College, University of Calcutta, Kolkata, India
e-mail: bhaswati173@gmail.com

Md. A. Abedin
Department of Soil Science, Bangladesh Agricultural University, Mymensingh, Bangladesh

R. Shaw
Graduate School of Media and Governance, Keio University, Fujisawa, Japan

E. Ying Yang Chan and R. Shaw (eds.), *Public Health and Disasters*,
Disaster Risk Reduction, https://doi.org/10.1007/978-981-15-0924-7_12

Keywords Community participation · Stakeholders · Health issues · Co-designing

12.1 Introduction

Water influences every aspect of human life and the growing human population. It plays a key role in the economic development of a nation and affects agricultural productivity and food systems, community health and hygiene, urban development and ecosystem functioning (Shomar 2013). Hence, water and sanitation have been recognized as a matter of fundamental human right. The United Nations General Assembly on 28 July 2010 recognized water and sanitation as a basic human right and called upon international organizations to provide financial and technological support to the less developed nations to enable the provisioning of safe, clean, accessible and affordable drinking water for all. The United Nations Human Rights Council through a resolution on 28 September 2011 invited good practices to safe drinking water and sanitation with an emphasis on the adoption of practical solutions. It is an attempt to secure human right to safe water and sanitation as well as for improved delivery options. The World Health Organization in the World Health Assembly in May 2011 resolved that member states must strive to ensure that national health strategies realize water and sanitation goals as set in the Millennium Development Goals and strengthen collaborations to increase access to safe drinking water and sanitation, thus improving the human right to water. The 2030 Agenda for Sustainable Development mentions water and sanitation among all recognized human rights in Goal 6, to ensure a holistic approach in tackling global water problems and water security issues for sustainable development.

The world is, however, facing a global water crisis with the capacity of hydrological cycles to supply water challenged by escalating demands, pollution of water resources and poor water management (Panos 1998). Fresh water available for use accounts for only 1% of all fresh water on earth and is distributed disproportionately across the globe (Shomar 2013). The unreliable and inequitable distribution of fresh water, resulting from various rainfall distribution systems spread across time and space, fundamentally impacts the environmental systems and infrastructures on which the life and livelihood of humans depend (Shomar 2013; Allan and Soden 2008). Around 1.2 billion or one-fifth of the world's population currently live in regions affected by physical water scarcity. Water resources are found to be inadequate to meet total demand, including the water needed to fulfil the demand of ecosystems to function effectively (Watkins et al. 2006). Water scarcity is accompanied by excessive withdrawal of groundwater. Underground aquifers that supply 35% of human water use are getting depleted. NASA satellites reveal that 21 of the world's 37 largest aquifers have exceeded their sustainability point (Frankel 2015). Thirteen aquifers declined at rates that forced them to be included in the most troubled or threatened categories and are located mostly in the poor and densely populated regions in South Asia and North Africa with limited alternatives to meet water scarcity. In spite of being declared the right to water and sanitation as basic human right, 884 million people do not have sufficient and reliable access to safe water while 2.6 billion are without basic sanitation facilities (Shomar 2013; Kirby 2009). It is predicted that in 2030, half of the world population will be living in high water-stressed areas (United Nations Educational Scientific Cultural Organization 2014). According to a report by McKinsey and Company, collective water demands will exceed predicted supply by 40%. (Shomar 2013) within two decades (Shomar 2013; Mckinsey and Company 2009) resulting in increased food prices, disrupted energy supply, the creation of environmental refugees and would eventually lead to the establishment of water-food-energy nexus as a policy tool for integrated development. Water is also a key element in political stability. The problem of physical scarcity of water is aggravated by climate change impacts on the global water cycle with potentially more frequent occurrences of floods and droughts, sea level rise and saline water intrusions into the groundwater system particularly in coastal areas. Being a universal solvent, water is a major source of pathogens, geogenic and chemical contaminants. Water storage and supply infrastructure like surface water reservoirs and desalinization plants in association with human water use behaviour and seasonal cycles in increased water demand also impact water availability and are matters of concern (Mehran et al. 2017; Vogel et al. 2015). Surface reservoirs control 20% of the global annual water discharged into streams and rivers (Mehran et al. 2017; Shiklomanov et al. 2000) and provide resilience against disaster during floods and droughts in addition to increasing energy production and addressing water resource management concerns (Mehran et al. 2017; Padowski et al. 2015; Madani 2010; Hallegatte 2009; Palmer et al. 2008).

Given the vital role of water in sustainable development and human well-being, it is important that the community takes up the responsibility of providing safe drinking water. The concept is not new. Examples of participatory management date back to over 1000 years in European rural communities (Razzaque 2009). The 1970s and 1980s evolved as the era for people-centred approach to economic development (Razzaque 2009; Spyke 1999). The 1990s saw a similar trend when stakeholder consultation and participation were considered essential for successful decision-making and good governance (Razzaque 2009; Kuhn 1999; McCormick 1995). Active stakeholder participation in water resource management including the local communities and the introduction of the traditional sustainable practices that have been discontinued could play a crucial role in ensuring sustainable improvements in water systems. Community-owned and community-managed innovative and smart solutions are needed to ensure that water and wastewater are managed efficiently in the era of rapid population growth and climate variability. The water management system must be sustainable and resilient to long term and consistent increase in water consumption levels and diminishing water availability. It should be equipped with the capacity to be monitored and networked with other critical systems to obtain more sophisticated information on their efficient management.

12.2 The Study Area

West Bengal in India suffers from both physical and economic scarcity of water. While physical scarcity of water is caused by over-exploitation of water resources, changing land use pattern interfering with infiltration and recharge, pollution of water bodies and severed by climate change, economic scarcity arises from lack of investment in the water sector and the inability to utilize available water resources because of lower affordability, ethnic and political conflicts. Almost all districts in West Bengal suffer from diminishing water reserves, discontinued traditional water harvesting systems, excessive withdrawal of groundwater for drinking and irrigation accompanied by exposure to geogenic and chemical contaminants present in the groundwater system, polluted surface water reserves and poor policy environment. Seasonal water scarcity is accentuated by variability in monsoon rains and depleting groundwater reserves. Eighty-three C. D. Blocks in 8 districts of West Bengal suffer from arsenic above permissible limit (0.05 mg/l according to Bureau of Indian Standards and 0. 01 mg/l according to the World Health Organization), and 43 C. D. Blocks in 7 districts suffer from fluoride above the permissible limit (1.5 mg/l according to the Bureau of Indian Standards, 1.0 mg/l according to World Health Organization). Salinity above permissible limit (chloride 250 mg/l, TDS 500 mg/l, Bureau of Indian Standards and World Health Organization) is seen in 57 C. D. Blocks across 4 districts. The state also suffers from inadequacy of drinking water sources with one stand post per 475 persons against a norm of one stand post for every 250 persons by the Working Group, Rural Domestic Water and Sanitation, Twelfth Plan 2012–17.

The C. D. Blocks of Baruipur, Sonarpur, Bhangar I, Bhangar II, Bishnupur I, Bishnupur II, Joynagar I, Magrahat II, Canning I and Canning II affected by arsenic, salinity and iron above permissible limit. C. D. Blocks Baruipur, Canning I and Canning II have been chosen for detailed study (Fig. 12.1). The groundwater system that supplies more than 90% of drinking water in these C. D. Blocks has iron, arsenic and salinity above permissible limit. Health impacts are also evident among the local communities.

12.3 Profile of the Study Area

The region is located in the southern part of the Ganga delta. It is a flat monotonous plain between 1 and 5 m above mean sea level, intercepted by paleo-channels and tidal creeks. The region is thus endowed with sufficient quantities of surface and groundwater. But the availability is hampered by sea water ingression, high soil salinity and contamination. It enjoys tropical monsoon type of climate with an annual average rainfall of 1762 mm. The rainfall distribution is highly skewed and concentrated in the four monsoon months from June to September (Bandyopadhyay and Basu 2017). Water scarcity is thus more pronounced in the pre-monsoon

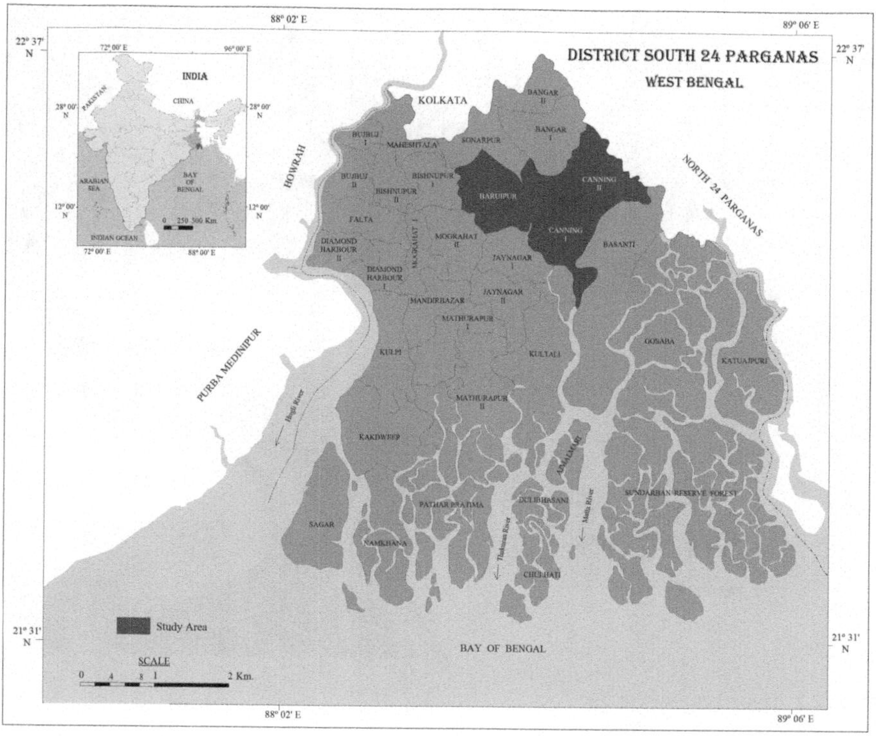

Fig. 12.1 Location of the study area

months when the temperature reaches up to nearly 40 °C. The presence of arsenic and salinity in the groundwater system further aggravates the scarcity. The level of arsenic in groundwater is also higher in the pre-monsoon months. Moreover, the incidence of arsenic in groundwater is erratic and difficult to predict because of its patchy discontinuous nature of distribution. High arsenic and low arsenic tube wells often occur side by side.

The aquifers belong to the unconsolidated sediments of quaternary and upper tertiary period deposited under fluvial and submarine to marine conditions. Groundwater occurs under confined conditions at a depth ranging from 20 to 400 m (Bandyopadhyay and Basu 2017). Freshwater aquifers are seen to occur at greater depths ranging from 160 to 360 m below ground level (Bandyopadhyay and Basu 2017). Shallower aquifers are, however, contaminated by the presence of arsenic and sodium above the permissible limit. Deeper aquifers at more than 300 m below ground level have not been found contaminated except in extremely rare cases. While even small traces of arsenic of 0.05 mg/l is harmful to the human body, amount of arsenic in groundwater reaches to more than 0.25 mg/l in parts of Baruipur and Canning C. D. Blocks along with Bhangar I, Bishnupur I, Jaynagar I, Magrahat II and Sonarpur at depths up to 200 m. Early clinical symptoms of arsenic

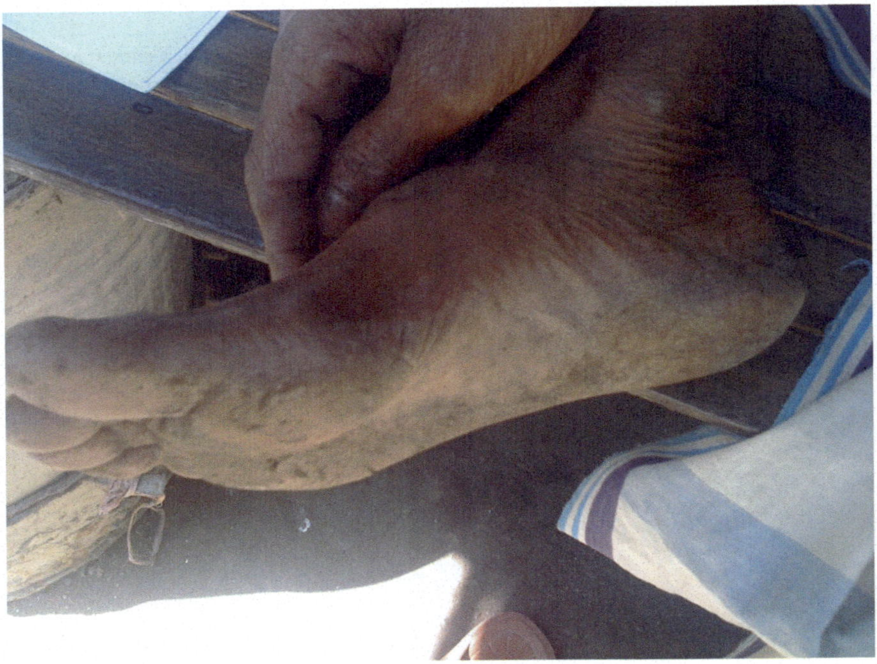

Fig. 12.2 Hyperkeratosis of feet

intoxication include abdominal pain, vomiting, diarrhoea, muscle pain and flushing of the skin. These symptoms are followed by numbness and tingling of the limbs, muscular cramping and the appearance of erythematic rash (Murphy et al. 1981). Within a month, burning paraesthesia and hyperkeratosis appear on palm and feet (Fig. 12.2) making walking and manual activities painful. There is also progressive deterioration in motor and sensory responses (Fennell and Stacy 1981; Murphy et al. 1981; Wesbey and Kunis 1981). Chronic exposure leads to dermal lesions, hyper- and hypo-pigmentation, neuropathy and a major carcinogen of skin, lungs and kidney. Arsenic ingestion is also often associated with increased risks for a number of vascular and cerebrovascular diseases, hypertension and coronary heart diseases (CHD) (Hopenhyan 2006). Arsenic may turn fatal at acute doses.

The upper aquifers at a depth of 20–160 m contain saline water, with the salinity level above the permissible limit of 600 ppm (Bandyopadhyay and Basu 2017). Saline intrusions are common in this region because of their coastal location and accentuated by global warming, sea level rise, unplanned embankment construction and increasing shrimp farming. Sodium chloride (NaCl) concentration in shallow aquifers is derived from sea water ingress and trapped salt water under marine conditions in late Pleistocene to early quaternary period when marine conditions prevailed in the Lower Bengal basin (Bandyopadhyay and Basu 2017). The region is also experiencing a decline in freshwater reserves due to the reduction of existing freshwater ponds, over-exploitation of sweet water aquifers and increased salinity in adjacent freshwater ponds and shallow aquifers through seepage of salt water brought in for wetland shrimp cultivation.

Sodium salts are not acutely toxic but accidental overdoses can turn fatal. Hypertension, coronary heart disease, vomiting, convulsions, muscular twitching and rigidity and cerebral and pulmonary oedema are the common effects of high salinity (Elton et al. 1963). Excessive salt intake seriously aggravates chronic congestive heart failure (World Health Organization 2003). High salt concentration in groundwater can damage crops and affect plant growth and pollute drinking water, thereby affecting economic development. Most salts do not degrade naturally and can be persistent in groundwater.

12.4 Drinking Water and Health Issues in Study Area: People's Perception

In order to assess people's perception on the health issues arising from contaminated drinking water, questionnaire survey was conducted in 200 households in March and April 2018, from each of the three C. D. Blocks in villages where drinking water was seen to be affected by arsenic contamination or salinity intrusions. Within the village, the method of random sampling was adopted. The respondents were all adults aged between 20 and 50 years and are educated at least up to the secondary level.

The source of drinking water for majority of the households constitutes groundwater. Tube wells account for 97.95% of drinking water sources at household level (Fig. 12.3), of which 16.76% continue to use shallow tube wells, making themselves vulnerable to the health impacts of excess arsenic and salinity. The

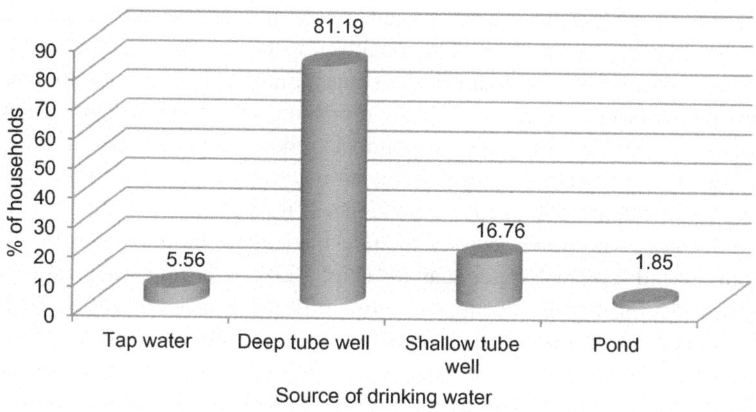

Fig. 12.3 Source of drinking water in the study area

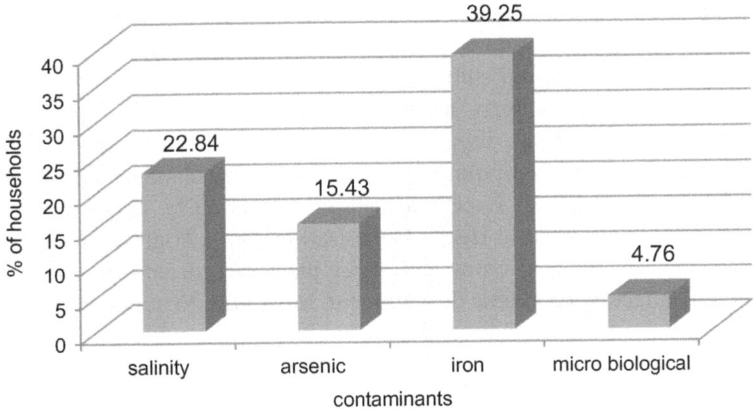

Fig. 12.4 Contaminants in drinking water in the study area

major contaminants identified by the local communities include iron, salinity, arsenic and microbiological. Arsenic, with the most damaging health impacts, has been identified by only 15.43% of the respondent households (Fig. 12.4). Salinity has been identified by 22.84% while iron as a major contaminant has been identified by 39.25%. Awareness level appears to be quite low. Iron has been identified by more than 39% of the households. The reasons were obvious. The presence of iron is easiest to detect by the distinct colour and smell, with red stains visible in the containers that store water. Salinity is still easy to detect by the taste while arsenic with no taste or smell is most difficult to detect until the disease symptoms become apparent. Identification of problems including contaminants was done based on disease symptoms and the characteristics of water, promoting citizen science and community participation. For identification of disease symptoms as well as the water scarcity issues by the local communities, the questions were of simple multiple-choice type where the identified symptoms were of diseases most commonly caused by each of these contaminants while the water scarcity issues address local environmental constraints. The health impacts easily discernible include indigestion, hyperacidity, and diarrhoea caused mostly by microbiological contaminants particularly during and after the monsoon season. These symptoms were reported by 20–35% of the total respondents. Skin problems like itching, pigmentation and hyperkeratosis caused by arsenic above permissible limit have been identified by less than 15% of the respondents (Fig. 12.5).

The other issues of water scarcity include inadequate number of safe drinking water sources and seasonal water scarcity (Fig. 12.6). A falling water table results in many sources turning dry during the summer or pre-monsoon season. However, there have been no efforts on the part of the local communities to address water issues because of their conviction that it is a state matter and the state government should be responsible for providing safe water to all (Fig. 12.7). Less than 10% of the households treat water or use water filters.

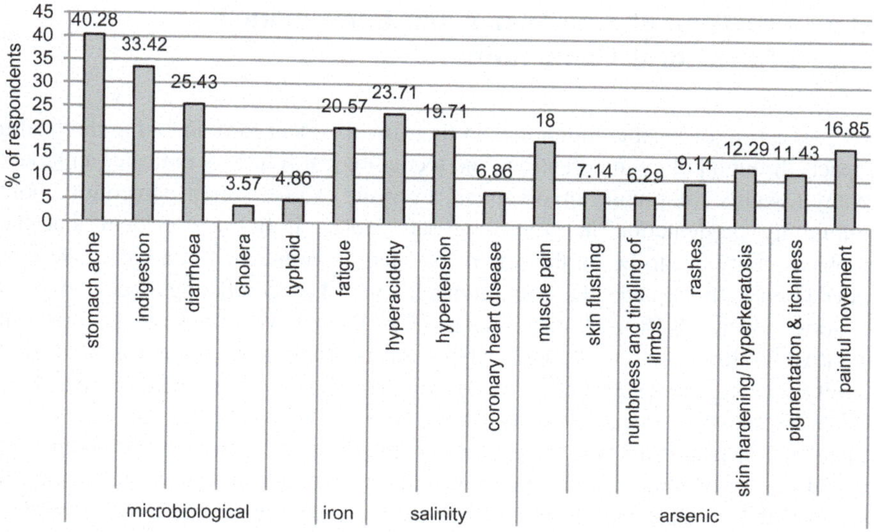

Fig. 12.5 Health impacts identified by the local communities in the study area

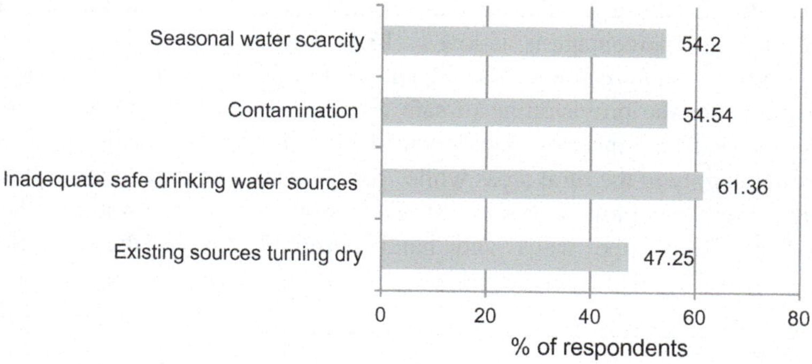

Fig. 12.6 Water scarcity issues identified by the local communities in the study area

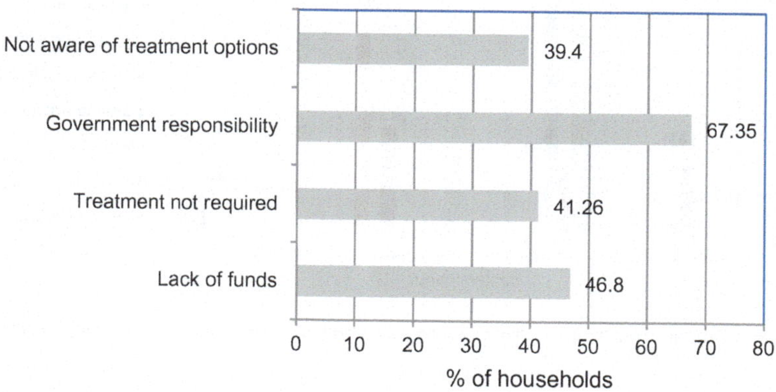

Fig. 12.7 Reasons for not using treatment options for drinking water in the study area

12.5 Provision of Safe Water and Participation of the Local Communities

The perception of local communities in their involvement for combating the health impacts through policy interventions and technological solutions indicates a marked apathy towards community involvement. In trying to assess their perception about community participation in promoting safe water, it became evident that the emphasis continues to be on the role of the state government to find alternate safe water sources to replace the contaminated ones. For combating excess iron in drinking water, 25.18% of respondents (Fig. 12.8) have opted for government initiative to find alternate source. The corresponding percentage for managing salinity is as high as 81.37% (Fig. 12.8). Another 51.11% have opted for iron filters which are mostly used at household level (Fig. 12.8). Rainwater harvesting as a viable option is still a taboo, and local communities are apprehensive about successful implementation of rainwater harvesting techniques. Rainwater harvesting if implemented at community level could be the most effective solution in this rainfall surplus district of West Bengal using low-cost sand filters and chlorine for treatment. But very few respondents showed preference for rainwater harvesting. Only 25.52% of respondents are in favour of rainwater harvesting and increased use of surface water for the iron affected point sources (Fig. 12.8), but for the salinity affected ones, the percentage is as low as 15.11% (Fig. 12.8).

For arsenic contamination also, local communities prefer to bestow the responsibility for the provisioning of safe drinking water to the local government and public health engineering departments. 59.14% of respondents prefer subsidized water supply in the rural areas while 75.45% want a change in the aquifer that is being currently exploited (Fig. 12.8). They want the shallow tube wells to be replaced by deep tube wells and hand pumps to avoid the arsenic- and

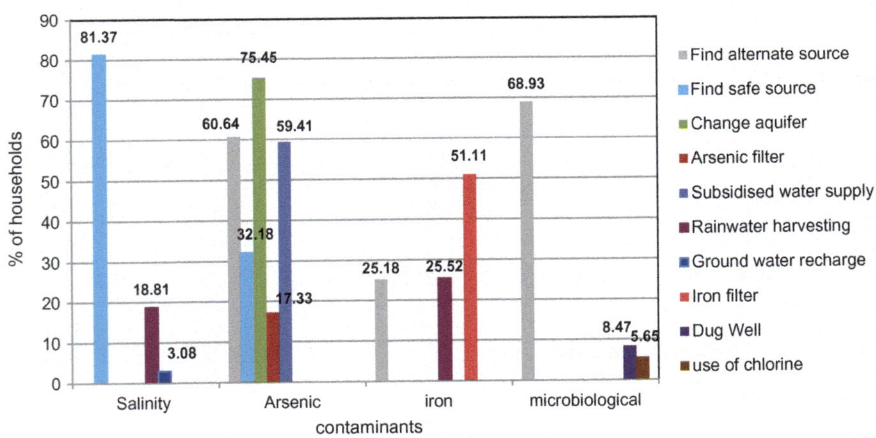

Fig. 12.8 Community initiatives for managing drinking water toxicity in the study area

salinity-contaminated aquifer layer. Deep tube wells would also take care of excess iron in water. More than 80% of respondents prefer the discontinuation of the contaminated sources. These sources need to be replaced by alternate safe sources. However, none have agreed to go in for rainwater harvesting though it is the safest and cheapest remedy for arsenic-contaminated groundwater. Pond filters also work well in arsenic-contaminated areas as surface water is free from these harmful chemical contaminants. But only 17.33% have voiced in favour of arsenic filters. There is thus a clear indication of a lack of involvement of the local communities in water resource management.

Assessment of the involvement of local communities in managing drinking water toxicity is also based on the use of multiple-choice type questions where alternative measures were suggested. For each contaminant, the best possible solutions were highlighted and explained to the local communities, and they were then asked to choose the best possible alternative. Such initiative was also undertaken to raise the level of awareness on the prevention and mitigation measures and to ensure their active involvement and participation. Any corrective needs to evolve from the local communities to ensure that such measures are effectively managed, a prerequisite for its long-term sustainability.

12.6 Stakeholder Mapping and Participatory Water Management for Safe Drinking Water

It is now an established fact that multi-stakeholder participatory management of water systems including the participation of local community is increasing throughout the world with an aim to aid collective decision-making and implementation of locally adapted water management strategies (Daniell et al. 2010). While community participation involves the citizens or non-technical persons in the designing and implementation of water projects, more commonly referred to as citizen science, stakeholder engagement encourages the involvement of different actors like the service providers, local government, investors and technical experts in addition to the civil society. Proper participatory management would enable the formulation of locally suited and acceptable water solutions and promote the use of local indigenous knowledge like traditional water harvesting systems for sustainable water management and effective water governance. Efforts to improve water governance explored by the United Nations, the World Bank and Organisation for Economic Co-operation and Development (OECD) in association with local government, academia, NGOs and the private sector (Groundwater Governance 2017; Transboundary Water Assessment Programme 2017; Organisation for Economic Co-operation and Development 2017) highlight the importance of shared vision and experiences and that of assessing alternatives in addressing water scarcity issues.

Stakeholder participation is being promoted and has been highlighted in World Water Forums since the Dublin Principles on Water and Environment 1992 that made stakeholder engagement an integral part of water resources management (IWRM). It was stated at the conference that development and management of water resources need a participatory approach and should involve users, planners and policymakers. The Hague Conference 2000 or the Second World Water Forum called for participatory approach in water management. The Third World Water Forum in Kyoto 2003 also prompted alliances and networking among various stakeholders while the next World Water Forum in Mexico 2006 was in favour of co-ordination across different levels of government. The critical role of vulnerable and marginalized groups was realized in the Fifth World Water Forum in Istanbul conference, 2009, and the need for multi-stakeholder initiatives was accepted. The increasing stakeholder engagement in the water sector indicates a paradigm shift in public management and water governance (Mollinga 2010). The Eight World Water Forum at Brasilia, Brazil, also strives to promote awareness building and initiate political commitment on solving critical water issues at all levels. It further envisages the coming together of all stakeholders including policymakers, scientists, private organizations and the civil society organizations. It also highlighted the need for innovations, transfer of technology, exchange of good practices and advanced research in promoting water security.

In many countries, stakeholder participation is practised for sustainable water resource management (Razzaque 2009). The National Water Policy of Bangladesh (1999) necessitates participation of local communities for water system management (Razzaque 2009; Akhter 2005). The Water Policy of Pakistan (2004) initiates stakeholder consultation and participation in all sectors of water management. The National Water Policy of India (2002) also encourages community participation in the management and development of water resources.

The first step to stakeholder participation and involvement is stakeholder mapping (Fig. 12.9) or the identification of stakeholders and the core stakeholder functions, their roles and responsibilities in the water sector and their impact on other relevant sectors. Promoters of stakeholder engagement appreciate the role of the stakeholders in taking up responsibilities arising out of a sense of ownership and a stake in the concerned system.

The fragmented nature of the water sector with governments, service providers, end users, private actors and financial institutions all claiming a stake further benefits from multi-stakeholder engagement. Involving the stakeholders would also ensure the participation of the scientific community to not only ensure proper management of existing water resources but also bring forward the mitigation measures that can be implemented to prevent and combat the health issues arising out of contamination of drinking water. Involving the local community in co-designing and managing water contamination to reduce the incidence of waterborne diseases would also ensure proper implementation because the local community would be more passionate as they are the worst affected. Safe water solutions like aquifer recharge or rainwater harvesting not only increases the aquifer potential and increase water availability, such measures would also help reduce the

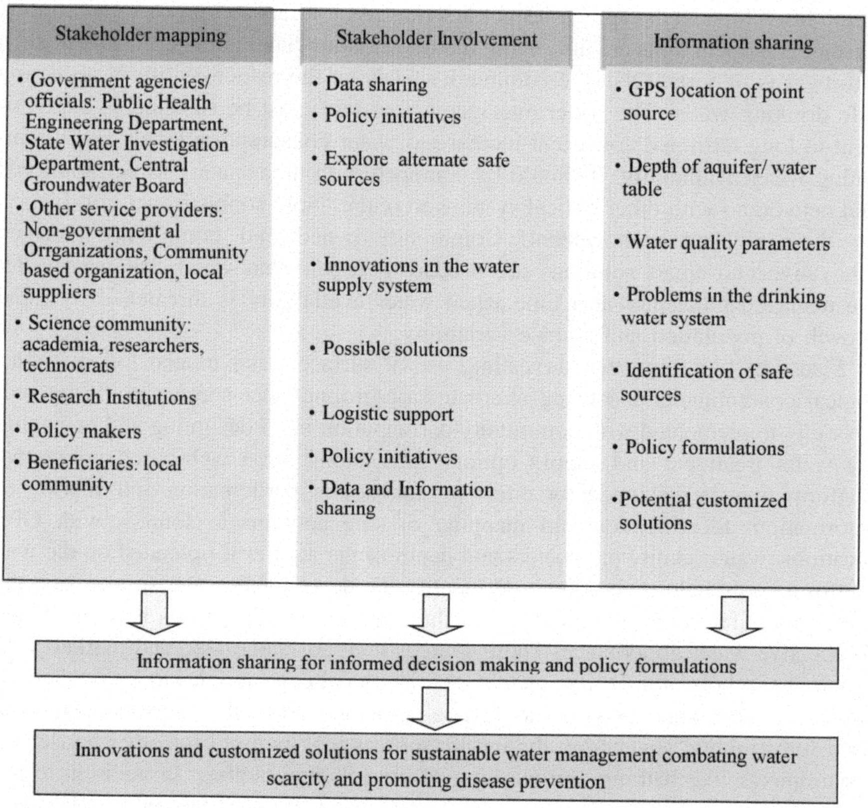

Fig. 12.9 Stakeholder mapping and involvement in the study area

threat of hyperkeratosis and cardio-vascular diseases caused by arsenic and salinity above permissible limit. Safe water solutions would thus ensure water availability and combat health issues.

However, stakeholder engagement and community participation are largely lacking in the study area as is evident from the fact that 67.35% of respondent households feel that provisioning of safe drinking water is a state responsibility. Thus, the only stakeholder seems to be the local government, and the local communities prefer to be at the receiving end. However, co-designing safe water solutions that are community owned and community managed with technical assistance from the science community is the need of the hour to ensure that the projects are resilient and sustainable. Local communities are still reluctant to take up the responsibility of safe water provisions as appeared from the primary survey.

Efforts are urgently needed to go in for stakeholder mapping and community participation. Public Health Engineering Department, Government of West Bengal responsible for rural water supply, has implemented different measures for supplying arsenic-free water to the affected communities through dug wells, hand pump fitted tube wells tapping deeper arsenic-free aquifers and piped water supply. Arsenic treatment units are also being fitted to existing hand pumps. In spite of such efforts, it is important that the community takes up the responsibility of accessing safe drinking water. The water management system must be sustainable and resilient to long term and consistent increase in water consumption levels and diminishing water availability. It should be equipped with the capacity to be monitored and networked with other critical systems to obtain more sophisticated information on their efficient management. Community-owned and community-managed innovative and smart solutions are needed to ensure that water and wastewater are managed efficiently at a time when water availability is threatened by rapid growth of population and climate variability.

Measures like rainwater harvesting, use of surface water treated for microbiological contaminants and using alternate safe groundwater sources could be successfully implemented with community participation in co-designing and managing the water treatment and supply options. Provisions for a web-based interactive platform may be designed for effective stakeholder participation and sharing of information. Identification and mapping of safe and unsafe sources with GPS locations, water quality parameters and depth of the aquifer if uploaded on the web platform could help each household identify the nearest safe water sources and the affected aquifer depth. Identification of the safe and unsafe sources is the first step to effective water governance. These sources may be categorized and marked with different symbols or colour so as to enable the local communities to know the safe sources. A dual water usage could then be promoted. While the safe sources may be used for drinking, water from the unsafe sources may be used for other household requirements like bathing and washing of utensils and clothes. In addition to the availability of water in adequate quantities to meet the multifaceted requirement, such measures would also be effective in addressing existing health issues and preventing future occurrences. Management tools involving the local communities can be easily customized to suit local conditions and would thus be sustainable and resilient. The dual water usage, however, is impossible to implement without community awareness and initiative. Similarly, rainwater harvesting may be adopted at household as well as at community level with appropriate storage structures including ponds and tanks to facilitate collection. The stored rainwater can be used for drinking to manage the water crisis in peak summers after a treatment by horizontal roughing filter (HRF) followed by slow sand filter (SSF) to remove waterborne pathogens (Gayen and Zaman 2013). Rainwater harvesting and aquifer recharge may also be adopted in high arsenic and high salinity areas to dilute their concentration in groundwater. Surface water-based schemes, desilting and protection of ponds before monsoon and treatment of pond water are alternate safe options.

12.7 Conclusion

Arsenic contamination, saltwater intrusion, depleting groundwater reserves and progressively falling water table as well as water-related hazards like flood and drought are the major obstacles in effective water use and water resource management for the rural water systems in the study area. Large parts of the area are threatened by groundwater contamination with grave impact on human health. In the absence of alternate safe drinking water sources, 90% of the rural population is dependent on groundwater sources, mostly shallow tube wells, thought to provide bacteria free safe and clean drinking water. The shallow aquifers in turn increase the vulnerability of the local communities to adverse health impacts.

Regarded as the responsibility of the state government, little has been done at community level to manage the water crisis. Simple measures like compulsory installation of rainwater harvesting systems in government offices and institutions, using surface water after treatment for microbiological contaminants and using alternate tube wells marked as safe are some of the policy interventions suggested. Support from public–private partnership may be encouraged to initiate and install the required infrastructure like rainwater harvesting reservoirs, pond sand filters and the like, but sustainable and resilient water systems shall emerge only through co-engineering water management systems through co-initiation, co-design and co-implementation phase.

References

Akhter BR (2005) Pollution and its management approach: integrated coastal zone management plan project. Government of Bangladesh, Dhaka

Allan RP, Soden BJ (2008) Atmospheric warming and the amplification of precipitation extremes. Science 321:1481–1484

Bandyopadhyay M, Basu R (2017) Crisis of fresh water in South 24 Parganas District, West Bengal: causes and consequences. IOSR J Hum Soc Sci (IOSR-JHSS) 22(6):4–15

Daniell KA, White I, Ferrand N, Ribarova IS, Coad P, Rougier JE, Hare M, Jones NA, Popova A, Rollin D, Perez P, Burn S (2010) Co-engineering participatory water management processes: theory and insights from Australian and Bulgarian interventions. Ecol Soc 15(4):1–37

Elton NW, Elton WJ, Narzareno JP (1963) Pathology of acute salt poisoning in infants. Am J Clin Pathol 39:252–264

Fennell JS, Stacy WK (1981) Electrocardiographic changes in acute arsenic poisoning. Ir J Med Sci 150:338–339

Frankel TC (2015) New NASA data show how the world is running out of water. The Washington Post, 16 June 2015

Gayen A, Zaman A (2013) Groundwater development and sustainable management in the delta front area of Sundarban Islands. Int J Emerg Technol Adv Eng 3(11):606–611

Groundwater Governance (2017) A global framework for action. http://www.Groundwatergovernance.org/home/en. Accessed 30 Jan 2017

Hallegatte S (2009) Strategies to adapt to an uncertain climate change. Glob Environ Change 19:240–247

Hopenhyan C (2006) Arsenic in drinking water: impact on human health. Elements 2:103–107

Kirby T (2009) Stephen Luby: promoting the importance of water for health. Lancet 374 (9706):1961. https://doi.org/10.1016/S0140-6736(09)62110-3

Kuhn S (1999) Expanding public participation is essential to environmental justice and the democratic decision making process. Ecol Law Quart 25:647–658

Madani K (2010) Game theory and water resources. J Hydrol 381:225–238

McCormick J (1995) The global environment movement. Wiley, Chichester

McKinsey and Company (2009) Charting our water future- economic frameworks to inform decision making. 2030 Water Resources Group, McKinsey and Company, New York

Mehran A, AghaKouchak A, Nakhjiri N, Stewardson MJ, Peel MC, Phillips TJ, Wada Y, Ravaliko JK (2017) Compounding impacts of human-induced water stress and ClimateChange on water availability. Sci Rep 7(6282). https://doi.org/10.1038/s41598-017-06765-0

Mollinga P (2010) Public engagement in water governance in the MENA region: a review. Deliverable 1 of the PEWM public engagement in water management consultancy for the Arab Water Council

Murphy MJ, Lyon LW, Taylor JW (1981) Sub acute arsenic neuropathy: clinical and electrophysiological observations. J Neurol Neurosurg Psychiatry 44:896–900

Organisation for Economic Co-operation and Development (2017) Water Governance Initiative. http://www.oecd.org/gov/regional-policy/watergovernance-initiative.htm. Accessed 30 Jan 2017

Padowski JC, Gorelick SM, Thompson BH, Rozelle S, Fendorf S (2015) Assessment of human–natural system characteristics influencing global freshwater supply vulnerability. Environ Res Lett 10:104014. https://doi.org/10.1088/1748-9326/10/10/104014

Palmer MA, Liermann CAR, Nilsson C, Florke M, Alcano J, Lake PS, Bond N (2008) Climate change and the world's river basins: anticipating management options. Front Ecol Environ 6:81–89

Panos (1998) Liquid assets: is water privatisation the answer to access? Panos Media Briefing No. 29, p 1. Retrieved from http://www.oneworld.org/panos/briefing/water.htm

Razzaque J (2009) Public participation in water governance. In: Dallepenna JW, Gupta J (eds) The evolution of the law and politics of water. Springer Science+Business Media B.V., England

Shiklomanov IA, Shiklomanov AI, Lammers RB, Peterson BJ, Vorosmarty CJ (2000) The dynamics of river water inflow to the Arctic Ocean. In: Lewis EL, Jones EP, Prowse TD, Wadhams P (eds) The freshwater budget of the Arctic Ocean, 1st edn. Springer, Netherlands

Shomar B (2013) Water resources, water quality and human health in regions of extreme stress: Middle East. J Earth Sci Clim Change 4:153. https://doi.org/10.4172/2157-7617.1000153

Spyke NP (1999) Public participation in environmental decision-making at the new millennium: structuring new spheres of public influence. Boston College Environ Affairs Law Rev 26:263–313

Transboundary Water Assessment Programme (2017) http://www.geftwap.org/. Accessed 30 Jan 2017

United Nations Educational Scientific Cultural Organisation (2014) International Hydrological Programme-Water in the post-2015 development agenda and sustainable development goals. Paper presented at the 21st session of the Intergovernmental Council of the International Hydrological Programme, UNESCO, Paris, 18–20 June 2014

Vogel RM, Lane M, Ravindiran RS, Kirshen P (2015) Storage reservoir behavior in the United States. J Water Resour Plann Manag 125:245–254

Watkins K, Carvajal L, Coppard D, Fuentes R, Ghosh A (2006) Beyond scarcity: power, poverty and the global water crisis. (Human Development Report 2006). United Nations Development Programme, New York

Wesbey G, Kunis A (1981) Arsenical neuropathy. Illinois Med J 150:396–398

World Health Organization (2003) Sodium in drinking water (background document for development of WHO guidelines for drinking-water quality). World Health Organization, Geneva, Switzerland

Chapter 13
External Support and Community Cooperation During Long-Term Sheltering—From the Case of the Great East Japan Earthquake

Shoko Miyagawa, Miho Satoh, Satoshi Furuya, Yoko Yamada and Rajib Shaw

Abstract At the time of large-scale disasters, people's life and property are at very high risk due to the building collapses, tsunamis, and fires. However, the health risks are not just those of the acute phase. On top of the trauma caused by the disaster, disaster-vulnerable people such as hospitalized patients, in-home care recipients, dialysis patients, pregnant women, infants, and people with disabilities also face high health risk in the long-term sheltering. Issues of health care in disaster situations tend to be discussed in the context of acute medical support. On the other hand, the health impact of long-term sheltering and living in the temporary housing has not been discussed enough. In this chapter, based on the cases that were seen in the Great East Japan Earthquake, we will discuss the health risks that stem from long-term sheltering and what kind of support is necessary to reduce such risks. In these cases, the early detection and response against the health risk of affected people were realized by a collaboration of those monitoring the residents and those managing the shelters. In addition, the importance of collaboration among healthcare and non-healthcare support teams as well as the use of ICT to smoothen the collaboration has been suggested.

Keywords Long-term sheltering · Health risk · Health support coordination

---

S. Miyagawa (✉) · R. Shaw
Keio University, Fujisawa, Japan
e-mail: miyagawa@sfc.keio.ac.jp

M. Satoh
Yokohama City University, Yokohama, Japan

S. Furuya
Makioka Hospital, Yamanashi, Japan

Y. Yamada
CANNUS, Ishinomaki, Japan

E. Ying Yang Chan and R. Shaw (eds.), *Public Health and Disasters*,
Disaster Risk Reduction, https://doi.org/10.1007/978-981-15-0924-7_13

13.1 Disaster Cycle and Health Risk

Disaster management is a cycling process of multiple stages that includes (1) mitigation, (2) preparedness, (3) response, and (4) recovery. Of these, this chapter discusses the response stage. The primary focus during the response stage, which occurs immediately after a disaster, is likely to be about various kinds of general disaster assistance that range from rebuilding living spaces to providing psychological support to help victims prepare for life in the recovery stage. The living conditions at the temporary quarters where the evacuees are sheltered lead to unexpected and sudden changes in people's living arrangements, and the environment of a shelter can exacerbate existing healthcare risks such as stress-related mental disorders and the need to manage long-term and chronic health conditions. The acute medical needs of patients' suffering from trauma and those who are critically ill are also similarly impacted.

In the event of a disaster, a region's healthcare system experiences heavy demands and may become dysfunctional. This is because the people and facilities that are responsible for the region's medical, health, and long-term care have also been struck by the disaster, impairing their ability to serve, while at the same time, the needs for these services increase exponentially in the aftermath of the disaster. Resolving these dysfunctions depends on receiving healthcare services from providers outside the region, understanding the health risks of individuals in the shelter and throughout the region, and having access to an information network that will link them with the appropriate support. Finally, as the recovery stage progresses, the services provided by external healthcare resources must be transitioned to local healthcare providers.

In this chapter, we focus on long-term sheltering during the recovery stage of disaster management cycle. The following section provides a literature review of the health risks associated with long-term sheltering, the support needed to resolve evacuees' health problems, and minimum guidelines for the management of these problems. The next section provides a case study timeline of the health risks experienced by evacuees during the months they spent in the Ishinomaki Watanoha elementary school shelter after the Great East Japan Earthquake (GEJE) in 2011. The fourth section summarizes the literature and case study observations to prioritize countermeasures that should be initiated to mitigate health risks associated with long-term sheltering. The fifth section explores the application of information communication technologies (ICTs) to overcome obstacles that hamper the effective delivery of a unified healthcare response following a disaster. The chapter concludes by introducing a three-pronged strategy for reducing long-term health risks in long-term sheltering.

13.2 Health Issues and Long-Term Shelters

13.2.1 Health Risks Associated with Long-Term Sheltering

The victims of disasters who are forced to live in shelters are exposed to extreme stress that can weaken their immune systems. Prior research suggests that disasters and the ensuing sheltering have a serious impact on victims' physical and mental health. With regard to mental health, as time passes after a disaster, many affected people are likely to suffer from a sense of loss, post-traumatic stress disorder (PTSD), fatigue, anxiety about the future, depression, or insomnia. Nagata et al. (2015) and Tanji et al. (2018) found that victims living in temporary housing for an extended time experienced serious depression and loneliness. Those who had spent long periods of time away from their residences after the disaster were shown to feel strong stress (Fussell and Lowe 2014). In addition, it is revealed that anxiety and depression related to long-term sheltering would induce increase in alcohol consumption (Ueda et al. 2016).

In Fukushima, a Health Management Survey was conducted after the GEJE for the purpose of understanding the health status among affected people. This research reported that in the span of a year, the stress associated with living in a shelter long-term and the reduction in physical activities would lead to rise in dyslipidemia, blood pressure, and blood sugar levels (Kawasaki et al. 2014) and increase in the body mass index (BMI) and obesity (Hashimoto et al. 2017; Ohira et al. 2016). Even four years after the GEJE, the occurrence of diabetes and dyslipidemia remained at increased levels (Ohira et al. 2016). It has also been suggested that the difficulties of having balanced daily meals at a shelter due to the shortage of cooking equipment and utilities as well as limited access to grocery stores may be the cause of the physical symptoms associated with lifestyle-related diseases (Zhang et al. 2017).

The number of patients with peptic ulcers after the GEJE increased by about 1.5 fold as compared with the number in the previous year, and additionally, cases of hemorrhagic ulcers increased by 2.16 fold (Kannno et al. 2013). Psychological stress has been identified as a contributing factor for causing peptic ulcers. The unusual living conditions in the shelter and the shelter's poor environment (such as cold exposure, food shortages, and lack of privacy) after the GEJE were theorized to be contributors to these increases (Iijima et al. 2014). Other physical health problems have been also revealed because of the stress experienced after the GEJE: the development of cerebrovascular disease (Omama et al. 2013), the onset of ischemic heart disease, arrhythmia (Suzuki et al. 2015), worsening of chronic renal failure (Hayashi et al. 2017), liver function deterioration (Takahashi et al. 2017), and polycythemia (Sakai et al. 2017).

In addition, living in an evacuation center would increase the risk of deep vein thrombosis and pulmonary embolisms. Previous studies reported that the decrease in physical activities associated with living in a shelter or sheltering in a car could cause deep vein thrombosis and increase risk of pulmonary embolisms, and the average age of these patients was approximately 70 years (Hanzawa 2011; Shibata et al. 2017). It was also observed that many elderly people exhibited disuse syndrome due to decrease in their physical activities (Motoya 2013; Ohkawa 2016).

Another serious health problem of sheltered living is infectious diseases (Kouadio et al. 2012; Kawano et al. 2016; Watson et al. 2007; Wilder-Smith 2005). In shelter situations, infectious diseases are likely to spread quickly due to the crammed living conditions. In the 1995 Great Hanshin Earthquake, among the patients with infectious diseases, around 40% were shelter residents, and this incident showed that people living in shelters were at risk of infection (Maeda et al. 1996; Mitani et al. 2014). Poor hygienic conditions also contribute to the spread of infection. Many, if not most, of the facilities utilized as evacuation centers in disasters are not equipped with adequate sanitation facilities such as toilets, washrooms, and kitchens. In addition, partitions in the facilities are insufficient, and free-standing partitions or curtains are often used to separate living spaces. At the time of the GEJE, the people who lived in shelters had a higher risk of getting digestive infections, because of inadequate sanitation caused by the shortage of hygienic supplies, high population density, and lack of heating during the deep winter season in snow country (Tokuda et al. 2014). In the aftermath of Hurricane Katrina, the lack of hygienic supplies and high population density caused delays in quarantining the infected as well as disinfecting the polluted areas, thereby leading to an outbreak of the norovirus (Yee et al. 2007).

13.2.2 Support Needed to Resolve Evacuees' Health Problems

Though an infectious diseases surveillance system should be used to track the transfer of infectious diseases in the shelters, current systems would be inadequate, because existing standardized systems and guidelines on the systems for each type of disaster are insufficient (Babaie et al. 2015). The effective prevention of infectious disease epidemics in shelters also depends on creating an environment for smooth coordination among shelter managers and health data collection personnel from the Health Department and the sharing of information between these two. To facilitate such a relationship, protocols must be developed for documenting the information that needs to be collected and the process that should be used to collect it (Ridpath et al. 2015).

A relationship of mutual trust must be established between the affected people and those who support them, in order to facilitate the rapid provisioning of appropriate assistance (Takumi et al. 2016). Iwamuro and Sasaki (2012) claimed that building on former peacetime relationships with the residents is essential in this regard. Additionally, major challenges involved in managing health problems included having public health and other health and welfare-related organizations share information and cooperate with one another when planning activities (Nohara 2011).

After the GEJE, individual volunteers, NGOs, and NPOs from outside the area made significant contributions to the affected areas through supporting activities and offering other assistance in relation to health problems. Nonetheless, Takumi et al. (2016) cited three challenges related to the disaster relief provided by NPOs and NGOs. First, NPOs are financially vulnerable in terms of providing long-term reconstruction assistance. Second, they can offer only a finite amount of assistance, because they are limited in the volume of information or content they can handle, in comparison with professionals. Third, the lack of diversity in skill set and/or mentality within an NPO creates a problem. Nonetheless, NPOs with different features and areas of expertise, individual volunteers, and local governments should cooperate with each other and build a framework for providing valuable assistance (Takumi et al. 2016).

13.2.3 Guidelines for Providing Health Assistance to Evacuees

In the "Minimum standards for Humanitarian Charter and humanitarian response" developed by the Sphere project—a group of humanitarian agencies working on a common theme that includes NGOs, the International Red Cross, and the Red Crescent Movement—victims whose social infrastructure has been destroyed by a disaster continue to have the right to a dignified life and the right to receive the assistance (The Sphere project 2011).

Sphere views the four following elements as essential for human survival: (1) access to water, sanitation, and hygiene promotion, (2) food security and nutrition, (3) shelter, settlement, and non-food items, and (4) health actions. It also specifies the minimum amount of water needed to sustain human life, the necessary nutritional value of food provided, the installation of toilets that meet specified standards, the number of shelters and settlements, and the minimum area and health services to be provided per person at a shelter. The Cabinet Office (2016) defines "shelter management guidelines" in Japan. These guidelines state that shelter residents' health management requires the cooperation of health professionals with expertise in medical care, public health and welfare, individual volunteers, and NPOs.

13.3 The Timeline of Changing Health Needs During Long-Term Sheltering After the Great East Japan Earthquake

13.3.1 The Ishinomaki Watanoha Elementary School Shelter

Given the number of houses damaged in the GEJE, many residents had to live in shelters for periods of time that ranged from three to seven months. Approximately 120,000 housing units had been completely destroyed, and 1,000,000 units had been partially destroyed, while the number of recorded survivors who evacuated to shelters rose to 470,000. As described in Sect. 13.2, such long-term sheltering presented a serious risk to the residents' health.

At that time, the Ishinomaki Watanoha elementary school shelter housed as many as 2000 people (Fig. 13.1). This section describes how their time in the shelter corresponded to a variety of health risks, from a shelter manager's perspective. The shelter manager, who summarized this case, is a female resident of the Ishinomaki Watanoha district, 43 year old at that time. After the earthquake, she—along with her family who had taken refuge in Watanoha elementary school—became involved in management of the shelter when she asked for medical assistance for a family member who is a dialysis patient. For seven months, until October 2011 when the Watanoha elementary school shelter was closed, she worked as the temporary shelter manager.

The health needs and responses in the shelter changed as time passes, from the acute response to psychological care (Fig. 13.2). This section describes the detail of the health needs along with the change of situation in the shelter.

Fig. 13.1 Location of Ishinomaki and the photo of Watanoha elementary school shelter (April 4, 2011). *Source* https://www.watanohasmile.jp/: Copyright Watanoha Smile

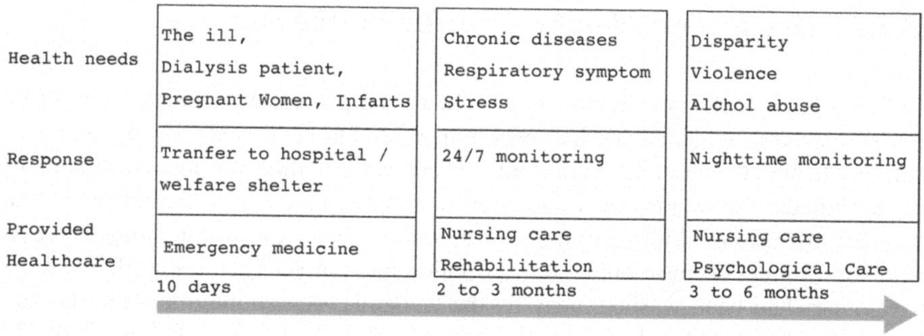

Fig. 13.2 Change of health needs and response in Watanoha elementary school shelter

13.3.2 Ten Days After the Earthquake

Rescue and relief activities did not reach the Watanoha district until March 13, two days after the earthquake, when the Self-Defense Forces and city officials arrived at the shelter. On the same day, two nurses who lived in a nearby shelter, and one nurse who had taken refuge in Watanoha elementary school, volunteered to help and began to triage the evacuees.

Many people in the shelter needed health assistance at that time, including dialysis patients, people with dementia, people requiring nursing care or medical assistance, pregnant or postpartum women, infants, and people with disabilities. However, because the road and communication infrastructure had been disrupted by the huge tsunami, they could not be transported to medical facilities by ambulance. Based on the triage by the volunteer nurses, patients with urgent needs were transported to a hospital when the Self-Defense Forces (SDF) team visited the shelter. Even after dealing with the needs of people with the most urgent medical needs, many other people remained in need of various healthcare assistance. Depending on each situation, patients were sent to the hospital by either ambulance or private car or were moved to the welfare shelter under the direction of the hospital, when routes had been cleared by the SDF. The coordination between hospitals and welfare shelters, however, was less than optimal, and sometimes patients were not accepted at the welfare shelter. By March 19, the Japanese Red Cross medical team and the Japanese Nursing Association disaster nurses team had started providing healthcare support in Watanoha elementary school shelter. The Japanese Red Cross medical team opened a clinic in a classroom where it provided consultations for ailing people in the shelter and the region. The Japanese Nursing Association's disaster relief nurses patrolled the residential area to locate people who needed help but had refrained from speaking out voluntarily.

13.3.3 One to Two Months After the Earthquake

By this time, a healthcare support system had been placed in the shelter, and those in urgent need of medical care had been transported to the hospital or moved to the welfare shelter. Concerning community health care, within one to two months of the earthquake the major roads had been restored, and some of the hospitals had resumed operations. Although signs of recovery began to appear, hospital visits were still limited, because public transportation had not yet been restored, and most private cars had been washed away by the tsunami. As a countermeasure, the city government provided a bus that ran between the shelter and the hospitals. In addition, taxis were used and volunteers provided transport assistance.

Once the urgent medical needs had been addressed and community healthcare service restored, the healthcare support team began to scale back their activity. By mid-April, the Japanese Nursing Association's disaster nurses team—and by mid-May the Red Cross medical team—had concluded their activities in the shelter. Meanwhile, there were still various kinds of health risks in the shelter, which came from prolonged living in the shelter. For people with chronic diseases such as diabetes and high blood pressure, the stress of shelter life, constipation from water deficiencies, nutritional imbalances, and loss of appetite due to a repetitive diet had made it difficult to control these underlying diseases. Respiratory symptom complaints caused by the sludge carried in by the tsunami also increased. It was around this time that people complaining of symptoms that could be attributed to the pain and bitterness felt due to the loss of loved ones or from the stress of living with so many others in the shelter began to increase.

At the urging of shelter management, a 24-h resident volunteer nursing team was instated to watch over residents' living quarters in the shelter. They monitored those with chronic diseases and oversaw emergency responses at night. The volunteer nurses also urged residents to wear masks to counter the respiratory symptoms at a shelter management meeting that was also attended by a delegation of residents. A mental care support team was also introduced.

A volunteer group of nurses organized a cooperative effort called "Cannus" in the Watanoha elementary school shelter beginning on March 20, nine days after the earthquake. Day-to-day living and acute health needs were both provided in the shelter, and workers sought to improve the health situation by improving the environment. Cannus was open 24 h a day, and over 300 volunteers, including nurses, physical therapists, occupational therapists, and other professionals, were involved through Cannus. The support activities provided in total were equivalent to about 5000 man-days. Cannus' activities were not limited to the shelter, but included the at-home evacuees who could not stay in the shelters for various reasons. Cannus also supported residents in temporary housing after the shelter closed, by providing medical and nursing care, long-term care such as bathing assistance, rehabilitation, and mental health care. It also provided goods, play and education supports for children, initiated housing repairs, located ICT devices, and provided ICT training for shelter staff and residents. These activities were loosely linked through the Cannus' support coordinator. In some cases, when the support needs could not be covered by individual support activities, support was provided cooperatively.

The activity of Cannus and other health-related teams was done in "autonomous" way, meaning they were not completely isolated from shelter management but they acted independently in their assessment and intervention activities. When a person who needed help from the nurses was identified, Cannus and other the medical professionals collaborated to design an intervention on their own. In cases where non-healthcare volunteers such as the child support team found a child and parent in need of mental support, Cannus and the medical professionals took the case. The shelter manager was informed of intervention in advance, but not details such as subject's personal information.

13.3.4 Three to Six Months After the Earthquake

By June 2011, most of the roads were passable, and healthcare services could again use road transport. At the same time, as the move to the temporary housing commenced, the shelter population began to decrease. While the need for the nurses' 24-h health center station declined, an emergency response capability was still required during the night. In addition, it was around this time that a divide was seen between those who were either able to move to temporary housing or who had a realistic road map to returning to a more normal life and those who could not. Anxiety, coupled with the frustration, associated with the uncertainty involved in reconstructing their lives and in some cases manifested as problems of violence and alcohol abuse, which were difficult for the shelter manager to deal with.

In response, the nurses' station continued to operate, but only at night. This arrangement remained in place until the Watanoha elementary school shelter was closed on October 11. The shelter manager then became the coordinator of the Cannus Disaster Relief Team and took on the dual role of watching over the temporary housing residents in Ishinomaki as well as managing the health assistance provided in other disaster-affected areas.

13.4 Mitigating the Health Risks of Long-Term Sheltering

When a disaster occurs, community care systems in the affected areas lose their ability to provide services to varying extents. Earthquakes, floods, landslides, and tsunamis may damage medical facilities or buildings. Similarly, day-to-day operations and decision-making for disaster response are also compromised due to both the staff and management being impacted by the disaster. In addition, damage to the transportation infrastructure such as roads, railways, and airports becomes an obstacle for ambulance transportation, patients' hospital visits, and health workers' home visits. This section summarizes the literature and case study observations to prioritize countermeasures that should be initiated to mitigate health risks caused by long-term sheltering. They include the following: (1) evaluating health risks quickly; (2) improving the capacity of the region's healthcare disaster preparedness; and (3) developing a coordinated and transparent approach for mitigating community healthcare shortages.

13.4.1 Evaluating Health Risks Quickly

As mentioned in Sect. 13.2, when disasters lead to long-term sheltering, increased health risks include infectious disease, acute stress reactions, increased blood pressure and blood sugar levels, decreased gastrointestinal function, cerebrovascular disease or ischemic heart disease, and deep vein thrombosis. By recognizing the symptoms associated with these conditions, and initiating appropriate responses in the early stage, health risks can be mitigated.

Undertaking appropriate treatment for the individual patients exposed to these health risks is only part of what is required. If these risks occur frequently in the population, or there are many people with high health risks, a different approach that will address the issues for the entire population is needed, as it has become a public health problem. One example of a public health approach is improving conditions in shelters. Sphere project's "Minimum standards for Humanitarian Charter and disaster assistance," which is mentioned in Sect. 13.2, provides specific guidance for the environmental improvement of shelters. According to these standards, an integrated approach for decreasing a population's health risk can be realized by providing drinking water, domestic water, toilets, food, living space and beds, clothing, garbage disposal, wastewater facilities, vaccinations for infectious diseases, and mental support in response to PTSD.

In the case of the Watanoha elementary school shelter, the medical triage performed by the volunteer nurses prior to the arrival of Self-Defense Force medical team helped in transporting people with urgent medical needs to hospitals. After that, disaster nurses watched and assessed the residents in the shelter. This enabled the early detection of health risks and the implementation of countermeasures.

13.4.2 Improving the Region's Healthcare Disaster Preparedness

Health emergency is caused when healthcare institutions that satisfy the regional health needs during peacetime can no longer manage the post-disaster situation due to the double problem of impaired serviceability from loss of staff and facilities and the sharp increase in patient numbers. In areas that struggle to meet the regional health needs even in peacetime, recovery of healthcare services after a disaster takes an especially long time and, as a result, people who are displaced for lengthy periods are not able to receive their usual healthcare services. Strengthening the peacetime regional healthcare system will ultimately reduce health risks in the event of a disaster.

In the case of Japan's comprehensive community care system, hospitals, clinics, and other various care providers collaboratively provide individual care in the regional community. Strengthening this collaboration will lead to providing robust and sustainable care not only in peacetime, but also during disaster situation. By replacing dysfunctional medical services with functioning one, the severe downgrading of medical service level can be avoided.

When performing such a re-arrangement of medical and nursing care, sharing information—that is, integrating the life support and medical needs to develop a plan for the delivery of service—is a big challenge. Key factors for sharing information include being aware of the providers that are offering services after a disaster occurs, coordinating these services so as to provide "normal" health care and arranging appropriate care services based on the assessments by healthcare professionals in the event new medical or nursing care needs arise.

13.4.3 Developing a Coordinated and Transparent Approach for Mitigating Community Healthcare Shortages

Even though people try to strengthen the community healthcare system, the impact of the disaster sometimes exceeds the resilience of it. In that case, to offset the lack of community healthcare resources, it is necessary to introduce external healthcare resources as an emergency countermeasure. What should be noted in these cases is that a lack of healthcare resources in the affected area is a temporary one. Considering that the whole community, including local medical services such as hospitals, clinics, and long-term care services, will eventually be restored, when bringing in external resources, it is important to include planning of the handover of the healthcare services to the original regional healthcare providers, although the time to recovery of these medical and long-term care services may vary from several months to several years. From this point of view, health needs stemming from the disaster which spike immediately after the strike and decrease as the community is restored should be taken care of by external health assistance, allowing them to exit after the restoration, while community healthcare resource should operate regular medical and care service which will remain needed on a continuing basis.

In order for these external healthcare resources to function effectively, it is important to clearly define their roles but, at the same time, allow for cooperation and collaboration with the community healthcare services. Shelter monitoring by public health nurses dispatched from the outside is an effective example of external health support that addresses health needs emerging from a disaster. The health needs that have been identified by monitoring should be forwarded to regional resources such as public health nurses from the community. Of course, if the region's healthcare resources are severely damaged, this aspect may also be managed from outside. To facilitate effective use of external resources, the method of cooperation between regional and external healthcare resources should be described in the disaster preparedness plan of each local government.

There is also another challenge for smooth cooperation between community and external healthcare resources. In general, although external assistance maintains continuity as a service, the individual members of the external assistance team changes frequently, every week or two. The transition of information regarding the shelter's situation, assessment check sheets, and the storing and sharing of health information is essential at the time of this change. Insufficient handover increases the risk that support needs will be overlooked. When coordinating external supporters, it is important to ensure a steady handover.

As mentioned in Sect. 13.2, the importance of cooperation in the disaster health management is pointed out (Nohara et al.) In the case of international rescue in disaster situations, the United Nations holds cluster meetings on humanitarian aid where they summarize information about activities, as well as the tasks and problems that should be exchanged among the rescue teams (UN OCHA). Another example of an effective information exchange between healthcare professionals and non-healthcare supporters occurred during a coordination meeting following the Kumamoto earthquake of 2016 in Japan, which led to cooperative environmental improvements to the shelter. However, this type of information exchange is not necessarily being carried out systematically, and previous research on methods and the effects of cooperation is not widely known. A mechanism to organize systematic coordination needs to be established.

In the case of the Watanoha elementary school shelter, it was reported that the need for psychological support for children and parents was identified through the activities of non-healthcare support teams. In these cases, the psychological support was provided through the cooperation between healthcare professionals and non-healthcare support teams. This became possible in this instance, because the volunteer organization that took the role of the shelter's health support also coordinated various kinds of assistance such as child support, home improvements, and ICT support. This kind of coordination was not conventional at the time of the GEJE. Currently as of 2018, the importance of coordination is gradually becoming understood, and efforts to strengthen this type of coordination are being carried out by both the government and the NPO sector.

Concerning the handover, in the case of Watanoha elementary school shelter, the records and the documents had not been integrated, and there was not sufficient information to facilitate long-term follow-up. A systematic support method that makes recording and transition easy is required.

13.5 Using Information Communication Technology (ICT) to Deliver a Unified Healthcare Response for Long-Term Sheltering

The types of problems identified in relation to providing effective and efficient healthcare services to long-term evacuees are as follows: (1) sharing and transferring information about specific a patient's health status, (2) being proactive in observing and evaluating health risk information in relation to a resident's health status, (3) coordinating the delivery of community health resources in changing and unpredictable circumstances, (4) establishing chains of command for overseeing the implementation of appropriate care, and (5) protecting a patient's records while ensuring care is not delayed. Although each of these problems tend to be discussed in relation to ICT capabilities, simply introducing an information system or a smart device will not resolve them.

Generally speaking, conventional information systems process data and provide information. Information can reduce uncertainty and help people's decision-making. However, information itself does not suggest what problems should be solved. Therefore, to utilize information processed from various kinds of data, it is important to know what decision-making is needed to countermeasure the situation. It is also important to develop a flow of information, so that decision makers can obtain appropriate information without delay at the time they have to make decisions. The standardization of the data format and organizational scheme of cooperation is as essential as the performance of the information system.

This can be seen by the following scenario which occurred often during the GEJE. Sharing and transferring information about specific patients' health status becomes problematic when a healthcare team working in a shelter changes to, or cooperate with another team that uses different formats for patient records, which results in a loss of data integrity. The chronologies of the patient's health records are subsequently lost, and/or data is not accurately transmitted between the two healthcare teams. The unification and standardization of recording formats are essential for promoting cooperation and a smooth transmission of patient information. Cooperation is necessary not only between healthcare teams, but also between healthcare and non-healthcare teams as it will enable the discovery and counteraction of health risks which lie behind the various difficulties of victims' daily life. A mechanism that includes information systems, standardized data format, and a cooperation scheme is also needed to promote healthcare–healthcare and healthcare–non-healthcare partnership.

Being proactive in observing and evaluating long-term evacuees' health risks depends on continuous population monitoring, which facilitates identifying people requiring health support and recognizing signs of infection. To obtain information about health risks from the data collected, it is essential that the recording formats described above be unified and standardized. When this requirement has been satisfied, ICT tools can be used to analyze data in a timely manner. In shelters with several hundred to several thousand people, it is very difficult to summarize and analyze data using paper-based monitoring. ICT can facilitate input-tabulations and summaries.

When community health resources are damaged and there is a need to re-arrange health services for people who cannot receive conventional care, it is important to be able to compare both the pre- and post-disaster community situations. Having access to a community's population dynamics, health risks during peacetime, and information about the residents' social networking within the community is also helpful. In addition, given that transportation is likely to be disrupted in the event of a disaster, information about mobility (available transportation, roads, and mobility support volunteers) is also required. At the individual level, being able to understand the existing chain of service for health cases, such as who the primary care doctor was or who the care manager was, is also crucial when rearranging care services.

Disaster scenarios are by definition chaotic and unpredictable. When existing community health resources are unable to fulfill regional health needs, and intervention by the external supporter is appropriate, the needed support must be found elsewhere. ICT can help document the health needs of individual residents, provide access to a list of available healthcare support and non-healthcare support, and facilitate the smooth introduction of external support. ICT can also be used to document and maintain a record of the chronological steps taken for each countermeasure. The information needed most in the recovery stage is a description of available support resources. Smooth coordination can be realized by aggregating information on the healthcare assistance provided by the local government, healthcare and health-related support provided by private support teams, and non-healthcare support such as supplies or soup kitchens. The aggregated 3 W (who, what, where) information can be viewed on the map by introducing a geospatial information system.

ICT can also facilitate the complex management and coordination that characterizes a disaster's recovery stage. In Japan, public medical support organizations such as the Disaster Medical Assistance Team (DMAT), Disaster Psychiatric Assistance Team (DPAT), Disaster Health Emergency Assistance Team (DHEAT), and the Nursing Association of Japan, as well as major NGOs, usually perform their healthcare activities under the direction of the local government's disaster response headquarters. There are, however, many disaster support teams from health care and healthcare-related fields that are not subject to the disaster response headquarters and go into the field directly. In the event of a large-scale disaster, with extensive damage to a community's health resources, it is essential that these "organized" support organizations and those that operate independently act cooperatively, to ensure the uninterrupted provision of health support. To realize such cooperation, a communication channel that enables the delivery of information from the non-healthcare teams to the healthcare teams must be established, allowing for an understanding of the various available support resources and coordination among supporters. Information sharing meetings, providing training in developing cooperation during peacetime, and collecting information about support activities (3 W) will contribute to establish the communication channel. When this information is accessible electronically, effective communication and oversight will be greatly enhanced.

Protecting patients' health records in disaster situations is a serious ethical concern. Even at the time of a disaster, sensitive information such as an individual's health condition should be protected as much as possible. On the other hand, to provide timely support—depending on each victim's health risks—information must often be shared. When a variety of supporters work together, managing personal information becomes a challenging issue. ICT can set various access controls to information, thereby making it available on a need-to-know basis, with access permissions granted appropriately. In this way, ICT can make information available, while protecting confidentiality.

13.6 Discussion

Long-term living in a shelter causes a variety of health problems in both mind and body. Impacts on vulnerable people including the elderly, people with illness, disabled people, pregnant women, and infants are a particularly serious concern. At the time of the GEJE, more than 3600 earthquake-related deaths were caused by physical and mental fatigue, due to living conditions in the shelter, the exacerbation of pre-existing illnesses at the shelter, and the interruption in the treatment of chronic diseases (The Reconstruction Agency of the Government of Japan 2018). Therefore, shelter management must assess the health risks of all evacuees to discern the high health risk patients and deliver appropriate care to them. With regard to constructing countermeasures, information about the health problems of displaced residents must be accumulated. With the exception of Japan, where some studies have focused on evacuees from the Fukushima Prefecture after the GEJE, the study of shelter residents in other regions or from other natural disasters has been insufficient.

Health management should improve the living environment and includes providing hygienic conditions, nutrition, preventing disease outbreaks, attending to victims' mental health, and providing assistance according to individuals' life stages and care needs. Guidelines from the Cabinet Office of the Japanese government are quite general, however, compared to Sphere standards, and it has been said that many of the shelters did not fully meet Sphere standards at the time of the GEJE. Although various organizations have produced guidelines for shelter management, health management, and disease prevention, most of them are premised on the guidelines developed by the Cabinet Office and lack both clarity and detail. The current guidelines should be improved to provide more details, by referring to Sphere standards. Another pressing task is strengthening the organizational system and the level of preparedness in such a way that shelter management evolves to meet the guidelines.

Handling of personal information is another challenge. Under Japan's Act on the Protection of Personal Information and its related laws and regulations, business entities, whether in the government or in private, are required to handle personal information properly, that is, to define the usage of personal information, to notify the purpose of use, and to control the usage in accordance with the purpose. Therefore, the provision of personal information to third parties without previous notification is basically prohibited. On the other hand, in an emergency like a disaster, the provision of personal information of vulnerable people to the rescue team may save their lives and properties. The act defines the exemption in the case that it is needed to protect human life, health, or property, and it is difficult to obtain the consent of the person. In the GEJE, however, some local governments were hesitant to apply for this exemption, and in some instances, this led to a delay in providing support for disabled people and their families (Fujii 2015). The sharing of information during long-term sheltering makes invoking this exemption even more difficult because it is not an acute "emergency" situation any longer. Moreover, the entities involved in supporting health are not limited to nationally qualified people such as physicians and nurses, who have a legally imposed duty of confidentiality. Cooperative management of personal information is very complicated, and, at the moment, no clear policy has been developed to address this problem. It will be necessary to organize these issues and decide on a countermeasure that can be widely deployed.

ICT provides an effective tool for addressing the information problems discussed above. It can be used to aggregate regional information (district diagnostic information, regional healthcare system), document the status of a community's healthcare services to facilitate the re-arrangement of delivery of health care, and provide geographically mapped information to external support team members who are not familiar with an area's geography. Unifying and standardizing the format for documenting assessments can be achieved by providing standard apps for mobile phones or tablets or bubble sheets. They also reduce human labor needed for data entry and facilitate rapid analysis.

Since it is difficult for users to start using a new tool when confronted by a disaster, efforts for disaster preparedness in peacetime should include continuous training or utilization of these tools for their daily business.

From the viewpoint of reliable storage and access, the use of cloud services is generally considered desirable. On the other hand, there are some concerns as to whether the cloud service itself is providing a service that satisfies the necessary protection of personal health information, or whether the Internet infrastructure will be continuously available in areas affected by a disaster.

Posts to social media are also currently becoming viewed as an important source of information in understanding the current situation. Applying artificial intelligence (AI) technologies such as natural language processing and machine learning will facilitate the analysis of various posts or tweets on social media and the extraction of potential health risks from them. It is also anticipated that information will ultimately be provided in multiple languages by applying automatic translation technologies.

The utilization of ICT for disaster management is just beginning, in both the public and private sectors. And it cannot be emphasized too much that ICT is not

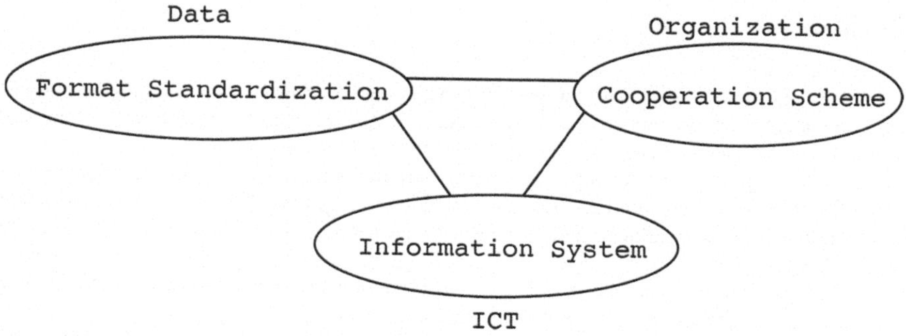

Fig. 13.3 Challenges related to the utilization of ICT

the "silver bullet" of smooth cooperation in emergency, as mentioned in the beginning of Sect. 5. There still remain big challenges such as unifying and standardizing the formatting of records and establishment of the scheme of cooperation that need to be solved before implementing ICT (Fig. 13.3).

Support teams that can provide ICT services are not sufficient in number compared to the rising expectation of ICT utilization in disaster support organizations. While strengthening efforts to clarify and systematize information issues, one objective should be developing a system that will be effective in any disaster.

13.7 Conclusion

This chapter focused on long-term sheltering associated with disasters. We discussed a framework of countermeasures that would mitigate the health risks that emerge when sheltering becomes long-term. The following steps should be adopted to mitigate the health risks of long-term evacuees living in a shelter:

1. Understand which health risks are likely to increase in the event of a disaster, detect and evaluate health risks at an early stage, and initiate appropriate healthcare responses
2. Develop the regional healthcare system's disaster preparedness capacity in peacetime
3. Coordinate the delivery of external support to address community healthcare shortages and various health problems caused by disaster
4. Clarify information challenges related to long-term evacuees' health needs, and investigate the technologies needed to address these challenges.

References

Babaie J, Ardalan A, Vatandoost H, Goya MM, Akbarisari A (2015) Performance assessment of communicable disease surveillance in disasters: a systematic review. PLoS Curr 24:7. https://doi.org/10.1371/currents.dis.c72864d9c7ee99ff8fbe9ea707fe4465

Fujii K (2015) The great east japan earthquake and persons with disabilities affected by the earthquake—why is the mortality rate so high? Interim Report on JDF Support Activities and Proposals. http://www.dinf.ne.jp/doc/english/resource/JDF_201503/1-1-1.html. Accessed 30 June 2018

Fussell E, Lowe S (2014) The impact of housing displacement on the mental health of low-income parents after hurricane Katrina. Soc Sci Med 113:137–144

Hanzawa K (2011) Deep Vein Thrombosis (DVT) after the GEJE and its problems. Jpn J Qual Saf Healthc 6:248–251 (in Japanese)

Hashimoto S, Nagai M, Fukuma S, Ohira T, Hosoya M, Yasumura S, Satoh H et al (2017) Influence of post-disaster evacuation on incidence of metabolic syndrome. J Atheroscler Thromb 24:327–337

Hayashi Y, Nagai M, Ohira T, Satoh H, Sakai A, Ohtsuru A, Hosoya M et al (2017) The impact of evacuation on the incidence of chronic kidney disease after the Great East Japan Earthquake: the Fukushima health management survey. Clin Exp Nephrol 21:995–1002

Iijima K, Kanno T, Koike T, Shimosegawa T (2014) Medical disease learned from the great East Japan Earthquake-feature, treatment and prevention-; 3) digestive diseases. Nihon Naika Gakkai Zasshi 103:557–560 (in Japanese)

Iwamuro S, Sasaki R (2012) Lessons from the Great East Japan Earthquake (Rikuzentakata). In: Kunii O (ed) Public health during disasters. Nanzando, Tokyo

Kannno T, IIjima K, Abe Y, Koike T, Shimada N, Hoshi T, Sano N, et al (2013) Peptic ulcers after the Great East Japan earthquake and tsunami: Possible existence of psychosocial stress ulcers in humans. J Gastroenterol 48:483–490

Kawano T, Tsugawa Y, Nishiyama K, Morita H, Yamamura O, Hasegawa K (2016) Shelter crowding and increased incidence of acute respiratory infection in evacuees following the Great Eastern Japan earthquake and tsunami. Epidemiol Infect 144:787–795

Kawasaki Y, Hosoya M, Yasumura S, Ohira T, Satoh H, Suzuki H, Sakai A et al (2014) The basic data for residents aged 16 years or older who received a comprehensive health check examinations in 2011–2012 as a part of the Fukushima Health Management Survey after the Great East Japan Earthquake. Fukushima J Med Sci 60:159–169

Kouadio IK, Aljunid S, Kamigaki T, Hammad K, Oshitani H (2012) Infectious diseases following natural disasters: prevention and control measures. Expert Rev Anti infect Ther 10:95–104

Maeda H, Nakagawa M, Yokoyama M (1996) Hospital admissions for respiratory diseases in the aftermath of the Great Hanshin Earthquake. Nihon Kyobu Shikkann Gakkai Zasshi 34:164–173 (in Japanese)

Mitani S, Murakami Y, Imamura Y (2014) Consideration of vulnerability concerning the elderly on disaster and disaster-related deaths. J Jpn Acad Health Behav Sci 29:2–11 (in Japanese)

Motoya R (2013) Health promotion for the victims and evacuees of the Great East Japan Earthquake. Japanese Soc Behav Med 19:68–74 (in Japanese)

Nagata S, Matsunaga A, Teramoto C (2015) Follow-up study of the general and mental health of people living in temporary housing at 10 and 20 months after the Great East Japan Earthquake. Jpn J Nurs Sci 12:162–165

Nohara M (2011) Impact of the great East Japan Earthquake and tsunami on health, medical care and public health systems in Iwate Prefecture, Japan, 2011. Western Pac Surveill Response J 2:24–30

Ohira T, Hosoya M, Yasumura S, Satoh H, Suzuki H, Sakai A, Ohtsuru A et al (2016) Effect of evacuation on body weight after the Great East Japan Earthquake. Am J Prev Med 50:553–560

Ohkawa Y (2016) Preventable disability as a new task at the time of disaster: the disuse syndrome in elderly people. Nihon Ronen Igakkai Zasshi 53:187–194 (in Japanese)

Omama S, Yoshida Y, Ogasawara K, Ogawa A, Ishibashi Y, Nakamura M, Tanno K (2013) Influence of the great east Japan earthquake and tsunami 2011 on occurrence of cerebrovascular diseases in Iwate, Japan. Stroke 44:1518–1524

Ridpath AD, Bregman B, Jones L, Reddy V, Waechter H, Balter S (2015) Challenges to implementing communicable disease surveillance in New York City evacuation shelters after hurricane Sandy, November 2012. Public Health Rep 130:48–53

Sakai A, Nakano H, Ohira T, Hosoya M, Yasumura S, Ohtsuru A, Satoh H et al (2017) Persistent prevalence of polycythemia among evacuees 4 years after the Great East Japan Earthquake: a follow-up study. Prev Med Rep 5:251–256

Shibata M, Chiba H, Sakaki K, Ueda S, Yamamura O, Hanzawa K (2017) The utility of on-site ultrasound screening in population at high-risk for deep venous thrombosis in temporary housing after the Great East Japan Earthquake. J Clin Ultrasound 45:566–574

Suzuki H, Ohira T, Takeishi Y, Hosoya M, Yasumura S, Satoh H, Kawasaki Y et al (2015) Increased prevalence of atrial fibrillation after the Great East Japan Earthquake: results from the Fukushima Health Management Survey. Int J Cardiol 198:102–105

Takahashi A, Ohira T, Hosoya M, Yasumura S, Nagai M, Ohira H, Hashimoto S et al (2017) Increased prevalence of atrial fibrillation after the Great East Japan Earthquake: results from the Fukushima Health Management Survey. J Epidemiol 27:180–185

Takumi S, Takumi Y, Otsuka Y (2016) The current trends and issues of exercise implement at the time of disaster area—the case of the great East Japan Earthquake of affected areas and exercise experts who live disaster areas. Hokkaido J Phy Edu Health Sport Sci 51:41–49 (in Japanese)

Tanji F, Tomata Y, Sekiguchi T, Tuji I (2018) Period of residence in prefabricated temporary housing and psychological distress after the Great East Japan Earthquake: a longitudinal study. BMJ Open 8. https://doi.org/10.1136/bmjopen-2017-018211

The Cabinet Office of the Government of Japan (2016) Shelter management guidelines. (in Japanese). http://www.bousai.go.jp/taisaku/hinanjo/pdf/1604hinanjo_guideline.pdf. Accessed 30 June 2018

The Reconstruction Agency of the Government of Japan (2018) The number of disaster related death in the Great East Japan Earthquake as of Mar. 31, 2018. http://www.reconstruction.go.jp/topics/main-cat2/sub-cat2-6/20140526131634.html. Accessed 30 June 2018 (In Japanese)

The Sphere project (2011) The sphere project handbook: Humanitarian charter and minimum standards in humanitarian response. http://www.ifrc.org/PageFiles/95530/The-Sphere-Project-Handbook-20111.pdf. Accessed 30 June 2018

Tokuda K, Kunishima H, Gu Y, Endo S, Hatta M, Kanamori H, Aoyagi T et al (2014) A survey conducted immediately after the 2011 Great East Japan Earthquake: evaluation of infectious risks associated with sanitary conditions in evacuation centers. J Infect Chemother 20:498–501

Ueda Y, Yabe H, Maeda M, Ohira T, Fujii S, Niwa S, Ohtsuru A et al (2016) Drinking behavior and mental illness among evacuees in Fukushima following the Great East Japan Earthquake: the Fukushima Health Management Survey. Alcohol Clini Exp Res 40:623–630

UN OCHA. Inter Cluster Coordination. https://www.humanitarianresponse.info/en/operations/afghanistan/inter-cluster-coordination. Accessed 30 June 2018

Watanoha Smile. https://www.watanohasmile.jp/. Accessed 1 July 2018

Watson JT, Gayer M, Connolly MA (2007) Epidemics after natural disasters. Emerg Infect Dis 13:1–5

Wilder-Smith A (2005) Tsunami in South Asia: what is the risk of post-disaster infectious disease outbreaks? Ann Acad Med Singapore 34:625–631

Yee E, Palacio H, Atmar R, Shah U, Kilbon C, Faul M, Gavagan T, Feigin R, Versalovic J et al (2007) Widespread outbreak of Norovirus gastroenteritis among evacuees of hurricane Katrina residing in a large "Megashelter" in Huston, Texas: lessons learned for prevention. Clin Infect Dis 44:1032–1039

Zhang W, Ohira T, Abe M, Kamiya K, Yamashita S, Yasumura S, Ohtsuru A, et al (2017) Evacuation after the Great East Japan Earthquake was associated with poor dietary intake: the Fukushima Health Management Survey. J Epidemiol 27:14–23

Chapter 14
Health Issues in the Aftermath of Nuclear Power Plant Accident in Fukushima

Takeshi Komino

Abstract The recovery from the nuclear power plant accident in Fukushima has just begun, and Japanese society is posed with a significant challenge of to what extent and how much the affected people's health should be monitored. There are broadly two types of health surveys Fukushima Prefectural government conduct. One is targeted for all Fukushima residents called Fukushima health management survey, and the other is more detailed screening which includes thyroid screening for those who were age 18 or younger at the time of the accident on 11 March 2011. Fukushima Medical University implements the thyroid ultrasound examination survey, comprehensive health check, mental health and lifestyle survey, and pregnancy and birth survey. Such comprehensive health screening efforts, Fukushima Prefecture is undertaking, are commendable, and it is in line with the spirit of Sendai Framework for DRR. However, as there are debates over to what extent such comprehensive health surveys should continue, it needs to be re-emphasized that the continuation of health surveys for early diagnosis and treatment is essential for any build back better efforts in Fukushima. From Japan's experience, the world can learn that disaster risk reduction strategies and contingency planning should be pre-planned with the involvement of wide range of stakeholders in society. This is critical particularly for those countries who possess nuclear power plant—safety myths should never be an option.

Keywords Fukushima · Nuclear power plant accident · East Japan Earthquake and Tsunami

T. Komino (✉)
CWS Japan, Japan CSO Coalition for DRR, Room 25, 2-3-18 Nishi-Waseda, Shinjuku-Ku, Tokyo 169-0051, Japan
e-mail: t.komino@cwsjapan.org

E. Ying Yang Chan and R. Shaw (eds.), *Public Health and Disasters*, Disaster Risk Reduction, https://doi.org/10.1007/978-981-15-0924-7_14

14.1 Introduction

Magnitude 9.0 earthquake in northeastern coast of Japan has triggered tsunami waves which recorded as high as 30 m in certain areas, and Fukushima Daiichi nuclear power plant was struck with 15 m high tsunami. The tsunami waves crippled emergency power supply for the power plant, which then resulted to one of the worst nuclear disasters in the world's history.

The loss of power supply meant pumping water into cool energy rods was no longer possible, and despite the national effort of carrying both fresh and sea water into the four reactor units at the power plant, the internal pressure rose to critical level, which resulted to hydrogen explosion which transmitted massive amount of radioactive material. The National Diet of Japan commissioned an official investigation, and the report (2012) indicates that the accident was clearly 'manmade' disaster.

Initial evacuation created chaotic situation as the official direction on evacuation kept on changing. On March 11th, evacuation order was issued within 3 km from the power plant, and on the same day, the order to stay indoors was issued within 10 km radius. The following day, on March 12th, the evacuation order was issued within a 10 km, and on the same day, this evacuation order was expanded to 20 km radius. As the radiated plume travelled by wind towards northwest of the power plant, in April 2011, areas outside of 20 km radius were also designated as either planned evacuation zone or emergency evacuation preparation zone.

Whereas the international dosage limit to non-natural radiation is set as 1 msv/year by International Commission on Radiological Protection (ICRP), there are measures being debated and recommended for cases where prolonged radiation exposure is inevitable. Thus, Japanese government made a decision to increase the limit from 1 to 20 msv/year after the accident, and the same new limit is being applied to determine whether evacuees are able to return or not. This is one of the most controversial areas when one considers how affected areas should recover. As shown in Fig. 14.1, after the Chernobyl nuclear power plant accident in 1986, areas that measured 5msv and above were classified as compulsory resettlement zone. Below chart shows the difference in category zoning for both Fukushima and Chernobyl. When and how to change the emergency dosage limitation need to be debated, not only by medical professionals and researchers, but with policymakers and stakeholders involved in disaster risks reduction and recovery process.

There were various radiation types that were discharged to air after the accident differs, such as Iodine-131, Cessium-134, and Cessium-137. Iodine-131 can accumulate in thyroid organ particularly among young children; thus, thyroid cancer was feared to spread. As per statistics from Fukushima Prefectural government, there are still almost 34,000 affected population reside outside of Fukushima prefecture and more than 13,000 people evacuated within Fukushima prefecture, both as of March 2018. At peak time in May 2012, the number of evacuees recorded 164,865 individuals.

Differences in evacuation segmentation between Fukushima and Chernobyl

Air radiation dose(annual)	Fukushima zoning	Chernobyl zoning
50mSv and above	"Difficult to return" zone	Forced evacuation zone
20 - less than 50mSv	Habitation restricted zone (temporary return possible)	Forced evacuation zone
Less than 20mSv	Zone being prepared for lifting of evacuation order	Forced evacuation zone
5mSv and above	No instructions	Compulsory resettlement zone
1 - less than 5mSv	No instructions	Right to resettlement zone
0.5 - less than 1mSv	No instructions	Radiation control zone

N.B. 1: Segments in red are in principle off-limits

N.B. 2: Zone designation in Chernobyl was carried out mainly according to soil contamination dose, and the method used for calculating annual exposure is also different in the case of Fukushima. References here are simplified for the purpose of general comparison.

Fig. 14.1 Difference in evacuation segmentation between Fukushima and Chernobyl (*credit* Fukushima Booklet Committee, http://fukushimalessons.jp/assets/content/doc/Fukushima10Lessons_ENG.pdf)

14.2 Health Surveys in Fukushima

There are broadly two types of health surveys Fukushima Prefectural government conducts. One is targeted for all Fukushima residents called Fukushima health management survey, and the other is more detailed screening which includes thyroid screening for those who were age 18 or younger at the time of the accident on 11 March 2011.

The survey targets everyone who was a Fukushima resident at the time of 11 March 2011, and it will be carried out for 30 years. Aside from basic personal information such as name, registered address, and contact number, the survey asks on what type of house (house or apartment, concrete or wooden, etc.) one lived at the time of the accident and detailed movements outside/indoors particularly for the first two weeks from the accident (11–25 March 2011). The survey also includes any periodic/concurrent movements to a particular place during 26 March to 11 July 2011 and how much of activities took place outdoors and indoors. On internal exposure to radiation, the survey asks questions on whether one had eaten self-grown vegetables or dairy products from his/her own farm during March 11 to end March 2011 and whether one had taken iodine tablets in the same period. The responsive rate to the survey remains around 27% according to Fukushima Prefecture.

Fukushima Medical University implements the survey commissioned by Fukushima Prefectural government, and it describes four detailed surveys conducted in addition to the basic survey above.

- Thyroid Ultrasound Examination Survey: The survey targets all residents who were age 0–18 at the time of the accident. Initial screening took place within the first three years from the accident, and complete thyroid examination followed since 2014 onwards.
- Comprehensive Health Check: The check targets around 210,000 former residents of evacuation zone. It aims to detect any early signs of disease and aims to prevent deterioration of such condition.
- Mental Health and Lifestyle Survey: The survey aims to detect any risks of post-traumatic stress disorder and other mental health issues.
- Pregnancy and Birth Survey: The survey aims to provide medical care and support for those pregnant mothers who were given maternal and child health book (Boshitecho in Japanese) between August 2010 and July 2011.

So far, there are 194 individuals who were diagnosed with thyroid cancer, and as per Fukushima Prefecture's information disclosure, 160 of them went through a surgery. Fukushima Prefecture discloses such information on the portal on constant basis, and the information can be obtained from https://www.pref.fukushima.lg.jp/sec/21045b/.

The survey, after the initial screening, groups the target population into below three categories:

- Category A: No nodule or cyst (A1) or nodule was below 5.0 mm, and cyst was below 20.00 mm (A2).
- Category B: Nodule was more than 5.1 mm and cyst 20.1 mm or more or those who were diagnosed as A2 but considered the requirement of further tests.
- Category C: those who require immediate second screening test based on the condition of thyroid.

Category B or C would go through second screening which would include detailed ultrasound test, blood test, and urine test (and when required, biopsy is also conducted). All medical cost is covered by the prefecture.

The debate of whether the number of thyroid cancer is high or low is currently going on. Whereas the authority expresses that there are no clear linkages between the survey results and the effect of the nuclear power plant accident, and it is likely due to screening surge which is unprecedented in Japanese history; Hiroko (2017) indicates 'the Fukushima Prefectural government and the organization charged with conducting the examination are trying to rationalize the results in many ways, without invoking the radiation impact of the reactor meltdowns. If this is indeed unrelated to the radiation from the damaged Fukushima nuclear power plants, a similarly high rate of thyroid cancer should be found all over Japan. The survey should be expanded in order to see whether that is indeed the case'. Similar perspective is shown by Tsuda et al. (2016) that an excess of thyroid cancer has been detected by ultrasound among children and adolescents in Fukushima Prefecture within four years of the release of the radiation, and the result is unlikely to be explained simply by a screening surge. Yasumura and Abe (2017) also question the validity of the current survey that indicates that 'internal radiation exposure estimated by wholebody counter has not yet been linked to external radiation dose, and assessment of total radiation exposure is needed to evaluate the extent of radiation to the body. Another problem is that internal doses to the thyroid gland were not accurately evaluated in the affected area'. The average figure shown by the National Cancer Center in Japan is that the prevalence rate of thyroid cancer for men is 6.8 persons/100,000 persons and for women 17.4 persons/100,000 persons. Different scientists use different studies to indicate whether the current prevalence rate is normal or abnormal, but the important point is that screening needs to continue in order to capture the actual impact of the disaster.

A not-for-profit organization called 3.11 Fund for Children with Thyroid Cancer issued a press release on 1 March 2018, raising concerns over an incident where a four-year-old at the time of the accident undergone the treatment, which contradicts the official communication that no under-five thyroid abnormality is observed. The NPO provides financial support to those who are going through treatment, detailed examination, and surgery for thyroid abnormality after the accident, and out of those for who the NPO provided funding support, over 70% of parents of the patients wish the continuation of detailed thyroid examination by the Fukushima Prefecture. Relatively, younger aged patients have tendency of faster development of cancer-related abnormalities, and there are numerous cases of re-surgery of patients (which is not shown in official record).

The United Nations Scientific Committee on the Effects of Atomic Radiation (UNSCEAR) Report (2013) concludes that 'because the exposure of the Japanese general public was low, so too were the risks to heath later in life', which is one of the bases of Japanese authority to deny any linkages between the health consequences and the nuclear power plant accident. However, the committee which undertook the study also indicates that there is 'theoretical possibility for infants and children of an increased risk of thyroid cancer and the situation should be monitored closely', which should be interpreted that anything 'unknown' should not be concluded as 'no risk'.

Sendai Framework for Disaster Risk Reduction (2015), a widely recognized international framework agreed and adopted by more than 190 states in 2015, sets out the following linkages between DRR and health.

- 19(c) Managing the risk of disasters is aimed at protecting persons and their property, health, livelihoods, and productive assets, as well as cultural and environmental assets, while promoting and protecting all human rights, including the right to development.
- 24(d) to systematically evaluate, record, share, and publicly account for disaster losses and understand the economic, social, health, education, environmental, and cultural heritage impacts, as appropriate, in the context of event-specific hazard—exposure and vulnerability information.
- 27(b) to adopt and implement national and local disaster risk reduction strategies and plans, across different timescales, with targets, indicators, and time frames, aimed at preventing the creation of risk, the reduction of existing risk and the strengthening of economic, social, health, and environmental resilience.

Furthermore, Priority 4b of Sendai Framework for DRR focuses on build back better in recovery, rehabilitation, and reconstruction, and UNISDR (2017) indicates the following tasks that are emphasized to enable this priority action:

- Develop all-stakeholder, national-level disaster recovery framework;
- Enable pre-disaster recovery planning among all stakeholders;
- Formalize processes and systems to enable effective assessment of post-disaster damages and needs in order to more accurately quantify and characterize recovery needs and to formulate broad recovery strategies;
- Institute or strengthen policies, laws, and programs that promote (incentivize), guide, ensure, and support build back better in recovery, rehabilitation, and reconstruction in both the public and private sectors and by individuals and households.

The pre-requisite to all these tasks is accurate understanding of disaster risks and impact and proactive attitude towards reducing the risks. Creation of safety myths, which was the biggest mistake Japan has made (also reaffirmed by Japanese government representatives at the Third World Conference on Disaster Risk Reduction in Sendai in 2015 that Japan will 'no longer create safety myths around nuclear power plant'), inevitably turns eyes away from all the pre-requisite steps for build back better; thus, it should be emphasized that the creation of any safety myths around nuclear power plants in the world should be strongly discouraged. The safety myth was backed up with a campaign led by the nuclear industry with slogans such as 'nuclear energy is the future energy of the world' and 'it is completely safe so no need to talk about if and when it goes wrong'.

Furthermore, as experienced in Fukushima, turning eyes away from the actual risks and possible disaster scenarios hinder the preparedness at all levels. If a state is serious about mainstreaming DRR into development, there needs to be multi-stakeholder dialogue on possible risks and possible mitigation measures. Only then, build back better can be discussed and be planned.

14.3 Conclusion and Recommendations

First of all, comprehensive health screening efforts Fukushima Prefecture is undertaking is commendable, and it is in line with the spirit of Sendai Framework for DRR. However, as there are debates over to what extent such comprehensive health surveys should continue, it needs to be re-emphasized that the continuation of health surveys for early diagnosis and treatment is essential for any build back better efforts in Fukushima. Health impacts from anxiety should be decreased as much as possible, and the residents' right to health needs to be respected. Reliable and continuous health examination support needs to exist for genuine relief of those residing in disaster-affected areas.

There are clearly difficulties for the residents to continuously take part in the annual health surveys, and the tailored support mechanism, particularly for those who are diagnosed with B or C category from the screening, should be considered to be budgeted and provided. Also, as there is no compensation support for those who miss work to get examined (transportation cost is out of pocket expense by the residents), the system requires enabling environment for the tests to be widely accepted and does not become a burden for those who need to be tested.

As a nation that hosted the Third World Conference on DRR where Sendai Framework for DRR was adopted, any measures Japan takes on in recovering from East Japan Earthquake and Tsunami in 2011 should carry forward the spirit of the framework. This would mean that uncertainty in potential impact from the unprecedented disaster should not be concluded as 'no risks' at the moment, and continuously pursue understanding of risks and investing into risk reduction should be encouraged and applied. And the process should be transparent and involves wide range of stakeholders so that it is well accepted by the society. This is pre-requisite if Japan wishes to build back better from one of the worst nuclear disasters in the world's history.

After all, the creation of safety myths around the nuclear power plant turned all eyes away from disaster risk identification and mitigation on nuclear power plant, and health management system we need to put in place is also one of the critical costs of the safety myths we have created. As Xiang and Zhu (2011) pointed out, such creation of safety myths is completely against the ethical principles of utilitarianism, no maleficence, beneficence, justice, disclosure, and autonomy. From Japan's experience, the world can learn that disaster risk reduction strategies and contingency planning should be pre-planned with the involvement of wide range of stakeholders in society. This is critical particularly for those countries who possess nuclear power plant—safety myths should never be an option.

In order for the world to be more prepared against nuclear risks, each state should be encouraged to think through crisis management process with proper simulation, taking both Chernobyl and Fukushima cases as examples, and see where the gap lies in the existing crisis management structure and capacity.

References

Fukushima Booklet Committee (2015) 10 Lessons from Fukushima reducing risks and protecting communities from nuclear disasters. http://fukushimalessons.jp/en-booklet.html

Hiroko A (2017) Follow up on thyroid cancer! Patient group voices opposition to scaling down the Fukushima prefectural health survey. Asia-Pac J Jpn Focus 15(2), 3 Jan 2017

National Cancer Center statistics. https://ganjoho.jp/reg_stat/statistics/stat/summary.html

Office of International Cooperation Radiation Medical Science Center Fukushima Medical University. Overview of the Fukushima Health Management Survey. http://fmu-global.jp/fukushima-health-management-survey/

The National Diet of Japan (2012) The official report of The Fukushima Nuclear Accident Independent Investigation Commission

Tsuda T, Tokinobu A, Yamamoto E, Suzuki E (2016) Thyroid Cancer Detection by ultrasound among residents ages 18 years and younger in Fukushima, Japan: 2011 to 2014. Epidemiology 27(3). May 2016

United Nations (2015) Sendai framework for DRR 2015–2030. https://www.preventionweb.net/files/43291_sendaiframeworkfordrren.pdf

United Nations Office for Disaster Risk Reduction (2017) Build back better in recovery, rehabilitation and reconstruction. https://www.unisdr.org/files/53213_bbb.pdf

United Nations Scientific Committee on the Effects of Atomic Radiation (UNSCEAR) Report (2013) Report to the General Assembly, Scientific Annex A

Xiang H, Zhu Y (2011) The ethics issues of nuclear energy: hard lessons learned from Chernobyl and Fukushima. Online J Health Ethics 7(2). November 2011

Yasumura S, Abe M (2017) Fukushima health management survey and related issues. Asia Pac J Public Health

Chapter 15
EpiNurse, Health Monitoring by Local Nurses on Nepal Earth Quake 2015

Sakiko Kanbara, Apsara Pandey, Maria Regina E. Estuar, Hyeon Ju Lee and Hiroyuki Miyazaki

Abstract The name, EpiNurse, refers to local nurses who perform epidemiological surveillance and care to ensure human security in and communicate with health authorities on the health status of communities. They collect and report epidemiological information by using easy-to-use technology in communities in which access to health information of the populace is hindered by catastrophic accidents or other geo-/socio-political reasons. EpiNurse Nepal Project (August 2015–April 2016) carried out epidemiological surveillances at 24 camps in 9 districts after the 2015 Nepal earthquake. By applying ICT and questionnaire, EpiNurse member nurses collected shelter-related data and information, such as location and timestamps, and geo-tagged photos. Identification of typical health behavior patterns and comparative information about differences between community and temporary shelters provided insights into the health security assessment. This initiative experiments how nurse should collect and deliver the health emergency information on their own local culture, lifestyle, and perceptions. The potential of EpiNurse concept lies not only in producing innovative research outcomes by improving or optimizing existing ICT application in health sector, but also in promoting research knowledge and exchange of ideas regarding social issues and challenges in the field of health emergency and community resilience. The most critical challenge in practice relates to collecting and storing data, which later would have been generated into reasonable health security index information to be used for predicting the likelihood of occurrence of health emergency events. It is necessary to apply

S. Kanbara (✉) · H. J. Lee
Faculty of Nursing, University of Kochi, 2751-1 Ike, Kochi 781-8515, Japan
e-mail: kanbara@cc.u-kochi.ac.jp

A. Pandey
Nursing Association of Nepal, Kathmandu, Nepal

M. R. E. Estuar
Ateneo de Manila University, Quezon City, Philippines

H. Miyazaki
University of Tokyo, Tokyo, Japan

E. Ying Yang Chan and R. Shaw (eds.), *Public Health and Disasters*,
Disaster Risk Reduction, https://doi.org/10.1007/978-981-15-0924-7_15

human behavior modeling using geospatial technology in order to create data transferring modules for first responders and civilian populations regarding DRR and behaviors.

Keywords Participatory monitoring · Information and communications technology · Nursing · Health security · Disaster risk reduction

15.1 Nursing and Disaster Risk Reduction

Nurses are practitioners in health care whose professional commitment is highly related to disaster risk reduction. Throughout history, nurses have called upon to respond to the needs of people. The International Council of Nurses (ICN) acknowledges the value of nurse involvement in disaster relief. Nurses with technical skills and knowledge of epidemiology, physiology, pharmacology, cultural-familial structures, and psychosocial issues can assist during actual disasters as well as in disaster preparedness programs. Following the Great Hanshin-Awaji earthquake and the Sarin attack in Japan in 1995, Japan Society for Disaster Nursing (JSDN) developed and described "disaster nursing" as "the systematic and flexible utilization of knowledge and skills specific to disaster-related nursing." (JSDN 1998) JSDN acknowledges nursing capacity in the promotion of a wide range of activities relevant to minimizing health hazards and life-threatening damages in disasters while collaborating with other specialized field experts. In order to respond effectively to disaster, nursing scholars in disaster nursing are educating students in order to produce human resources that can respond to the emergency events, manage facilities, and other aid equipment, and establish an emergency response training network.

15.2 Lessons Learned from the Great East Japan Earthquake

Despite concerted efforts and preparation, natural disasters continue to pose a threat to the health and human security of the Japanese people. In the aftermath of the Great East Japan Earthquake that occurred on March 11, 2011, it has become clear that trained leadership with comprehensive understanding of disaster-related issues and decision-making skills in the field of disaster nursing is needed. The disaster management cycle-based approach to disaster relief efforts posed problems in coordination due to vertically structured management system that did not allow flexible cooperation and coordination among disaster responders.

15.3 EpiNurse: Local Nurses as Responders, Informants, and Monitors

EpiNurse concepts envision assessing and meeting health care needs through the collaboration of local nurses in epidemiological surveillance in order to protect and promote health and safety of people and communities during disasters. EpiNurse members report epidemiological, human security information by using available technology to provide actionable information. It is expected that these nurses would share rapid-assessment data and shelter positioning with local responders and those in other public services and functions of relief. This would ensure the sufficient running of health system in promoting health, preventing illness, and caring for the ill and disabled within the shelter environment.

Governments and affected populations have to manage a complex, unpredictable interplay of environmental, social and economic shocks, and disturbances in this setting. Hence, it is more cost-effective to adopt comprehensive multiple disaster risk approaches to protect and enhance lives, livelihoods, and assets in non-disaster times. Maintaining the livable environment in times of disaster is central to disaster risk reduction efforts, especially in terms of securing health (see Fig. 15.1).

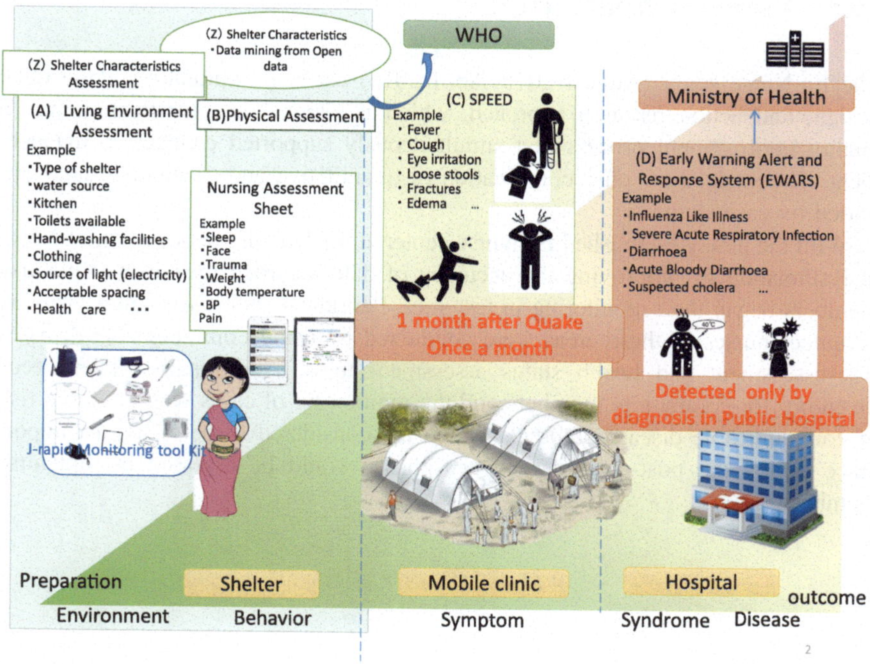

Fig. 15.1 Pathogenesis and surveillance system

15.4 Response to the 2015 Nepal Earthquake

Just after Nepal earthquake occurred in 2015, members of the global EpiNurse team reviewed various reports on emergency health situations which often occur as a consequence of a primary disaster of great magnitude due to its impact on living environments. An outbreak of communicable diseases, for example, is common among displaced populations in the aftermath of a large scale disaster due to poor access to basic supplies of adequate shelter, clean water and sanitation, sufficient nutrition, and primary healthcare services. In addition, the relationship between the occurrence of a disaster and an outbreak of communicable diseases is reported to be associated with the size and characteristics of the displaced populations (Sphere, 2014). In case of the 2015 Nepal earthquake, the lack of adequate infrastructure and resources coupled with an inexistence of disaster preparedness programs seemed to be among factors that contribute to increased risk of epidemics following a humanitarian crisis, according to this review (Kanbara et al. 2018). Critical attention should be paid to hygiene control in shelters to maintain minimum health safety level. In this context, monitoring the environment and health status of shelters during temporary settlement period after disaster becomes crucial for maintenance of health.

15.5 EpiNurse Approach

The EpiNurse study (August 2015–April 2016) used an exploratory, descriptive design, and action research approach, which it allowed an investigation into a quality improvement process that simultaneously supported changes in management. The study drew on a combination of quantitative and qualitative data supported by case studies.

With the newly established EpiNurse center at the Nursing Association of Nepal in Kathmandu for improving the accuracy of data recorded by nurses, EpiNurse members began their monitoring in test sites. Periodical shelter-to-shelter visits by volunteer nurses in these areas were conducted by using communicable diseases risk assessment and health status assessment tools. Additionally, relief needs inventory for community members and the monitoring of cases with potential risk of communicable disease outbreak, which are considered to be of utmost importance during the post-disaster recovery phase, would be assessed by EpiNurse members.

Fig. 15.2 Monitoring sites on the map and example pictures of shelters

15.6 Settings

Monitoring was conducted by trained local nurses in Kathmandu for 4 months at 24 Camps in 10 affected districts (Kathmandu, Gorkha, Dolakha, Sindhupalchok, Kavrepalanchok, Nuwakot, Rasuwa, Bhaktapur, Laltipur, and Dhading). Figure 15.2 shows the example of the photo the demographic infromation taken by EpiNurse on the same map.

15.7 Enabling Open Framework for Application Programming Interface (API) and Integration with Other Public Health Platform Systems

Monitoring by nurses can help improve community resilience and assist in disaster risk reduction. As the system is dynamic and measurements are ongoing, the proposed project may also facilitate the behavioral changes among the shelter residents.

The monitoring toolkit items used by EpiNurse members include living environment assessment and Surveillance in Post Extreme Emergencies and Disasters (SPEED) items. SPEED was conceptualized and developed by WHO/WPRO to provide real-time health information reporting after a disaster. The assessment items were also taken from the Early Warning Alert and Response System (EWARS), which have been used by ministries of health in various countries.

The toolkit was developed to create an application programming interface (API) for data collection and sharing with other health sectors and national surveillance by ministry of health and population (MOHP) and WHO (see Tables 15.1 and 15.2).

15.7.1 Pilot Observation

Health is highly related to local population characteristics and emergency responses in disaster situations. When the Nepal earthquake occurred, the Japanese team reviewed situation reports by UN and other disaster health responders and developed the environmental and behavior-specific indicators in health care, derived from global standards identified in the literature and lessons learned from the past disasters. The latest Nepali national census in 2012 was used as site-specific thresholds of pre-disaster situation. An onsite inspection and fieldwork by the EpiNurse team in Dhading district made the researchers realize that a number of disaster-affected areas have not been properly provided with relief health care services. Evacuees with health problems did not benefit from any medical assistance, and, in areas without a rural primary health center, they rarely visited health care providers who were based on health facilities. Moreover, pregnant evacuees delivered babies in village shelters built by villagers themselves. Such health and security issues occurring in disaster-affected communities in remote areas were not being monitored, and there were looming threats of outbreaks of community cases among local population and neighboring districts. The communication tool for qualitative information and unexpected event report was required. The follow-up network established with WHO and MOHP, in case of detection of sign of outbreak (Fig. 15.3).

15.7.2 Initial Training

A two-day workshop was conducted for representative nurses from the disaster-affected areas on collecting information about shelters, specifications of the survey toolkit. During the workshop, nurses also received skills training on

Table 15.1 Monitoring indicators

Living environment assessment	Monitoring tool kit	Surveillance in post extreme emergencies and disasters (SPEED)	Early warning alert and response system (EWARS)
Estimate population Estimate population under 5 Years Type of shelter Security/safety assured Type of drinking water source Adequate water supply Safe and clean food items provision Kitchen Appropriate waste storage Toilets available Adequate number of toilets Distance toilet—dwelling place Hand-washing facilities Soap available Clothing Non-food item Source of light (electricity) Acceptable spacing Acceptable cleanliness Blanket or other items Unusual symptom Unusual disease/event, outbreak, etc.	Bag T-shirts Stethoscope Sphygmomanometer-B.P Towel Basal thermometer Alcohol Torch Globes Mask Weight machine Soap Thermostat Measuring tape Tablet	1. Fever (FEV) 2. Cough, colds or sore throat with or without fever (ARI) 3. Fever with rash (MEA) 4. Fever with spontaneous bleeding (i.e., nose bleeding, gum bleeding) (AHF) 5. 12 months and over: sudden onset of fever (>38 °C) with severe headache and stiff neck; <12 months: fever (>38 °C) with bulging fontanel, or refusal to suckle (MEN) 6. Fever with headache, muscle pains, and any of the following: eye irritation, jaundice, skin rash, scanty urination (LEP) 7. Yellow eyes or skin with or without fever (AJS) 8. Fever with other symptoms not listed above (FOS) 9. Loose stools, three or more in the past 24 h with or without dehydration (AWD) 10. Loose stools with visible blood (ABD) 11. Open wounds and bruises/burns (WBS) 12. Fractures (FRS) 13. Skin disease (SDS) 14. Animal bites (ANB) 15. Eye itchiness, redness with or without discharge (CON) 16. Spasms of neck and jaw (lock jaw) (TET) 17. High blood pressure >140/90 (HBP) 18. Known diabetes (KDM) 19. Difficulty in breathing and wheezing (AAA)	1. Influenza like illness 2. Severe acute respiratory infection 3. Acute watery diarrhea 4. Acute bloody diarrhea 5. Suspected cholera 6. Fever with Jaundice 7. Fever with rash 8. Fever without rash and jaundice

(continued)

Table 15.1 (continued)

Living environment assessment	Monitoring tool kit	Surveillance in post extreme emergencies and disasters (SPEED)	Early warning alert and response system (EWARS)
Health care services on site Psychological support/ counseling Nursing care Characteristics of this shelter		20. Floppy paralysis of the limbs which occurred recently in a child <15 years who is previously normal (AFP) 21. Visible wasting with or without bipedal pitting edema (AMN)	

*Source for SPEED: Department of Health, Republic of the Philippines (2011) SPEED operations manual for managers. Retrieved from http://www.wpro.who.int/philippines/publications/speed_operations_manual.pdf?ua=1

Table 15.2 Characteristics of EpiNurse

Age in years	Average	40	21–54
SMS	Yes	22	91.7%
	No	2	8.3%
Line	Yes	0	0.0%
	No	24	100.0%
Viber	Yes	16	66.7%
	No	8	33.3%
Skype	Yes	6	25.0%
	No	18	75.0%
ICT user	User	12	50.0%
	Can but do not use	8	33.3%
	Cannot use	4	16.7%
Status of working	Staff nurse	9	37.5%
	Auxiliary nurse midwife (ANM)	15	62.5%
Past experience in years	Average	16.8	1.5–33

monitoring and rapid assessment so that they pay more attention to the hygiene control to maintain minimum health safety during monitoring. As a result, they could make the consensus described in Table 15.3.

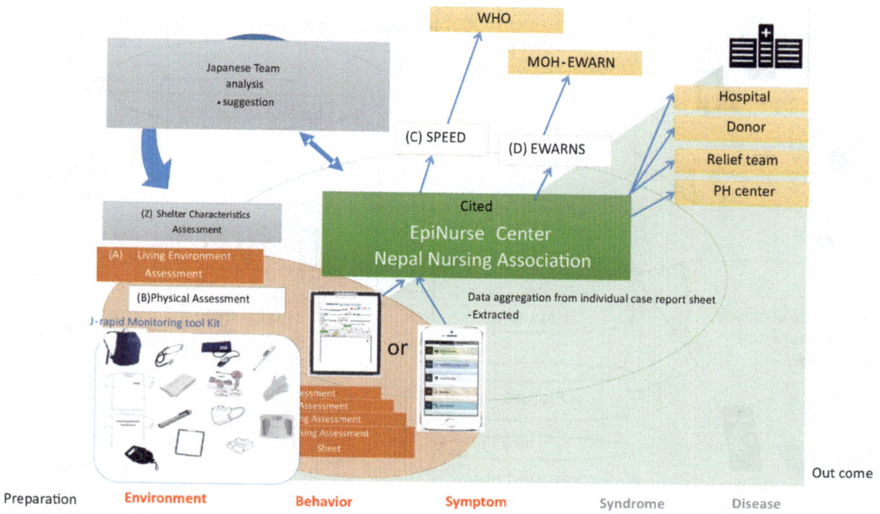

Fig. 15.3 Framework of the study

Table 15.3 EpiNurse perception for surveillance

Main health issue	Starvation Sanitation, communicable disease (diarrhea, measles, dysentery) Psychosocial problem Family crisis, over crowding, safety and security
How to improve?	Mobilize community leaders, FCHV health survey, data analysis health action as per priority implementation, health education Coordinate with DPHO, INGO, NGO, local leaders, social workers, and so on
Where to report?	Health center, Local leaders, DPHO, NAN Epidemiology Department/MOH
What kind of information feedback needed?	Human resources, money Communication materials, monthly supervision Medicines, referral format
How to report?	SMS, Email, Phone, Facebook Messenger

15.7.3 *Continuous Monitoring*

Traditional paper and telephone method for data collection took more time than the smartphone application due to accessibility to towers and power lines. To solve this problem, we developed new application collaborated with the local NPO as in Fig. 15.4.

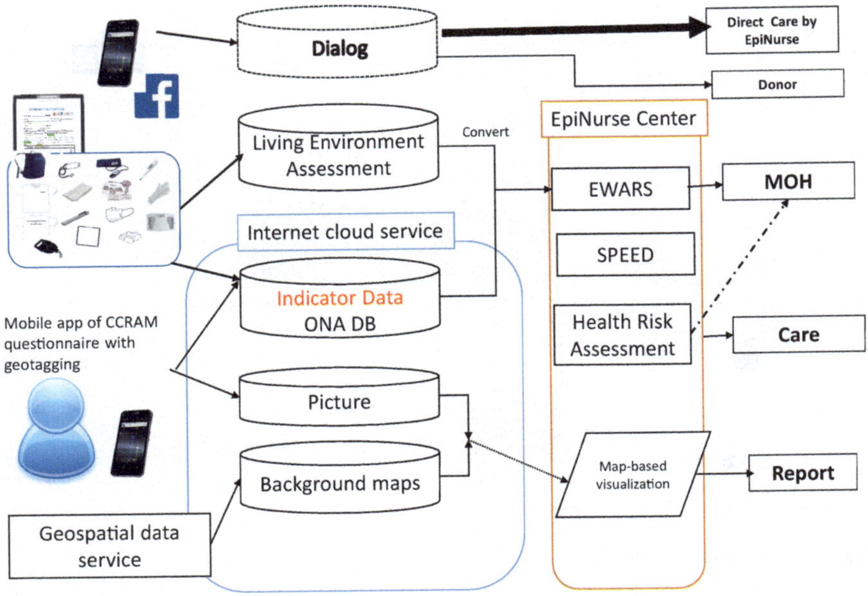

Fig. 15.4 Overview of the proposed system

Through engaging practitioners, the care improved, the quality of reporting changed, nurses' responses became flexible, and nursing association was provided with useful information. This study showed that identifying shared information and engaging domestic community nurses and stakeholders in practical activity can bring about sustainable improvement in community health. The partnering government and donors of the study now have a basis to apply the study results to reconstruction. Collected data can be consolidated and analyzed for further developing the surveillance system. Meanwhile, this study shows the need for securing domestic health professionals like community health nurses to take urgent steps to make intelligent communication from information and develop a sustainable network as social capital to ensure that no one is left behind in disaster risk reduction.

Domestic health professionals know that to bring about societal behavior reforms through institutional policies and procedures must co-evolve with health environment changes in relation to people's beliefs, norms, and values that become motivating forces behind particular types of health behavior (Kanbara et al. 2017).

An EpiNurse Member Profile: A Staff Nurse in the Public Health Center

Education: Bachelor of Nursing.

Work experience: 16 years, TRIAGE, how to make network for relief, how to utilize and flow the information and system. Has not had any training or drill about disaster while she was working. She had managed and experienced other disaster situations like thunderstorm and accident in the hospital. She participated in a WHO training under WASH one year before the earthquake. She also thought about how to give service outside of the hospital, like using tents, supply of water, and where to make toilet, and how to clean toilet and places.

Experience with working in disaster: CMA and nurses were so busy that they did not have time even to drink water. There were no foods to cook. Later, the army helicopter arrived and gave them dry food and water to drink. The army and local people helped them a lot too. She thought PHC knew about the disaster management but they were not prepared, people did not know who to coordinate with PHC as we have not done any awareness. They have very little manpower, one staff nurse myself, one ANM, one officer from National Planning Commission, three doctors. According to the government policy three ANM, one staff nurse, two CMA, one HA, and one doctor, we have requested different government for manpower as this is one of the most busy PHC before earthquake too. Daily there are about 150 patients and four to five critical patients to bed. They have about 45 delivery cases in a month and cover wide area too and other hospitals are difficult to access due to geographical reason. She reported using the apps as Attachment 2, 3, and Table 15.4 and Fig. 15.5 show the result of the continuous monitoring. The quality and completeness of the recording of important information are improved. Living environment data which had not been sufficiently well recorded previously, were absent from the mother's medical records, and were now being checked with picture she sent which depend on cultural perspective of the nurses as local residents. Most of this data could be found in one place in the record expect from the direct dialog and telephone communication though they used most. Results from the apps record showed the possibility of sustained improvement. Finally, she said that "I will continue going to shelters because I can understand that survivor become motivated to rebuild own house when I give mental health consultation."

Melamchi Case Study Report by an EpiNurse Member

15.7.4 Monitoring Results

Overall detection of epidemiological items assorted from collected data through the smartphone application is shown in Table 15.5 and Fig. 15.5. It was available for EpiNurse members to collect shelter information, including the shelter location,

Table 15.4 Total number of symptom suspected on site

Health issues	Total cases: under 5 years	Total cases: above 5 years
Fever (FEV)	352	536
High blood pressure >140/90 (HBP)	0	242
Difficulty in breathing and wheezing (AAA)	135	212
Fever with headache, muscle pains and any of the following: eye irritation, jaundice, skin rash, scanty urination (LEP)	38	122
Loose stools with visible blood (ABD)	39	52
Yellow eyes or skin with or without fever (AJS)	15	15
Known diabetes (KDM)	0	83
Fever with spontaneous bleeding (i.e. nose bleeding, gum bleeding) (AHF)	6	15
12 months and over: sudden onset of fever (>38 °C) with severe headache and stiff neck; <12 months: fever (>38 °C) with bulging fontanel, or refusal to suckle (MEN)	4	0
Animal bites (ANB)	7	9
Spasms of neck and jaw (lock jaw) (TET)	2	2
Loose stools, 3 or more in the past 24 h with or without dehydration (AWD)	206	270
Fractures (FRS)	6	22
Eye itchiness, redness with or without discharge (CON)	104	247
Skin disease (SDS)	136	261
Open wounds and bruises/burns (WBS)	36	142
Fever with other symptoms not listed above (FOS)	13	36
Fever with rash (MEA)	44	64

timestamps, photos, and questionnaire with location information by the GPS of smartphone and sight check. We found that initially, there was a relatively low commitment among the local nurses to continue monitoring. But the use of the smartphone application increased the nurses' sense of responsibility toward the care they provided. It was revealed that the cloud service should be used as the shared data storage and analysis and open data to other services, and visualization of the latest shelter information. The cloud service works as a one-stop platform at which all of the data and information of the epidemiological data can be accessed. The cloud infrastructure should be prepared with flexible extensibility and developed platform as daily use for community resilience.

The legends in Table 15.4 should be interpreted as follows: the red sign indicates that there were no such cases. EpiNurse members have included nose bleeding in AHF and fever with muscle pain in eye irritation, jaundice, skin rash, and scanty urination These data may not be reliable as they cannot exactly diagnose.

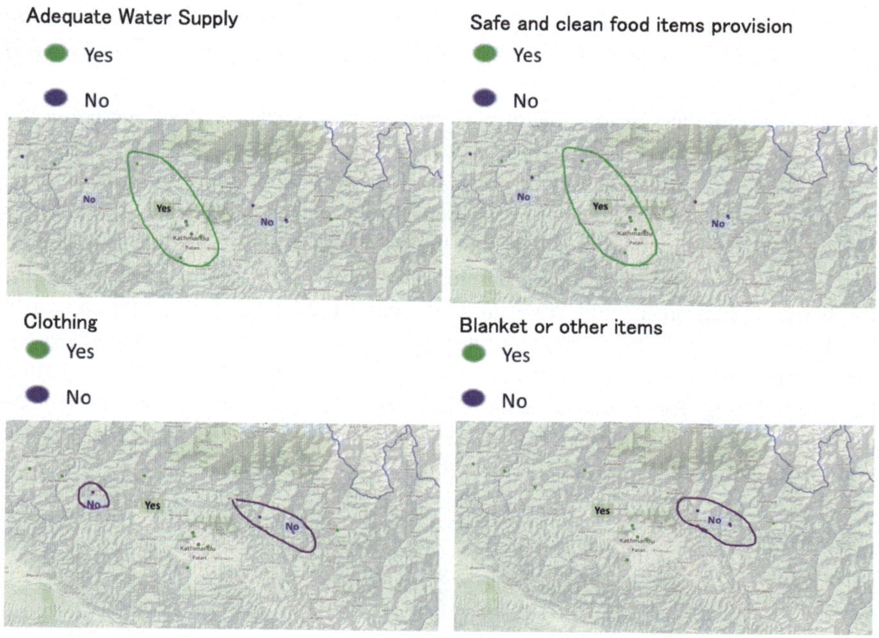

Fig. 15.5 Example geospatial aspect of 1 year after the Nepal Earthquake (April 2016)

Table 15.5 EpiNurse perceived barriers of health management after disaster

Problem	Proposed solution
Lack of communication and media	Promotion of media through Radio, Television Mobile, Magazine
Geographical condition	Mobilize locally available resources –Facility: Stretcher, Ambulance –Human resource –Road
Not proper system meeting	Stake holder meeting
Lack of disaster training	Disaster management training
Lack of monitoring and evaluation	Recording, reporting regular supervision appreciation and rewards
Not proper working environment	For health insurance and basic needs
Not proper supply and logistics	Adequate supply of goods and stocks
Post-disaster monitoring	Case finding by social capital

15.7.5 Barriers to Health Management After Disaster

At the focus group discussion in April 2016, participating EpiNurse members generated a list of barriers to health management after disaster. It is important for scientist to collate in working through the barriers to health management after disaster, as can be seen in Table 15.5, from the focus group discussion in April 2016.

EpiNurse members at this meeting expressed their wishes to continue using the smartphone application as a way to reduce the difficulties of monitoring because using the device allowed easier monitoring, took less time, and effort to complete the monitoring, and provided valuable data that assisted their decision-making in shelters. One EpiNurse member said that the information about conditions in shelter after disaster was similar to a black hole. There was also a report about needs for health risk reduction through a sustainable network as a form of social capital in disaster mitigation. Using simple technology like crowdsourcing application could help obtain human security information and generate actionable information from daily monitoring for future disaster response. Also feasibility development by utilizing the Plan-Do-Check-Act (PDCA) problem-solving model as well as routine monitoring would be helpful.

15.7.6 Community Monitoring Challenges

At the beginning, we faced some difficulties with reporting by nurses due to biases toward the monitoring among people seeking cure and required care in the disastrous situation. Geographic information of tentative migration points did not have correct addresses, only GPS on picture in smartphone enables to identify the shelters. With absent IT infrastructure, it took much more time to update data. In this research, eight of the participating EpiNurse members did not have smartphones initially, and no electricity or telephone services were available after disaster in some areas as shown in Table 15.2.

15.7.7 Conclusions and Future Prospects

This study is one of the practical, solution-oriented research which challenges the conventional public health security monitoring systems which require near real time, population-based, and statistical alarms to alert to unusual activity. EpiNurse approach also includes various key issues in developing the next methodological model of health monitoring by offering a strong interactive network, which crosses cultural and societal differences, geographic regions, and generations. The potential of this work is not only in its possibility for generating innovative research outcomes involving the improvement or optimization of services utilizing ICT, but also in promoting research knowledge and idea exchange regarding social issues and challenges in the fields of emergency preparedness and response.

Rapid assessment as risk reduction could become empowered by applying location information from smartphones used for data collection. Identification of typical behavior patterns could also provide useful insights to the health security assessment compare differences between community and temporary shelter site hourly using the brief EpiNurse communication.

It is urgent to apply geospatial technology for people's behavior modeling in disaster situation and suggest to explore the potential use of social networks to enhance the government's understanding of people's health security level. Development of trans-community preparation for comparative decision-making for prompt disaster response is also an urgent task for global health security. Our study showed that identifying a shared goal and engaging Nepali and Japanese researchers in this project to achieve this goal can bring suggestion to Nepali government for build back better reconstruction.

15.7.8 Routine Health Security Monitoring and Emergency Report Based on Public Health

Resources are often limited and do not allow monitoring tool to incorporate all of the reporting sources. It is required for government to coordinate the interoperability of rapid assessment with other data services for further cooperation with other sectors, such as infrastructure, energy, transport, and water, and close partnerships with local volunteer. This team responsible for rapid assessing each emergency should exist in sustainable community for health security. Routine monitoring and communication are essential in ensuring to adapt over time to changing social context, environments systems, and technology. At a minimum communication and regular social capital evaluation should be undertaken. This trajectory may later result in a mode of transferring guidance to first responders and civilian populations regarding DRR and behavior and also as a way to reach large populations.

15.7.9 Capacity Building Including Human and Technology in Sustainable Community

With formal education and drilling, competency can be developed in public health. It may enable local nurses to make decision to create a system that would allow smooth cooperation among stakeholders. In local communities, nurses collect information about and mitigate communicable diseases risks. The data produced and their improved quality confirm the claim by others that this information can assist with allocation of resources, establishment of budgetary and long-term planning, and productivity measurement and allotment of the work depending on literacy and skills of the community. It is crucial that capacity building for participants to coordinate the effective actions and interventions as well as periodical onsite risk assessment. Effective leadership might be also necessary to manage improvement in reconstruction settings. By position, being member of every part of the health care system, they can become critically needed leaders in emergency management and disaster preparedness. Nurses may manage community setting as well as hospital ward. They can play a vital role in the restoration of public health under disaster condition (water, sanitation, food, and shelter) and identification of high risk and vulnerable population including the unique needs of children in disaster.

This project has explored the ways in which the data collected by local nurses transform into the basis of the daily community health information system. This could then potentially be developed as a monitoring model and benchmark for a leap forward toward sustainable disaster risk reduction from the point of view of nursing science.

References

Japan Society of Disaster Nursing (1998) Prospectus on the establishment of the society. Accessible at: http://www.jsdn.gr.jp/english/about/index02.html

Kanbara S, Nlandu N, Pokhrel T, Pandey A, Sharma C, Lee H, Miyagawa S, Miyazaki H, Nojima S (2018) The 2015 Nepal Earthquake Disaster: is the threat of occurrence of communicable disease epidemic over? Int J Indones Natl Nurs Assoc 1(1):105–110

Kanbara S, Yamamoto Y, Sugishita T, Nakasa T, Moriguchi I (2017) Japanese experience of evolving nurses' roles in changing social contexts. Nurs Rev 64(2):181–186. https://doi.org/10.1111/inr.12365

Chapter 16
An Assessment of Primary Health Care Facilities and Their Preparedness Level in Khyber Pakhtunkhwa Province of Pakistan: Strengths, Weaknesses, Opportunities, and Threats (SWOT)

Ashfaq Ahmad Shah, Muhammad Abid and Jingzhong Ye

Abstract Pakistan is one of the countries that are most vulnerable to climate-induced natural disasters such as floods. For this reason, it is essential that emergency relief services are in a state of preparedness in order to reduce causalities and other damages. Rural health centers could play an important role in this regard, but there is little information available about their current and potential capacities. This chapter assesses the level of flood disaster preparedness of rural health centers in six flood-affected districts of Khyber Pakhtunkhwa (KP), Pakistan. For this purpose, we collected data on three components of rural health centers' disaster preparedness, using structured questionnaires from 48 respondents from 19 rural health centers. The study findings show that the facilities at rural health centers are currently insufficient to meet the needs of local communities at times of disaster and that there are significant gaps in all three components of their disaster preparedness. In particular, the staff at rural health centers lack training, and the buildings lack the space, beds, and equipment required for relief work. A SWOT analysis revealed that these centers have much potential to fulfill this role if they were provided with proper infrastructure, trained human resources, and financial resources.

A. A. Shah (✉)
School of Management Science and Engineering, Nanjing University of Information Science and Technology (NUIST), No. 219, Ningliu Road, Nanjing, Jiangsu 210044, People's Republic of China
e-mail: ashfaqahmadshah@s3h.nust.edu.pk; ahmad.ashfaq1986@gmail.com
Shahaa@cau.edu.cn

A. A. Shah · Jingzhong Ye
College of Humanities and Development Studies (COHD),
China Agricultural University, Beijing, People's Republic of China
e-mail: yejz@cau.edu.cn

M. Abid
Centre for Climate Research and Development (CCRD), COMSATS University,
Park Road, Tarlai Kalan, Islamabad 45550, Pakistan
e-mail: muhammad.abid@comsats.edu.pk

E. Ying Yang Chan and R. Shaw (eds.), *Public Health and Disasters*,
Disaster Risk Reduction, https://doi.org/10.1007/978-981-15-0924-7_16

Keywords Flood preparedness · Disaster risk reduction · Health facilities · Rural health centers · Pakistan

16.1 Introduction

Flood disasters are the most frequent type of natural disasters (Rattanakanlaya et al. 2016). Between 2005 and 2014, there have been a total of 1751 flood disasters (46% of the total of natural disasters globally) which killed 59, 0592 people and caused 342,836 million USD of economic losses (IFRC 2015). Floods can become catastrophic disasters in vulnerable places when they happen unexpectedly and/or with unpredicted intensity. Evidence shows that the frequency and intensity of floods (or hydro-meteorological disasters) have increased in Southeastern Asia countries, including Pakistan, India, and Bangladesh, over recent decades (Shah et al. 2017, 2018a; Hirabayashi et al. 2013). Pakistan experienced its most catastrophic flood in 2010, which affected more than 24 million people, damaged more than 2 million hectares of crops and 515 health care centers, and caused an economic loss of around 10 billion US dollars (Shah et al. 2017; Shabir 2013). Projections suggest that the frequency and magnitude of floods and their impacts on populations and property will increase in the future (GOP 2013) due to a combination of climate change and rapid and unplanned urbanization (Tariq et al. 2014).

Extreme weather-related events, such as floods, have dramatic effects on the environment and human health and put enormous pressure on health system. The Lancet Commission on climate change identifies that health systems in developing countries lack the capacity to respond effectively to the adverse effects of climate change due to their disorganization, inefficiency, and inadequate resourcing (Costello et al. 2009). One of the United Nations Sustainable Development Goals (SDG) is to build resilience and adaptive capacity so that societies can respond better to extreme weather-related events. As the first line of contact for disaster victims, health facilities are an essential component of disaster preparedness and response. Preparedness refers to the knowledge and capacities of individuals, including professionals, communities, and governmental and non-governmental organizations (NGOs) (UNISDR 2009). As such, it is important to understand the preparedness ability and functional capacity of health facilities to respond effectively to hydro-meteorological disasters (Farley et al. 2017).

Different health organizations and disaster epidemiologists define the concepts of risk and vulnerability in public health in different ways. The Society for Risk Analysis (SRA) defines risk as the negative impacts that arise as a result of risks on human lives, property, and health and estimates risk based on the probability and timing (Thompson et al. 2005). In public health, there is a dynamic relationship between the numbers of people and their distribution within a given area, their vulnerability and the skills of, and resources available to, medical and emergency personnel (Gillam et al. 2007). The 'vulnerable population' can be defined as those people who are more prone risks: those who live in flood-prone areas and/or have inadequate housing, who are economically insecure, have a poor health status, and/ or lack access to health care facilities (Weathers et al. 2004). The impacts of disasters can be complex. In addition to the loss of life and damage to property, infrastructure crops and livestock, they can damage the capacity of the health care system and other essential public services (AbouZahr and Boerma 2005; Greenough et al. 2001) and be the cause of acute stress (Waelde et al. 2001), depression (Tapsell et al. 2002; Reacher et al. 2004), anxiety, and posttraumatic stress disorder (PSTD). These impacts will depend on the nature and scale of flood disasters, the ability to return to a normal way of life, the presence of environmental contaminants, the evacuation procedures and preparedness measures in place, and assistance received (Tapsell and Tunstall 2006; Adeola 2003; Galea et al. 2005).

In Pakistan, health services can be classified into three categories: primary health care facilities (including basic health units, rural health centers, and civil dispensaries); secondary health facilities (district and tehsil headquarters hospitals), and tertiary health care facilities, which are mainly located in the major cities and affiliated with research and educational institutions (WHO 2013). When severe or prolonged emergencies occur in rural areas, the resources of primary health care facilities are quickly exhausted. According to the Federal Office of Rural Health Policy's document 'Rural Communities and Emergency Preparedness,' rural communities are highly dependent on the preparedness of emergency medical services. While such health facilities do exist, they are often unable to respond quickly and efficiently to flood disasters as they have limited funds and staff. In spite of the advances made by science and technology in the health sector, the majority of the rural population continue to be exposed to flood disasters. This can be attributed to the differential distribution of critical health facilities in the rural areas where the rural community are likely to use rural health facilities because such services are accessible to them and less costly. Therefore, it is worth examining the preparedness level and resilience of such facilities to disasters, as they can make a significant contribution to the socioeconomic and psychological recovery from catastrophic flood disasters (RHIhub 2017).

A considerable number of academic studies have been done on the preparedness for disasters of hospitals in different various countries. These studies have focused on the linkages, status, and capacity of hospitals' in responding to different types of disasters (Greenberg et al. 2002; Braun et al. 2004; Kaji and Lewis 2006). However, these studies have mostly focused on tertiary health facilities such as major hospitals in the cities, and little work has so far been done on the preparedness level of primary health facilities (Rural Health Centers (RHCs) in South Asia and there is a distinct lack of in-depth assessments of the preparedness of health facilities for flood disasters (Phalkey et al. 2012; Abaya et al. 2009).

In Pakistan, rural areas are often under-resourced in health care terms and lag behind in terms of the provision of health care facilities, with the rural poor lacking access to secondary and tertiary health facilities (Akram and Khan 2007). The only studies that have been done to date are limited to the preparedness level of secondary and tertiary level health care facilities (Khan et al. 2017; Ullah et al. 2017). There have (to the authors' knowledge) been no studies focusing on the preparedness level of primary health facilities. Like many developing countries in South Asia, Pakistan faces challenges with its preparedness level for dealing with flood disasters. The United Nation Development Program (UNDP 2004) stated that Pakistan lacks integrated disaster management policies or a system for disaster preparedness. This highlights the need for effective structures, strategies, and policies for disaster management. This study hopes to go some way to fill the existing research gap with respect to the current preparedness level of primary health care facilities in flood-prone districts of KP province (and by extension the rest of rural Pakistan). It also applies a SWOT analysis, to evaluate the strengths, weaknesses, opportunities, and threats of the rural primary health care system in terms of preparedness for disasters.

16.2 Conceptual Framework

The conceptual framework of this study first describes the role that health care facilities can play in enhancing preparedness for, and in enhancing societal resilience to, floods disasters. The framework starts with flood disasters (Fig. 16.1), which negatively affect societies in different ways, having economic, social, physical, and psychological impacts. These impacts can be reduced by societal

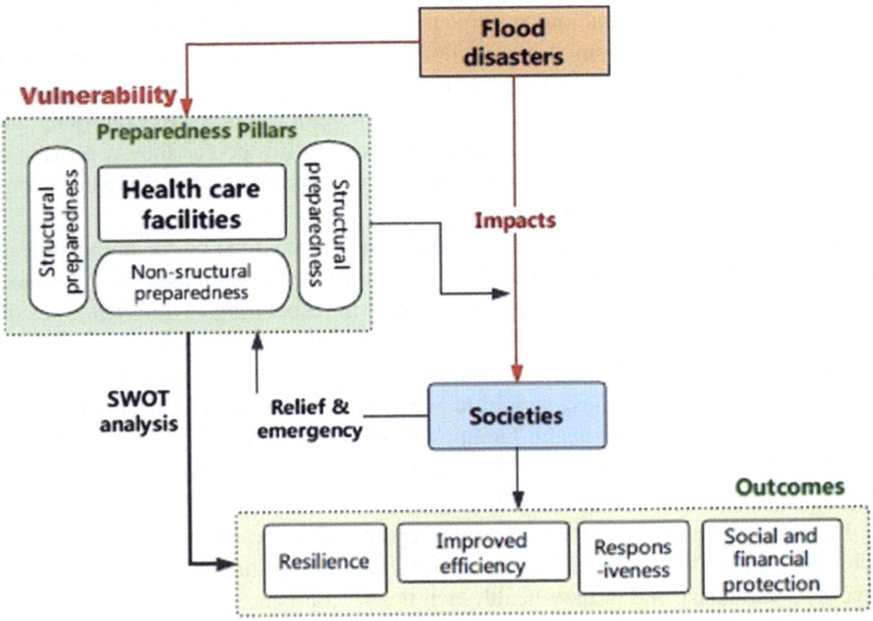

Fig. 16.1 The conceptual framework of the study (authors own construction)

adaptive capacity. One measure that can enhance societal adaptive capacity is to provide the necessary infrastructure to avert disasters or reduce their impacts. Here, basic health care facilities can play a key role in providing the necessary relief and emergency services at times of disaster. To do this, a health facility needs to be equipped with all the necessary infrastructure, structural, non-structural and to be functionally prepared for disasters. Without these tools, health care facilities are as vulnerable as the society that they are designed to serve. In order to assess the preparedness level and vulnerability of health care centers in Khyber Pakhtunkhwa (KP), this study considered three pillars of preparedness pillars: structural preparedness, non-structural preparedness, and functional preparedness. On the basis of this, we develop a SWOT analysis to assess the strengths, weaknesses, opportunities, and threats to health care facilities when serving communities during flood disasters.

16.3 Materials and Methods

16.3.1 Study Area and Sampling Method Description

The study was conducted in Khyber Pakhtunkhwa (KP) province to assess the preparedness level of health facilities across six districts in the flood-prone areas of KP. The province was selected purposively as it is one of the poorer provinces in the country, plagued by different natural disasters, especially floods and has a low adaptive capacity for coping with natural disasters (Shah et al. 2017, 2019). KP, which is divided into seven divisions and 25 districts, has 1583 primary health facilities including rural health centers (RHCs), basic health units (BHUs), and civil dispensaries (CDs); 125 secondary health care facilities, including district and tehsil headquarter hospitals and civil hospitals; and 11 tertiary care hospitals (IMU 2016). This research only focuses on the primary health care facilities, RHCs, which are open 24 h a day seven days a week across six districts of the province. For the current study, we have selected nineteen sample RHCs randomly from a list shared by the Khyber Pakhtunkhwa Health Department. The cross-sectional quantitative survey was done in July and August 2017. The questionnaire was pretested and corrected before the collection of final primary field data. Binary type questions were asked from the representatives of RHCs along with various cross-questions to assess the preparedness level of RHCs. Forty-eight respondents were selected randomly (including medical doctors, senior technicians, and lady health visitors) and interviewed for an in-depth assessment of their RHCs preparedness level. Table 16.1 provides the number of RHCs visited in study districts during field research.

Table 16.1 Total number of RHCs, sampled RHCs, and sampled respondents in the selected RHCs

District	# of RHCs	Sampled RHCs	Selected RHC	Sampled respondents
Charsadda	4	3	RHC Sherpao, RHC Jamal Abad, and RHCBattagram	07
Nowshera	5	4	RHC Kheshki, RHC Akbar Pura, RHC Pirpai, RHC Mankisharif, and RHC Dakismail Khel	12
Peshawar	5	4	RHC Gara Tajik, RHC Badaber, RHC Nahqi, and RHC Takht Abad	10
Mardan	4	4	RHC Palo Dheri, RHC Shergarh, RHC Gumbat, and RHC Manga	8
Swat	3	2	RHC Devlai, and RHC Chuprial	5
D.I.Khan	3	2	RHC Sidalian at Kot Joi, and RHC Karri Shamozi	6
Total	24	19		48

Source Corresponded from personal meeting with IMU Director, 2017

16.3.2 Three Pillars of Rural Health Centers' Vulnerability to Natural Disasters

In the study, we looked at three pillars of preparedness (structural preparedness, non-structural preparedness, and functional preparedness) to assess the vulnerability of RHCs to hazardous events, such as flooding. We discuss each of these pillars below.

- **Structural Preparedness**: Structural preparedness involves physical construction which will avoid or reduce the possible negative impacts of hazards, the adoption of engineering solutions and systems or structures that enhance resistance to hazards. Here, we primarily considered the construction of buildings and the infrastructure available for providing health and emergency evacuation facilities in case of floods or other hazards
- **Non-structural Preparedness**: Non-structural measures may be defined as non-structural short-term and small-scale measures that can reduce the health and societal impact of flood disasters. Here, we considered equipment and facilities, looking at the availability of medical gas facilities, sprinkler systems (in the case of fire), emergency exits, medical equipment, the safety of fixtures and equipment, and the availability of supplies in the case of flooding events.
- **Functional Preparedness**: This includes aspects such as accessibility, equipment, emergency supplies, communication and transportation systems, human resources, and the capacities of medical and auxiliary staff.

16.4 Results and Discussion

16.4.1 The Impacts of Disasters on Health Facilities and the Health Care System

The impact of flooding on the health sector brings a secondary, knock-on, effect, if health care facilities including primary level health care facilities (RHC) are damaged. The extent of such damage will depend on the nature and severity of the catastrophe and the vulnerability and resilience of existing health facilities (Mulyasari et al. 2013). The redevelopment of such facilities involves substantial financial investments for reconstruction and rehabilitation, which in turn will place be enormous pressure on the regional or national government's finances. In addition, the failure of health facilities to withstand flood disasters has indirect costs and social implications. People's morale can suffer as a result of the loss of loved ones, particularly elders and children (who are the most vulnerable to flood events), especially if this is the result of a failure of emergency services when they are the most needed (UN 2009). This can spill over and ignite political dissatisfaction (PAHO/WHO 2003).

The delivery of efficient emergency medical services to the people in a disaster situation is also highly dependent on having an efficient and effective coordination system among different levels of health facilities and with other emergency services. For instance, a lack of coordination among the primary, secondary, and tertiary levels of health facilities will lead to a loss of time, waste of resources, and overlapping of services. Coordination among all relevant stakeholders during times of emergency is considered to be an integral part of providing medical services (ADPC 2009). There are various literatures available on global initiatives to address the risks facing the health sector, including hospitals and another type of facilities. Bissell et al. (2004) examined the effectiveness of health sector preparedness in disaster response. Similarly, Khan et al. (2017) conducted a study on knowledge about disaster preparedness in tertiary level health care facilities in Lahore (Pakistan) and found that expertise regarding disaster preparedness is essential to improve the ability of the health system to perform efficiently in the time of disaster. Ullah et al. (2017) conducted a study on the challenges facing tertiary health facilities (hospitals) in terms of their preparedness for disasters (natural and man-made) in Quetta (Pakistan) and emphasized the need for the establishment of a crisis central command to coordinate emergency responses that can provide efficient and timely medical services.

Several global initiatives have been taken to enhance the preparedness level of health sectors for disasters. These include the Hyogo Framework for Action (2005–2015) by the United Nations International Strategy for Disaster Risk Reduction (UNISDR 2005), the World Disaster Reduction Campaign 2008–2009 by the United Nations (UN 2009), and the Asian Disaster Preparedness Centers (APDC 2009). Despite these efforts, which aim to ensure the safety of different levels of health facilities, there are still some countries where health care facilities from tertiary (hospitals) to rural health centers are built in the vicinity of disaster-prone areas (UN 2009). In Pakistan, the 2010 floods had a massive impact on health facilities. Almost three thousand (out of 9721) of Pakistan's health facilities are situated in flood-affected districts. In KP province, 10.9% of the health facilities were damaged, either partially or entirely destroyed, by the 2010 flood. Most of the facilities affected were located in rural areas and while the secondary and tertiary level of health facilities were not overly affected by the flood, the severe disruption caused to primary health facilities led to secondary and tertiary institutions being overwhelmed by the demand for medical services (Shabir 2013).

16.4.2 *Indicators for Assessing Rural Health Centers' Preparedness for Dealing with Flood Disasters*

16.4.2.1 Structural Preparedness

The structural preparedness elements for the primary health care facilities (RHCs) include having structures that are resilient enough to withstand the impacts of floods or other hazardous events and enough space to accommodate people who are need of evacuation during an emergency. These structural preparedness elements should be appropriate to the location of health facility, the population which it serves and the frequency and severity of hazardous events. The location of health facilities may, in itself, make it vulnerable to hazardous events, particularly floods. Choosing sites away from the risk of inundation should be a priority as should the adoption of effective mitigation measures that will minimize damage to facilities' buildings and infrastructure. Table 16.2 shows RHCs' preparedness under the first pillar that relates to the vulnerability of buildings and infrastructure. It shows that just 36% of the RHCs that we sampled are adequately prepared in this respect. The low structural preparedness level of health facilities requires serious attention from local health departments which should take adequate measures for their buildings to be able to withstand the adverse impacts of floods and be able to continue to function and deliver medical services at times of flood disaster. These findings are in line with those of Hatami et al. (2017) who reported a similar level (44%) of resilient health facility structures in Iran. The literature that examines this phenomenon identifies the main reason for the low structural preparedness of health facilities as being a lack of adequate supervision by the relevant agencies during different phases of construction, in which guidance about structural mitigation mechanisms is often ignored (Seyedin et al. 2011).

Another important aspect of the structural preparedness and vulnerability element used in this study is the availability of enough space for safe evacuation during a flood emergency. Table 16.2 shows that, on average, 41% of the primary health facilities do not have any, or sufficient, space for safe evacuation within their health facilities. This is in strong contrast with Japan, where Mulyasari et al. (2013)

Table 16.2 Structural preparedness indicators

Preparedness pillar	Vulnerability indicators (%)	Charsadda	Nowshera	Peshawar	Mardan	Swat	D. I. Khan
Structural preparedness	*Building and infrastructure*						
	Resilient structure	34	33	37	40	40	33
	Space available for flood emergency evacuation	33	38	44	58	40	57

Source Derived from field survey data, 2017

found that great majority (80%) of health facilities have enough space available for emergency evacuation of the evacuees during disasters. The lack of availability of space for evacuees implies a low level of preparedness for such eventualities. In addition, the low number of primary health facilities that reported having a resilient structure (36%) (one of the structural preparedness vulnerability elements) could compromise the safety of people at risk of flooding events. Local government and the administrations of health facilities should be aware of the protocols and building codes in their jurisdictions and ensure that these protocols are correctly followed and implemented. The use of substandard materials is a particular problem, which together with the lack of availability of sufficient space for evacuees, limits the ability of many primary health care facilities to provide services during flood emergencies and could lead to a major tragedy.

16.4.2.2 Non-structural Preparedness

Non-structural preparedness, the second preparedness pillar of RHC facilities, covers vulnerability elements such as the availability of medical gas and fire suppression systems, the provision of emergency exits, the availability of medical equipment, and the safety of fittings, fixtures, equipment, and supplies (Table 16.3). These issues are all very crucial for the efficient operation of primary health care facilities, especially in times of disaster. Only 29% of RHCs have adequate storage for their *medical gas systems* and only 20% an adequately safe distribution system. Concerning the *fire suppression system*, 51% of the facilities sampled reported having an automatic fire alarm system and 45% said that their primary health facility has fire exits that are easily accessible in case of an emergency situation and an evacuation plan. However, only 20% of facilities have a portable fire extinguisher in each room, largely due to a shortage of funds to purchase such equipment. *Emergency exit systems* can play a significant role in reducing physical injury to patients and personnel during a flood or other emergencies. In total, 49% of the sample reported that the local health department in each district provided luminous directional exit signs to make it easier to evacuate the building during an emergency situation. These make it easier for people to see and follow the glowing signs to

Table 16.3 Non-structural preparedness indicators

Preparedness pillar	Vulnerability indicators (%)	Charsadda	Nowshera	Peshawar	Mardan	Swat	D. I. Khan
Non-structural preparedness	*Medical gas system*						
	Safe and appropriate location for storage of oxygen gases	30	28	31	22	37	25
	Safety of oxygen gas distribution system (valves, pipes, connections) is ensured	21	25	19	17	23	15
	Fire suppression system						
	Automatic fire alarm system	55	61	44	37	51	57
	Each room provided with portable fire extinguishers	11	17	10	13	17	15
	Fire exit and evacuation plan	45	51	33	42	50	47
	Emergency exit system						
	Provided luminous directional exit signs located	51	50	41	51	55	47
	Size of signs—plainly legible letters	67	73	64	59	70	55
	Medical equipment						
	Safe from flooding	33	41	67	83	81	77
	Adequate power supply	22	29	44	53	57	32
	Clean and orderly, free of dirt and infectious materials	56	61	55	64	71	45
	Safety of fixtures, equipment, and supplies						
	Proper segregation and storage of hazardous materials and chemicals	33	29	30	27	39	22
	Material safety data sheets	14	25	20	13	60	17
	Stocking available	29	33	30	38	40	21
	Contingency supplies available	10	16	9	13	22	11

Source Derived from field survey data, 2017

safely leave the building during an emergency. Sixty-five percent of the survey respondents reported that the size of signs was appropriate and the lettering was plainly visible. This would help patients to go in the right direction in case of an emergency.

The primary health care facilities in our sample generally fared much better in relation to *the availability of medical equipment*. Almost two-thirds of facilities (64%) kept their medical equipment safe from flooding (64%) and more than half (59%) kept it clean free from dirt and infectious materials. Some respondents volunteered that their facilities had adopted specific protection measures, such as wooden foundations for heavy equipment to protect it from damage. However, almost 40% of the facilities lack an adequate power supply to run such machines and equipment in times of emergency, thereby limiting the provision of emergency health care when it is needed the most.

Safety issues regarding fixtures and fittings, medical equipment, and supplies are crucial and merit separate consideration, for instance, improper handling of hazardous chemicals may people cause serious injury. When the respondents were asked about safety issues, just 30% of respondents reported that their RHC has a facility for storing hazardous chemicals and materials. It is also important that people receive appropriate training required for handling such chemicals. One-quarter of the total survey respondents said that they used material safety data sheets and encouraged the dissemination of important information about hazardous chemicals among medical practitioners, other emergency response providers, and sometimes the general public. Slightly under one-third (32%) of facilities in the sample reported having adequate stock of medicines for the patients, but in case of emergencies, only 14% of facilities said that they have contingency medical supplies sufficient to last for a maximum two or three days at a time of emergency.

16.4.2.3 Functional Preparedness

When considering the functional preparedness of RHCs, the aspects of vulnerability that we considered in this study include: accessibility, equipment, and supplies for emergency, communication and transportation systems, human resources, and the capacities of medical emergency staff. The functional preparedness of primary health care facilities can play a significant role in the provision of medical services during emergencies or disasters when emergency medical services are most needed. *The accessibility of rural health centers* (RHCs) is an essential factor in determining functional preparedness and the level of vulnerability. A large majority of the sampled facilities (79%) reported no obstructions on the roads that lead to their facility. The majority (percentage) also have alternative access routes that can be used in case of flooding or other disasters. This could also help to ensure the safe evacuation of people from inundated areas. The location of health facilities is another vital issue and it is imperative that facilities are built close to roads that can provide adequate transportation and preferably in proximity to other educational research institutes. And it goes without saying that facilities should not be constructed in flood-prone areas as this will make them inaccessible at times of need and pose a significant threat to structural safety.

Table 16.4 also shows *the availability of equipment and supplies for emergency situations*. It shows that only 35% of primary health facilities in the sample have tents available in cases of emergency and only 45% have an in-house generator. These are both critical resources, especially emergency generators which are essential to provide electricity to the critical life-saving medical equipment in times of power outages (OSPHD-Electrical Requirement for Health Care Facilities 2011). Other supplies required for an emergency include a supply of clean drinking water inside the particular facility (which only 64% of facilities have), emergency food supplies (kept by 37%), folding beds for times when there are more patients than beds (54%), and wheelchairs for the aged and disabled people (54%). The large majority of primary health facilities (89%) lack triage facilities. Onsite triage of the patients and coordinated transfer will avoid the mismanagement of an influx of non-triaged patients into the trauma centers of health facilities (Ullah et al. 2017). All these types of equipment and supplies are essential for primary health centers to function effectively during emergencies, yet are lacking in many centers (PAHO 2000).

Communication and transportation are two other vital aspects when considering the vulnerability of primary health care facilities in times of emergency. Only 46% of the facilities sampled have a functioning emergency management information system (EMIS), although this varies enormously between the different districts. Facilities fare rather better in terms of having backup communication tools, such as mobile phones, or walkie-talkies (possessed by 80% of facilities); a vital back up should be the main communication tool break down (Aitken and Leggat 2012). Similarly, the higher average responses on the availability of ambulances to shift causalities from the field to nearby hospitals (72%), a list of all running ambulances (75%), and medical equipments and medical supplies details (81%) show that the primary health facilities in the sample are relatively well-prepared for emergencies in terms of communications and transportation systems, an important aspect of disaster preparedness (Mulyasari et al. 2013). *Human resources and capacity building* are the final aspects of functional preparedness for disasters that we considered. Those who are actively engaged (whether as medical officers, disaster risk management staff, or community members) in the provision of emergency medical services at times of disaster should be adequately prepared. By consideration of human resources preparedness, the respondents were asked their opinion about the Disaster Relief Management (DRM) coordination units at their health facilities. Only 36% of RHCs actually had a dedicated DRM unit and only 9% had staff with a specific responsibility for DRM activities. Twenty-four percent of the facilities have established a DRM health committee, which implies a worrying lack of community participation in health-related DRM (Chan 2014). This needs special consideration Alexander et al. (2015) showed that a lack of community engagement was one of the most important factors responsible for the scale and duration of the Ebola outbreak in West Africa. Another worrying feature is that less than one in ten (6%) of RHCs have a DRM health strategic plan, the lack of which often results in

Table 16.4 Structural preparedness vulnerability elements in the sampled area

Preparedness pillar	Vulnerability indicators (%)	Charsadda	Nowshera	Peshawar	Mardan	Swat	D. I. Khan
Functional preparedness	*Accessibility*						
	No obstructions on the roads leading to the RHC	73	85	87	71	69	89
	Access to more than one road (alternative routes)	80	87	81	67	57	90
	Equipment and supplies for emergency						
	Tents	29	33	40	38	40	33
	In-house generator	14	50	40	75	60	33
	Clean drinking water	70	68	61	70	80	33
	Food	29	58	30	38	20	50
	Folded beds	53	59	43	39	61	71
	Triage tags	4	17	6	9	13	17
	Wheelchairs	43	50	50	50	80	50
	Communications and transportation systems						
	EMIS	71	67	40	25	40	33
	Backup communications	81	83	85	89	71	69
	Ambulances for transporting casualties	61	71	73	69	79	81
	List of identified available and capable ambulances for use during emergencies	76	77	70	65	80	84
	Lists of available equipment, medical supplies, and emergency drugs	77	81	82	73	87	83
	Human resources						
	DRM coordination units among RHC	40	43	29	37	32	33
	DRM focal staff within RHC	5	17	10	0	20	0
	DRM health committees established	29	33	20	25	40	0
	DRM health strategic plans available	11	16	0	0	14	0
	Capacity building						
	Training availability for emergency medical staff	43	25	40	25	40	0
	Disaster drill for emergency medical staff conducted on an annual basis	29	33	30	38	60	50
	Disaster drill for patients conducted on an annual basis	7	9	2	0	3	0

Source Derived from field survey data, 2017

the fragmentation of health-related DRM activities and situations where RHCs may have more sources of command, sending out conflicting messages, and no concerted efforts to address the emergency (Olu et al. 2016). It is essential that all health facilities should have a defined command structure and be able to rely on it and clear-cut definitions of responsibility when a disaster does strike (Mulyasari et al. 2013).

There is also an essential need for staff to be trained in and prepared to deal with emergency situations. Yet in our sample, only 29% of emergency medical staff received such training. This is in line with the findings by Hsu et al. (2006), Tachibanai et al. (2005), and Bagatell and Wiese (2008) who all found a low level of preparedness and expertise among health professionals who are expected to actively engage in the provision of emergency medical services during times of disaster. Such gaps should be bridged through the provision of an advanced level of disaster preparedness training to emergency medical staff which defines specific activities and responsibilities to ensure an efficient response in times of disaster (Ullah et al. 2017). Other important aspects of training include disaster drills and exercises for emergency medical staff that should be carried out on an annual basis. Yet 60% of the sample did not carry out such drills for their staff and only 4% carried out such drills involving patients. Disaster drills are a critical component of functional preparedness to test the response of health facilities in simulated real time situations. The evaluation of these activities is essential to know the strengths and weaknesses of an institution's response to risk (Ahrq 2011).

16.5 SWOT Analysis of Primary Health Care Facilities (RHCs)

The key findings of all three-preparedness pillar of primary health care facilities can be categorized as internal (strengths and weaknesses) and external (opportunities and threats) (Shah et al. 2018b; Noordin et al. 2011) which can be analyzed through a SWOT analysis. A SWOT analysis examines and evaluates an organization's internal strengths and weaknesses, the opportunities for growth and improvement, and the threats that the external environment poses (Terzic et al. 2010). A SWOT analysis is often an important component of the planning process and can be practically applied in the field or to projects to estimate the level and extent of strengths, weaknesses, opportunities, and threats for achieving specific objectives. A SWOT analysis can also undertake a risk analysis to help organizations to recognize dangers, devise or adopt strategies for minimizing the risks, and identify coping mechanisms. As this study is focused on the flood disaster preparedness level of the health sector in Pakistan, following Table 16.5 presents a SWOT analysis for the health facilities preparedness for flood risks.

Table 16.5 SWOT analysis for Rural Health Centers in the study area

Strengths (S)	Weaknesses (W)
1. All the RHCs surveyed in this study are functional and provide primary health care services to their local communities	1. The RHCs are located in union councils and serve the community providing primary health services (treatment and medication). There are no secondary or tertiary health care services available. Patients requiring such services are referred to as Tehsil or District HQ hospitals
2. The local communities are able to avail themselves of the services of specialized emergency medical officers who provide primary health services to the rural population for	2. Rural communities are highly dependent on emergency medical services, and the existing health facilities do not have sufficient funds or resources to meet the health emergency needs of these rural communities. This is a serious weakness
3. The administrative and management responsibilities for all the health facilities covered by this study rest with government and the local community cannot interfere in their management and administration	3. Much equipment used in the rural health facilities is either outdated or out of order. This creates gaps in the delivery of health services
4. The government deploys financial and human resources, which if they were assigned to the community, could give rise confusion and opportunities for favoritism and nepotism in deploying resources	4. In Pakistan, there are 17 health service delivery and 17 health service management standards. The health facilities located in rural areas are rarely succeed in meeting these standards, which further contributes to poor health service delivery
5. Most of the health facilities in the study have alternate access routes that could be readily evacuated in times of emergency. These alternate access routes could also allow essential equipment to be removed from the premises (to be used elsewhere) if there was risk of flooding	5. There are no proper mechanisms of social accountability of the duty holders
6. Rural populations have easy access to health facilities to get treatment and medication as well as medical consultations at low or no cost on their doorsteps	6. There are no appropriate monitoring systems to monitor the duty hours', process, services delivery and use of resources
	7. Poor communication can lead to a waste of resources, especially in times of emergency
	8. The lack of disaster risk management mechanisms and health committees are further significant weaknesses of the health facilities
Opportunities (O)	Threats (T)
1. There is an opportunity for public and private partnerships in the vicinity of the health facilities. Specialized human resources	1. Many of the health facilities in our sample are located in flood-prone areas and are vulnerable to damage from flooding

(continued)

Table 16.5 (continued)

Strengths (S)	Weaknesses (W)
and experts could be deployed or encouraged to provide services at nominal costs, and the community could benefit	
2. Humanitarian organizations are already working in the flood-affected areas/districts of KP and other provinces of Pakistan. This provides an opportunity for local and district governments to engage these organizations in providing missing facilities, capacity building and meeting health care standards in the health facilities	2. Many of the structures are very old and, at the time of construction, no elements were incorporated to make them safe and resilient to flooding
3. Services provided at rural health facilities are either less costly or free which is an opportunity for the impoverished local community	3. As the structures are old and constructed with traditional materials and methods they are unlikely to be resilient to natural hazards such as floods
4. There is an opportunity for external training to build up the capacity of existing medical staff to provide deliver higher quality emergency health services	4. There is a lack of social awareness among rural communities and of a sense of ownership and responsibility on the part of the community to maintain a healthy and clean environment in these facilities
5. The existence of humanitarian organizations able to identify the gaps in health service delivery and to address such gaps and needs	5. Social inequalities regarding the provision of primary health services to community members which are often disbursed through favoritism and nepotism, which further increases social disparities
6. There are good opportunities for new construction as the many/most health facilities have enough space to permit the construction of flood protection measures and to provide additional (currently missing) facilities	6. The existing health facilities have inadequate provisions for maintenance due to a lack of funds and resources. This poses a severe threat to their sustainability
	7. Other than health standards, most facilities do not have access to standby ambulance services and only provide primary health care to patients. For patients with a severe condition, the lack of ambulance services for moving them to a secondary or tertiary hospital may be life threatening
	8. Lack of coordination between the primary health centers and the secondary, tertiary facilities are a threat to human and material resources

Source Authors own construction based on the field survey findings, 2017

16.6 Conclusion and Policy Implications

The general expectations of the people in flood-prone areas are that primary health care services are always prepared and ready to deal with flood emergency situations in an efficient manner. The primary purpose of this research was to assess the preparedness level and evaluation of rural health centers in flood-prone districts of KP province. The current study used three preparedness pillars (structural, non-structural, and functional preparedness). Structural vulnerability elements are crucial if primary health care facilities are to withstand the adverse effects of flood disasters, whereas the non-structural and functional indicators are essential for them to continue their routine operations. The findings of this research reveal that rural health care centers in flood-prone districts of KP province are at risk with low performance on all three indicators. At times of flood, or other, disasters, primary health care facilities should be the main organizations able to offer an effective response. Their preparedness and ability to deliver timely emergency medical services should play a significant role in reducing death tolls. Yet this study's findings show that the preparedness level of the majority of the sampled primary health care facilities in the flood-prone districts of KP province is inadequate. To be prepared to meet the challenges of providing an efficient and adequate emergency response at times of disaster, it is imperative that a fundamental review of the preparedness of RHCs in KP (and other parts of rural Pakistan) is undertaken and that measures taken, and resources devoted, to ensure that these essential public services are properly equipped so that they can be both safe havens and fully operational during disasters.

References

Abaya SW, Mandere N, Ewald G (2009) Floods and health in Gambella region. Ethiopia: a qualitative assessment of the strengths and weaknesses of coping mechanisms. Global Health Action 2. http://dx.doi.org/10.3402/gha.v2i0.2019

AbouZahr C, Boerma T (2005) Health information systems: the foundations of public health. Bull World Health Organ 78–583

Adeola FO (2003) Flood hazard vulnerability: a study of Tropical Storm Allison (TSA) flood impacts and adaptation modes in Louisiana. Citeseer

ADPC (Asian Disaster Preparedness Center) (2009) Safe hospital: the key to deliver effective emergency medical services. Public Health in Emergency Team, Asian Disaster Preparedness Center. http://www.adpc.net/v2007/ikm/ONLINE%20DOCUMENTS/downloads/ADUMP/SafeHospitalKeytodelivereffectiveEMS.pdf

Ahrq Archive Tool for Evaluating Core Elements of Hospital Disaster Drills (2011) Retrieved on 20 Jul 2013. Retrieved from archive.ahrq.gov/prep/drill elements/drillelements 1.htm

Aitken P, Leggat P (2012) Considerations in mass casualty and disaster management. In: Michael B (ed) Emergency Medicine: An International Perspective, pp 145–182. http://cdn.intechopen.com/pdfs/31947/InTechConsiderations_in_mass_casualty_and_disaster_management.pdf

Akram M, Khan FJ (2007) Health care services and government spending in Pakistan. PIDE (Pakistan Institute of Development Economics) Working Papers: 32

Alexander KA, Sanderson CE, Marathe M, Lewis BL, Rivers CM, Shaman J et al (2015) What factors might have led to the emergence of Ebola in West Africa? PLoS Neg l Trop Dis. https://doi.org/10.1371/journal.pntd.0003652

Bagatell S, Wiese J (2008) The elite code grey team: a new model for residency preparedness and training in advance of a disaster. Am J Med Sci 336(2):174–178

Bissell RA, Pinet L, Nelson N, Levy M (2004) Evidence of the effectiveness of health sector preparedness in disaster response: the example of four earthquakes. Family Commun Health 27 (3):193–203

Braun BI, Darcy L, Divi C, Robertson J, Fishbeck J (2004) Hospital bioterrorism preparedness linkages with the community: improvements over time. Am J Infect Control 32(6):317–326. https://doi.org/10.1016/j.ajic.2004.01.003

Chan M (2014) Ebola virus disease in West Africa—no early end to the outbreak. N Engl J Med 371:1183–1184

Costello A, Abbas M, Allen A, Ball S, Bell S, Bellamy R, Lee M (2009) Managing the health effects of climate change. Lancet 373(9676):1693–1733

Farley JM, Suraweera I, Perera WL, Hess J, Ebi KL (2017) Evaluation of flood preparedness in government healthcare facilities in Eastern Province, Sri Lanka. Glob health action 1;10 (1):1331539

Galea S, Nandi A, Vlahov D (2005) The epidemiology of post-traumatic stress disorder after disasters. Epidemiol Rev 27:78–91

Gillam S, Yates J, Badrinath P (2007) Essential public health: theory and practice. Cambridge University Press, Cambridge, UK

Government of Pakistan (GOP) (2013) National disaster risk reduction policy. National Disaster Management Authority, Islamabad

Greenberg MI, Jurgens MS, Gracely EdJ (2002) Emergency department preparedness for the evaluation and treatment of victims of biological or chemical terrorist attack. J Emerg Med 22 (3):273–278. https://doi.org/10.1016/S0736-4679(02)00427-4

Greenough G, McGeehin M, Bernard S, Trtanj J, Riad J, Engelberg D (2001) The potential impacts of climate variability and change on health impacts of extreme weather events in the United States. Environ Health Perspect 109(2):191–198. https://doi.org/10.2307/3435009

Hatami H, Neisi A, Kooti M (2017) Functional, structural and non-structural preparedness of Ahvaz Health Centers against disasters in 2014–2015. Jundishapur J Health Sci 9(1):e34239

Hirabayashi Y, Mahendran R, Koirala S, Konoshima L, Yamazaki D et al (2013) Global flood risk under climate change. Nat Clim Change 3:816

Hsu EB, Thomas TL, Bass E B, Whyne D, Kelen GD, Green GB (2006) Healthcare worker competencies for disaster training. BMC Med Educ 6(19). http://dx.doi.org/10.1186/1472-6920-6-19

IFRC (International Federation of Red Cross and Red Crescent Societies) (2015) World Disasters Report 2015: Focus on local actors, the key to humanitarian effectiveness. Available at: https://ifrc-media.org/interactive/wp-content/uploads/2015/09/1293600-World-Disasters-Report-2015en.pdf. Accessed 08 Apr 16

IMU-H (Independent Monitoring Unit Health) (2016). Monitoring report on KP's health care facilities. Government Of Khyber Pakhtunkwa. Available at http://www.healthkp.gov.pk/index.php/publications/

Kaji AH, Lewis RJ (2006) Hospital disaster preparedness in Los Angeles County. Acad Emerg Med 13(11):1198–1203. https://doi.org/10.1197/j.aem.2006.05.007

Khan S, Kausar S, Ghani M (2017) Knowledge of disaster preparedness among nurses at two tertiary care hospitals in Lahore. Biomedica 33(1)

Mulyasari F, Inoue S, Prashar S, Isayama K, Basu M, Srivastava N, Shaw R (2013) Disaster preparedness: looking through the Lens of Hospitals in Japan. Int J Disaster Risk Sci 4(2):89–100. https://doi.org/10.1007/s13753-013-0010-1

Noordin MF, Othman R, Zakaria NA. (2011) Peopleware and heartware—The philosophy of knowledge management, research and innovation in information systems (ICRIIS). International Conference on 23–24 Nov

Olu O, Usman A, Manga L, Anyangwe S, Kalambay K, Solomon Woldetsadik NN, Hampton C, Nguessan F, Benson A (2016) Strengthening health disaster risk management in Africa: multi-sectoral and people-centred approaches are required in the post-Hyogo Framework of Action era. BMC Public Health 16:691. https://doi.org/10.1186/s12889-016-3390-5

OSPHD-Electrical Requirement for Health Care Facilities (2011) Retrieved on 20 July 2013. Retrieved from http://www.oshpd.ca.gov/FDD/Plan-Review/electrical.Pdf

PAHO (Pan American Health Organization) (2000) Principles of disaster mitigation in health facilities. Disaster mitigation series. http://www.paho.org/english/dd/PED/mit3-intro.pdf

PAHO/WHO (Pan American Health Organization/World Health Organization) (2003) Protecting new health facilities from natural disasters: guidelines for the promotion of disaster mitigation. http://www.preventionweb.net/files/629_10343.pdf

Phalkey R, Dash SR, Mukhopadhyay A, Runge-Ranzinger S, Marx M (2012) Prepared to react? Assessing the functional capacity of the primary health care system in rural Orissa, India to respond to the devastating flood of September 2008. Glob Health Action 5

Rattanakanlaya K, Sukonthasarn A, Wangsrikhun S, Chanprasit C (2016) A survey of flood disaster preparedness among hospitals in the central region of Thailand. Aust Emerg Nurs J 19:191–197

Reacher M, McKenzie K, Lane C, Nichols T, Kedge I et al (2004) Health impacts of flooding in Lewes: a comparison of reported gastrointestinal and other illness and mental health in flooded and non-flooded households. Commun Dis Public Health 7:39–46

RHIhub (Rural Health Information Hub) (2017). Rural emergency preparedness and response. Last reviewed on 29 June 2017. Retrieved from https://www.ruralhealthinfo.org/topics/emergency-preparedness-and-response

Seyedin H, Maher S, Mohammadi Yeganeh S, Asadi Lari M (2011) Qualitative and quantitative performance of equipment and non-structural vulnerability in selected public hospitals in tehran city during an earthquake. Q Sci J Rescue Relief 3(2)-0-0. http://jorar.ir/article-1-90-en.html

Shabir O (2013) A summary case report on the health impacts and response to the Pakistan floods of 2010. PLoS Curr 5

Shah AA, Ye J, Abid M, Ullah R (2017) Determinants of flood risk mitigation strategies at household level: a case of Khyber Pakhtunkhwa (KP) Province, Pakistan. Nat Hazards 88:415–430

Shah AA, Ye J, Abid M, Khan J, Amir SM (2018a) Flood hazards: household vulnerability and resilience in disaster-prone districts of Khyber Pakhtunkhwa Province, Pakistan. Nat Hazards 1–19

Shah AA, Ye J, Pan L, Ullah R, Shah SIA, Fahad S, Naz S (2018b) Schools' flood emergency preparedness in Khyber Pakhtunkhwa Province, Pakistan. Int J Disaster Risk Sci 9:181–194. https://doi.org/10.1007/s13753-018-0175-8

Shah AA, Shaw R, Ye J, Abid M, Amir SA, Pervez AKMK, Naz S (2019) Current capacities, preparedness and needs of local institutions in dealing with disaster risk reduction in Khyber Pakhtunkhwa, Pakistan. Int J Disaster Risk Reduct 34:165–172

Tachibanai T, Takemura S, Sone T, Segami K, Kato N (2005) Competence necessary for Japanese public health center directors in responding to public health emergencies. Nippon Koshu Eisei Zasshi 52(11):943–956

Tapsell S, Tunstall S (2006) The mental health aspects of floods: evidence from England and Wales. Flood hazards and health: responding to present and future risks. Earthscan, London

Tapsell SM, Penning-Rowsell EC, Tunstall SM, Wilson TL (2002) Vulnerability to flooding: health and social dimensions. Philos Trans R Soc London A: Math Phys Eng Sci 360:1511–1525

Tariq MAUR, Hoes OAC, Van de Giesen NC (2014) Development of a risk based framework to integrate flood insurance. J Flood Risk Manag 7(4):291–307. https://doi.org/10.1111/jfr3.12056

Terzic Z, Vukasinovic Z, Bjegovic-Mikanovi V, Jovanovic V, Janicic R (2010) "SWOT analysis: the analytical method in the process of planning and its application in the development of orthopedic hospital department. Srpski Arhiv Za Celokupno Lekarstvo—Journals 138:473–479

Thompson KM, Deisler PF Jr, Schwing RC (2005) Interdisciplinary vision: the first 25 years of the Society for Risk Analysis (SRA), 1980–2005. Risk Anal 25(6):1333–1386. https://doi.org/10.1111/j.1539-6924.2005.00702.x

Ullah S, Latif N, Alam AN, Zaman T (2017) Challenges of hospital preparedness in disasters in Balochistan. Pak J Public Health 7(1)

UN (United Nations) (2009) Hospitals safe from disasters. 2008–2009 world disaster reduction campaign. Technical Guidance, United Nations. http://www.unisdr.org/2009/campaign/pdf/wdrc-2008-2009-information-kit.pdf

UNDP (United Nations Development Program) (2004) Pakistan: national disaster management program. [Database, on internet, cited 2010 October 10]. Available from: http://www.undp.org/cpr/disred/documents/publications/corporatereport/asia/pakistan1.pdf

UNISDR (United Nations International Strategy for Disaster Risk Reduction) (2005) Hyogo framework for action 2005–2015: building the resilience of nations and communities to disasters. http://www.unisdr.org/files/1037_hyogoframeworkforactionenglish.pdf

UNISDR (United Nations International Strategy for Disaster Risk Reduction) (2009) UNISDR terminology on disaster risk reduction. UNISDR, Geneva

Waelde LC, Koopman C, Rierdan J, vid Spiegel D (2001) Symptoms of acute stress disorder and posttraumatic stress disorder following exposure to disastrous flooding. J Trauma Dissociation 2:37–52

Weathers A, Minkovitz C, O'Campo P, Diener-West M (2004) Access to care for children of migratory agricultural workers: factors associated with unmet need for medical care. Pediatrics 113(4):e276–e282

WHO (World Health Organization) 2013 Annex 9: PAKISTAN case study. Available at http://www.who.int/workforcealliance/knowledge/resources/MLHWCountryCaseStudies_annex9_Pakistan.pdf

Chapter 17
Public Health and Disaster Risk Reduction: Experiences from Vietnam

Tran Nu Quy Linh and Tran Thi Tuyet Hanh

Abstract Vietnam is located in Southeastern Asia, along with the Indochinese Peninsula. With a long coastline, diversified topography, and high percent of the population living in low-lying coastal regions, Vietnam is one of the most disaster-prone nations in the world. The country is frequently exposed to a wide range of recurring natural hazards such as tropical storms, floods, droughts, landslides, and forest fires. Typhoons and floods are the two most frequent and dangerous occurrences, causing numerous fatalities, economic loss and infrastructure damages as well as other short and long-term public health problems. With the rising in both frequency and intensity of natural disasters in the last decades under the impacts of climate change, the vulnerability of local communities has been more remarkable. Therefore, Vietnam has actively taken a number of measures to prevent, respond, and mitigate disasters' consequences. There are both valuable experiences and challenges in national disaster reduction efforts. This chapter describes climate characteristics and common natural disasters of Vietnam; analyzes the vulnerability of communities in disasters; synthesizes and assesses policies related to natural disaster reduction; as well as discusses valuable experiences and existing problems in disaster adaptation strategies of Vietnam.

Keywords Disasters · Risk reduction · Flood · Typhoon · Vietnam

T. N. Q. Linh
Ha Tinh, Center for Disease Control and Prevention, Ha Tinh, Hatinh Province, Vietnam
e-mail: linhtrannu@gmail.com

T. T. T. Hanh (✉)
Hanoi University of Public Health, Hanoi City, Vietnam
e-mail: tth2@huph.edu.vn

E. Ying Yang Chan and R. Shaw (eds.), *Public Health and Disasters*,
Disaster Risk Reduction, https://doi.org/10.1007/978-981-15-0924-7_17

17.1 Introduction

In Vietnam, natural disasters are a significant barrier to poverty reduction, economic growth, health improvement, and social stability. The Globa Climate Risk Index 2017 ranked Vietnam as the world's eighth most affected by climate variability and extreme climate events over the period from 1996 to 2015 (Kreft et al. 2016). The spreading of unplanned urban development, economic activities, and ecosystem degradation in Vietnam has been increasing the risks, vulnerabilities and consequences of natural disasters. The increases in magnitude and frequency of extreme climate events such as floods, storms, droughts, and flash floods lead to high numbers of deaths, injuries, diseases, and other significant public health issues. For example, over ten years (from 2001 to 2010), storms, floods, landslides, and droughts claimed 9500 deaths and caused asset damage equivalent to 1.5% annual GDP lost in Vietnam (Socialist Republic of Viet Nam 2011). In addition, extreme weather events may lead to enormous impacts to public health such as environmental pollution, spreading infectious diseases and worsening chronic conditions, as well as causing unemployment, poverty, and other social and political issues (The World Bank and GFDRR 2011). Health problems caused by natural disasters often are direct, such as drownings, missing and injuries, or indirect such as vector-borne diseases, waterborne diseases, chronic health conditions, mental health problems, effects of exposure to pollutants in floodwater, and food shortages (Alexander 2017). Damage assessment system in Vietnam has not sufficiently evaluated indirect or long-term consequences of natural disasters due to the lack of a consistent standard assessment method among localities. Most of damage assessment reports focused on direct human losses and economic damage while other public health issues directly or indirectly affected due to disasters were not paid enough attention (Navrud et al. 2012). While health consequences due to disasters are remarkable, the capacity of health care system in Vietnam, especially in rural and mountainous areas, is not adequate for responding (Van Minh et al. 2014). During storms, floods, and landslides, health facilities can be destroyed, and the healthcare services may be interrupted; therefore, the health system has been facing more challenges in disasters and extreme weather events due to the impacts of climate change and global warming. The adverse consequences of natural disasters in Vietnam are projected to be more serious in the next decades. This is a threat to poverty reduction and a barrier to achieving the Millennium Development Goals, including health goals.

17.2 Geographical Location and Climate Characteristics of Vietnam

The complex topography and climate characteristics of Vietnam make it one of the world's most disaster-prone nations. Vietnam is located in Southeastern Asia with 331,114 km^2 of the mainland, one million km^2 of the territorial sea and 3200 km of coastline, which is along the tropical monsoon belt. The climate of Vietnam, in general, is hot and humid. In addition, about 59% of total land and approximately 71% of Vietnam's population are vulnerable to floods and storms (The World Bank and GFDRR 2011). The population of Vietnam reached around 95 million in 2017 (The World Bank 2018), and the population density is around 232 people per square, ranked the second in Southeast Asia (after Singapore); especially in the Northern Delta, the population density is up to 1000 people per square (Aid Irish 2017). These features contribute to increasing the risks of deaths, injuries, and diseases as well as to exaggerate economic losses under the effects of extreme weather events and natural hazards.

The average annual temperature in Vietnam has been increasing 0.4 °C since 1960 and especially in the dry season in the southern region, the rate of increase was more rapid. Moreover, the number of high-temperature days and nights each year has increased significantly (Field et al. 2014). Although the number of cold days has decreased, cold weather spells have been more complex, difficult to forecast and significantly fluctuated year by year. Particularly in recent years, ice and snow occur more frequently in the high mountains (Tran et al. 2015). Extreme rainfall events have increased during recent decades, while annual rainfall total has shown an upward trend in the southern regions and a downward trend in the northern parts (Field et al. 2014). Rainfall in Vietnam is mainly due to monsoon circulations; thus, in the North and South areas, heavy rains often occur from May to October while in the central region, it appears from September to January (Aid Irish 2017). According to reports of the water level gauging stations along the coast and island of Vietnam, the water levels had slightly increased during the recent 50 years, with average 2.45 and 3.34 mm per year for sea-water and river-water level, respectively (Ministry of Natural Resources and Environment 2016). This would bring slow-onset climate change risks for many provinces along coastal line in Vietnam, which can have adverse impacts on the health sector at different levels. Health facilities may need to be relocated to higher areas and the health system will need to respond to health risks associated with sea-level rise, such as water-related diseases due to reduced drinking water supplies, mental health impacts due to relocated of houses and crops lost, and malnutrition among children. Recent projections have issued threats that Vietnam can face due to climate change in the future. According to USAID (2017), the annual mean temperature is projected to increase by 1–2 °C, and annual rainfall is projected to rise by 2–7% by the 2050s. The number of days and nights considered as "hot" is expected to increase in the future. Extreme precipitation variability is more likely to be wider between the rainy and dry seasons. By 2090, it is projected that total rainfall during heavy events, especially in northern areas will increase from 2 to 14%. Also, sea levels are projected to rise by 28–33 cm (The World Bank and GFDRR 2011). Projections of future changes in climate indicate that Vietnam is more likely to be suffering from extreme weather events and natural disasters in the next decades.

17.3 Main Types of Natural Disasters in Vietnam and Their Public Health Impacts

Vietnam is devastated by natural disasters that cause severe damages to lives and properties, as well as leads to serious public health issues. Typhoons and floods are the two most frequent and noteworthy natural hazards in Vietnam. Figure 17.1 indicated that floods and storms caused around 92% (8330 million USD) of total economic losses from four main natural disasters during the period from 1990 to 2012. Also, the data from the Japan International Cooperation Agency (2015) confirmed that storms and floods were two most dangerous hazards that caused 92% of the total affected people, 98% of total deaths, and 92% of economic losses.

17.3.1 Typhoons

With 3200 km long coastline, tropical storms and typhoons have become the most common disaster hazards in Vietnam, killing hundreds of people every year and seriously affect livelihood. Annually, there are average 6–8 typhoons depressions in Vietnam (Nguyen Van Thang 2010). In the period of 1954–2016, there were more than 550 typhoons and tropical affecting Vietnam, in which 36% of storms hit the Central region, 31% to the North and 33% to the South Central and the Southern areas (Japan International Cooperation Agency 2015). Typhoons often result in floods because they usually accompany heavy rains, strong winds, and high tides. Around 80–90% of Vietnam's population can be affected by storms (Japan International Cooperation Agency 2015). The consequences of storms are remarkable, including infrastructural destruction, land loss, mangrove forest, erosion of sea dykes, and saltwater intrusion. There is an increase in the number of annual occurring storms from five main storms (level 6–12) per year in the period of 1980s and 1990s to over seven major storms annually occurring in the 2000s

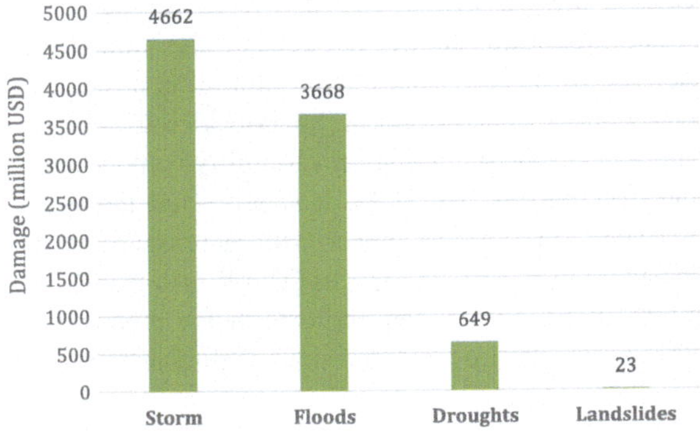

Fig. 17.1 Total economic losses due to four main natural hazards in Vietnam (1990–2012). *Source* Tran et al. (2015)

(National Centre for Hydro-Meteorological Forecasting 2015). Table 17.1 describes remarkable typhoons during the past 20 years (1997–2017), which caused high numbers of live losses, destroyed houses and economic damages. The health

Table 17.1 Some major typhoon from 1997 to 2017

Events	Year	Description	Consequences
Typhoon Linda	1997 (2–3, November)	– It was the strongest recorded in southern Vietnam for the last 100 years – Winds of 65 mph (100 km/h) – Strong rains (100–150 mm of rainfall)	– 464 deaths, 857 injuries, and 3218 missing – Damage totaled USD 385 million. 200,000 houses destroyed, 383,000 people homeless (UN Department of Human Affairs 1997)
Typhoon Xangsane	2006 (1 October)	– Heavy rains, sustained winds of 150–160 km/h – Heavy rains (400 and 846 mm of rainfall) – Affected 12 provinces of the central region	– 71 people died, 525 people injured, 1.3 million people were affected – The estimated loss was USD 443.2 million (International Federation of Red Cross and Red Crescent Societies 2006)
Typhoon Ketsana	2009 (29 September)	– Strong winds, rapid movement of up to 167 km/h, hit the coast of central provinces – Rapidly increasing water levels in rivers and resulting floods and landslides	– 163 people died, 11 people missing, over 258,000 houses were damaged – Economic loss: USD 785 million (both due to flooding and storm) (JICA 2009)
Tropical storm Son-Tinh (Storm No. 8)	2012 (28 October)	– Rapid and unpredictable movement – Strong winds, heavy rainfall and landed to 11 central and northern coastal provinces – Affected northern and central coastal provinces – Landslides occurred in northern provinces	– 4 people died, 6 others missing, 7 people injured, thousands of houses were damaged – Economic loss USD 473.1 million (Reliefweb 2012)
Typhoon Haiyan (locally named storm No 14)	2013 (10–11 November)	– Strong winds, heavy rains and flooding – Directly affected Quang Ninh and Hai Phong provinces, 13 other provinces were impacted	– Killed thousands and left widespread devastation in the Philippines but it weakened when landing in Vietnam – 13 people died, 81 people injured and near 800000 people were evacuated (UN Country Team in Viet Nam 2013)

(continued)

Table 17.1 (continued)

Events	Year	Description	Consequences
Tropical storm Mirinae	2016 (29 July)	– Strong winds, rapid movement – Directly affected Northern Vietnam	– 5 people died, 5 others missing, 1425 houses damaged – Estimated economic lost: USD 289 million (Reliefweb 2016)

Sources UN Department of Human Affairs (1997), International Federation of Red Cross and Red Crescent Societies (2006), JICA (2009), Reliefweb (2012, 2016), UN Country Team in Viet Nam (2013)

and economic consequences described in Table 17.1 only based on the data provided in the initial rapid assessments, while the long-term effects of these typhoons could be more serious.

17.3.2 Floods

Floods are the second devastating extreme weather events in Vietnam after storms. Over recent decades (2006–2016), there were more than 400 flood events in Vietnam, occurring mainly in the Central region (National Center for Hydro-Meteorological Forecasting 2017). Floods with major impact in the Mekong Delta occur every four to six years (Tran et al. 2015). Flooding is resulted from increasing water level in rivers and streams during and after heavy rains, storms, or typhoons in coastal areas. The annual flood season happens in different locations with the different time scales. For example, in the northern region, floods often occur from June to October, in South and Central Highlands from June to November, while in North Central of Vietnam from July to November (Tran et al. 2015). Due to varies in topography conditions, the flood risks in the North, the Central, and the South of Vietnam are different. In the North, thank to five hydropower plants on the Da River Basin, floods are significantly controlled. In the South region, heavy rains, storms, sea-level rise, and tides are main reasons for floods. The Central area is at the highest risk of floods because of narrow and steep terrain and fragmented by various rivers (Luu et al. 2017). The monsoon tropical climatic features combine to a long coastline (around 3200 km), a diversified river system with over 3450 rivers and streams, and high population density contributes to increasing the flood risk in Vietnam (Chau et al. 2014).

The health impacts related to floods may be direct during the events such as drownings, injuries, outbreaks of gastroenteritis, or indirect consequences after floods such as water-borne diseases, vector-borne diseases, chronic diseases, mental health problems, and malnutrition. The long-term health impacts of flooding events need to be more concerned than other injuries or illnesses because their consequences often continue for long time periods after floods. Table 17.2 shows typical

Table 17.2 List of severe floods in Vietnam from 1995 to 2014

Year	Affected regions	Main features	Consequences
September 1996	Northern Delta	The third biggest flood in Red River system of the last 100 years	– 60 deaths, most of the local dikes were broken – Estimated economic losses: USD 31.3 million
September 1999	Central of Vietnam (Quang Binh to Khanh Hoa provinces)	The highest flood level recorded so far	– 718 deaths – Estimated economic losses: USD 141.5 million
September 2000	Mekong Delta	The most serious historical flood (from 1900 to 2000) in Mekong Delta	– 448 deaths, 5 million people affected – Estimated losses: USD 134.7 million
August 2001	Mekong Delta	One-month flooding, leading to serious inundation, affected 10 provinces	– 539 deaths, 219 injured, over 6 million people evacuated – Estimated economic damages: USD 154.4 million
October 2010	North Central of Vietnam (from Nghe An Province to Hue city)	Heavy rains, deep inundation	– 143 deaths, 800 thousands people affected, thousands of houses flooded
October 2011	Mekong Delta	The flood remained for one month, leading to deep inundation	– 29 deaths, millions of people evacuated – Economic damages: USD 42.8 million
September 2013	Central of Vietnam (from Hue to Phu Yen Province)	The most serious flood from 1999 to now in the central region	– 45 deaths and missing. Over 200 thousand people affected

Source Modified from Tran et al. (2015) Copyright 2015 Vietnam Publishing House of Natural Resources, Environment and Cartography

floods in Vietnam for 20 years (1995–2014). Damages from these floods were not sufficient as they mainly based on quick assessment directly after floods.

17.4 Vulnerable Groups in Natural Disasters in Vietnam

The vulnerability of the community to a health risk often depends on population density, economic sustainability, food and water availability, household income, local environmental conditions, personal health status, as well as the quality and availability of health care system. However, in Vietnam, the lack of long-term datasets combining to the lack of stakeholders' awareness about the impacts climate change on health made the vulnerability assessment more challenges (Tran Thi Tuyet Hanh et al. 2018). Natural disasters affect communities in different levels from disrupting daily life to causing devastation, injuries and spreading diseases, also, some groups are more vulnerable than others. In general, poor people, ethnic minorities, elders, children, and pregnant women often are more vulnerable to the impacts of extreme weather events and disasters. Low-income households living in river banks, coastal zones, and low-lying regions tend to suffer the most from storms and floods (OXFAM and Actionaid 2011). Natural disasters exacerbate the gaps between poor and rich families as the rich decide to move away from disaster-prone areas and build a new life, while the poor often have no choice and they are left behind. Children also are at high risk of natural disasters. For example, from 1994 to 2006, 74% of fatalities in floods in Dong Thap Province in the South of Vietnam were children under six years old, and the majority of these lost came from low-income households because their parents had to work far from their families and did not have time to take a good care of the children (Birkmann et al. 2012). In addition, natural disasters contribute to extra gender inequalities because they create additional workloads for women who often have responsibilities for domestic duties (UN Vietnam and Oxfarm 2009). Ethnic minority communities in Vietnam also are considered as the vulnerable group in disasters because they may do not speak the King language fluently—the Vietnamese official language. Therefore, they may face difficulties in accessing early warning and communication information and they are often among the most neglected in relief programs. In terms of diseases in disasters, especially in floods, children under five years old and pregnant women are at high risk from water-borne diseases such as cholera, diarrhea, and shigellosis (UN Vietnam and Oxfarm 2009). Furthermore, farmers are more likely to be impacted by floods, storms, and droughts because the livelihood depends mainly on the natural environment (Tran et al. 2015). In Vietnam, moreover, the adaptive capacity of the health sector is an important determinant of disaster vulnerability for the community. The first and most updated assessment using a list of 33 indicators to assess the *level of exposure* to climate change, *level of health sensitivity*, and *level of adaptation capacity* of the Vietnam's health system showed that the level of vulnerability and adaptation to climate change during 2013–2017 were "*High risk*" and "*Very high risk*" (overall scored from 3.7 to 4.1 out of 5.0) (Tran Thi Tuyet Hanh et al. 2018). Health system in Vietnam, especially the primary health care, is facing many difficulties in responding to extreme weather events. Commune health stations, for example, often do not have enough equipment and medicine for first aid or treatment (especially water-borne disease outbreaks) during and after heavy rainfall and floods. Even in some extreme weather events, these facilities were severely damaged, isolated, and inaccessible. Therefore, adaptation solutions to minimize the vulnerability of the Vietnamese health system to climate change are urgently required in the coming decades.

17.5 Current Policies and Legislation Related to Disaster Risk Reduction and Public Health Programs in Vietnam

The Vietnamese Government recognizes the urgent need of adaptation strategies to national disasters, in which the government plays a central role in all these preventive programs while non-governmental organizations and communities play supportive roles. Disaster risk reduction has been integrated into national development strategies as it has been highlighted in the poverty reduction strategy and the 10-year socioeconomic development strategy. The National Committee on Climate Change, which was established in 2011, is responsible for maintaining and monitoring the progress in the implementation of legislation and adaptive activities at the national level, while the Ministry of Natural Resources and Environment is the Secretariat (Tran et al. 2015). Recently, in December 2018, the Vietnam Ministry of Health also approved the Climate Change Response Action Plan for the Health Sector for the 2019–2030 period, with the vision to 2050 (Bộ Y tế 2018).

The Vietnamese Government has issued a range of laws, legislation, and strategies related to natural disaster adaptation and mitigation. Some main policies include Decision No. 46/2006/QĐTTg about Master Planning for Search and Rescue to 2015, with vision up to 2020; Decision No. 172/2007/QD-TTg about National Strategy for Natural Disaster Prevention, Response and Mitigation to 2020; Law on Disaster Control and Prevention, 2013 (Law No. 33/2013/QH13); and Decision No. 1002/2009/QD-TTg about National Programme on Community-based Disaster Risk Management. In addition, there are policies related to climate change and sea-level rise adaptation including National Target Program to Respond to Climate Change; National Action Plan on Climate Change; Directive No. 35/2005/CT-TTg on implementation of the Kyoto Protocol of the UN Framework on Climate Change; the Climate Change Adaptation Action Program Framework for agriculture and rural development sector in 2008–2020. In addition, the government has developed other laws to create a legal corridor for natural disaster adaptation. These laws include Dykes Management Law (2006), Law on Environmental Protection (2013), Water Resources Law (2012), Fisheries Law (2003), Law on the Protection and Development of Forests, etc., and other legislation such as the Ordinance of Flood and Storm Control, the Ordinance of Exploitation and Protection of Water Irrigation Work (2001), and the Ordinance of Protection of Hydro-Meteorology.

These laws and legislations highlight principles of national disaster risk reduction, such as being active preparedness, expeditious response and efficient recovery. Active prevention is a key strategy for the community to prevent and mitigate different consequences including health lost caused by disasters. The regulation system on disaster management and reduction in Vietnam are quite comprehensive with laws, decrees, and decisions, which are implemented from national to local levels. However, the large numbers of laws and legislation related to national disaster reduction lead to difficulties in the implementation. For example, more than

150 laws, regulations and guidelines related to flood prevention cause overlapping in responded activities (The Ministry of Agriculture and Rural Development 2012). In addition, these laws and legislation are still separated by different types of hazards. Also, there are currently no regulations for some other important hazards such as freezing, damaging cold, and heat waves. Another limitation is that the law and legislation are not adequate nor enough detail to address specific social problems and concerns related to national disaster reduction. Therefore, local government and communities face challenges when these events occur. Another limitation in the policy system is that there is a lack of systematic coordination for adaptation to concurrent disasters (e.g., cyclones combine to floods, flash floods, and landslide).

17.6 Experiences from Disaster Risk Reduction and Health Adaptations in Vietnam

Recognizing current and future challenges under the impacts of climate change, the Vietnamese Government, non-governmental organizations and communities have been implementing and developing robust strategies to improve capacity for disaster risk reduction. This section introduces some experiences from disaster risk reduction and adaptation in the health sector in Vietnam.

The "Four on-the-Spot" Motto

In natural disaster reduction, the Vietnamese Government and communities have learnt various valuable lessons from practical experiences. The most current effective solution in natural disaster reduction is the "Four on-the-spot" motto, which has been formed and developed since the early 1970s (JANI—The Joint Advocacy Network Initiative 2010). The motto aims to encourage the participation of governmental agencies at all levels, non-governmental organizations, political and social organizations, as well as the communities. The phrase "on-the-spot" is often understood as relying mainly on the capacity and resources of the local authority and community to address local issues. The "four on-the-spot" motto includes *leadership on-the-spot, human resource on-the-spot, means and materials on-the-spot, and logistics on-the-spot* (JANI—The Joint Advocacy Network Initiative 2010).

- *Leadership on-the-spot* includes leadership from the local government, the political and social organizations, experienced people in disaster management and heads of households.
- *Human resources on-the-spot* include local rescue teams, military in the local area, political and social organizations, and healthy family members. These groups understand geographical characteristics and local context better than anyone else; therefore, when disasters occur, they are available to help local

people actively prepare, immediately respond to and quickly search and rescue victims.

- *Materials and means on-the-spot* include available equipment, facilities, and other resources such as moto-boats, canoes, rafts, and floating devices are used to respond to extreme weather events. For example, military, members of Women Union and Farmers' Association, and local people can prepare sand-bags, bamboo, and soil to proactively protect dykes or handle in the case if the dykes are broken (JANI 2010).
- *Logistics on-the-spot* include materials such as drinking water, food, energy for impacted communities during and after the natural disasters, especially prolonged floods where impacted areas are isolated from the outside (JANI 2010). For example, Typhoon Ketsana (26/9/2009) combined to long-days floods, some districts in Quang Nam Province was isolated for many days. However, the local community still had enough drinking water and food to survive without calling for external aid.

The motto has been highlighted in legal documents from the national level to local communes and households. It is clear that, when disasters caused by storms, floods, and landslides occurred, rescue actions can face lots of difficulties because impacted areas often are isolated due to transport system is destroyed, electronic is lost, and communication is not working. Therefore, the motto seems to be the most effective solution for disaster response (Tran et al. 2015). It can be implemented efficiently before, during and after disasters. Before a disaster, the motto shows its own effectiveness in proactive preparation such as trimming plants, strengthening houses, moving vulnerable people to safe places, preparing foods, drinking water, and some essential medicines. When disasters occur, the motto is manifested in emergency response, rescue affected people, and protecting the properties. After the disaster, the motto is necessary for a quicker recovery and stabilization.

Although "four on-the-spot" is one of the successful solutions for disaster reduction, it is mainly applied to storm and flood control. In addition, community-level decisions in disasters sometimes based on experiences rather than following guidelines (Luu et al. 2018). In order to effective adaptation to other types of disasters, it is necessary to have detail plans with specific guidelines for every type of disasters in critical regions based on specific contexts. In addition, local authorities and communities should be trained in rescuing skills and practiced those skills in emergency situations.

Safe Hospitals in Emergencies and Disasters

In the past decades, the health sector in Vietnam has been greatly affected by natural disasters. It is very important to maintain safe hospitals, especially during emergencies and natural disasters as these health facilities must be able to withstand and ensure fully operation at such time. WHO built a list of indicators to assess the vulnerability of hospitals in emergencies and disasters, in order to increase the resilience and safety as well as to ensure functional activities of these facilities. These indicators include structural, nonstructural, and functional components

(World Health Organization 2010). Structural indicators such as location, design, and structures need to be considered in order for buildings to withstand extreme weather events. Nonstructural elements include building plans, architectural indicators (windows, ceilings, and doors), medical and laboratory equipment, safety and security issues. They are the indispensable elements to maintain medical examinations and treatments. The functionality of hospitals such as site and accessibility, equipment and supplies, logistics system, transportation and communication system, and human resource, evaluation is crucial. This element helps the health services continue to work effectively in disasters and provide lifesaving medical care. For example, if the hospital is located near a major roadway with an adequate mean of transportation, close to political, cultural and economic centers, it would be more accessible in the emergency (World Health Organization 2010).

The Vietnamese Government and Minister of Health have been raising attention to the safe hospital. A number of training courses about hospital preparedness for the emergency have been implemented in order to improve hospitals responding capacity in disasters and extreme weather events. There are several assessments about the capacity of health facilities and hospitals to respond to emergency situations (Nhu and Trang 2011). Some strengths were highlighted from these assessments, for example, most hospitals had emergency management policies, procedures and guidelines; had the alternative source of electrical supply in emergencies, and the structural buildings were able to resist strong winds. However, many weaknesses of existing hospitals in Vietnam were reported; for instance, there were many major cracks on ceilings and walls in observed hospitals; the ceiling and door materials were not fired resistance, and there was a lack of health emergency planning. Identifying vulnerability and existing problems of health facilities in disasters are the initial step for the public health sector in Vietnam to better prepare and response to natural disasters as well as to improve the resilience of healthcare system in emergency situations.

17.7 Existing Problems from Disaster and Health Risk Reduction in Vietnam

The early warning system plays an important role in reducing human losses and economic damages cause by disasters. This system supplies information and warning about the change of extreme weather events from hours, days, weeks, months, or seasons, in order to help authorities and communities give decisions on immediate responsive actions as well as future preparedness (Brunet et al. 2010). It is clear that the early warning system plays a crucial role in disaster management, especially in the context of climate change. Vietnamese Government recently has more concern for early warning system of extreme weather events. The Law No. 33/2013/QH13 on Disaster Prevention and Control has highlighted the role of early warning in natural disaster management and reduction. Effective early warning significant contributes to reducing deaths, injuries, economic and livelihood damages due to disasters. However, this system requires modern technology, suitable policies and the comprehensive coordination of different levels of government. An early warning system should include four main elements: hazard identification, monitoring and prediction; risk assessment; the timely warning communication; preparedness and emergency response plan. In practice, although Vietnam has achieved considerable successes in implementing early warning in disaster management, this system still has some limitations and especially the development of early health warning system responding to extreme weather events and climate change is still at its early stage.

At present, the early warning system mostly based on public weather forecasting bulletins, which normally provide short-term warnings such as for heavy rainfall, cyclone, or landslide. The lag time for early warning weather and climate events is normally from 24 to 48 h. Such short-range warning often only enough to deploy emergency responses such as moving communities to safe places, rescue affected people, moving goods, and strengthening house. Natural resource and climate-dependent livelihoods such as forestry, agriculture, aquaculture, and fishing are still significantly damaged because they require a longer time to implement protect activities (Hansen et al. 2011). Although the government has paid attention to develop monitoring and forecasting system, the accuracy of weather event predictions still has some limitations due to uncertainties in the collected data and a lack of advanced technology (Tran et al. 2015). It is a fact that annual and seasonal forecasts often show averages and probability for a specific zone in a certain period; thus, they may not forecast weather events that occur unusually or unpredictably. For the health sector, the health vulnerability and adaptation capacity assessment was recently implemented under a project funded by ADB and based on this, the National Health Adaptation Plan was developed and approved on 24th December 2018 to respond to climate change in the country until 2030, with the vision to 2050. One of the adaptation options would be developing early health warning system based on the climate and health databases. In addition, building public health capacity to actively respond to climate change and extreme weather events is important for Vietnam as the country is among the most vulnerable countries being negatively impacted by climate change and sea-level rise.

17.8 Conclusion

Vietnam has been extensively impacted by natural disasters such as storms, floods, landslides, droughts, and earthquakes. It is an estimated that 71% of the population and 59% of total land are at risk of disasters. Extreme weather events are likely to be more frequent in the next decades, causing more damages to communities' health and livelihood, as well as impacting significantly on the economic and social development. Floods and storms are the two most devastating disaster hazards in Vietnam, leading to a huge number of deaths, injuries, diseases, economic damages, and other public health problems. Vietnamese Government has consolidated the legislative systems and implemented a number of measures to prevent, respond, and mitigate climate change's impacts. Although achieving some remarkable successes in natural disaster reduction, Vietnam still has some limitations in implementing strategies. Among strategies to strengthen the prevention and mitigation of disaster risk, "Four on-the-spot" motto and "Safe hospitals" are highlighted as effective solutions. These strategies promote the participation of local authorities, health facilities, social organizations, and local communities. However, Vietnam still faces the number of challenges such as developing and implementing effective early health warning system (e.g., in responding to dengue hemorrhagic fever, stroke, heat stress, injuries, etc. in extreme climate events and under the impacts of climate change). The government should pay more attention to consolidating legislation system, developing an effective community-based approach, and applying technology development in forecasting.

Addressing the negative impacts of climate change and extreme weather events to human health is especially challenging, but certain adverse health effects can probably be avoided if decisions are made in advance, such as identification of vulnerable populations and ensured access to appropriate public health preventive measures. In recent years, there have been several studies exploring the impacts of climate change and natural disasters on different health issues in certain areas in Vietnam. However, the need for scientific evidence on which to base health-related policies in responding to climate change and extreme weather events is still critical. Climate change health vulnerability and adaptation assessment was comprehensively implemented to provide evidence to inform the development of National Health Adaptation Plan for climate change, which identified different measures that the health sector, related stakeholders and the communities can address anticipated current and future climate threats to public health, especially in extreme weather events and natural disasters.

References

Aid Irish (2017) Vietnam climate action report for 2016. https://www.irishaid.ie/media/irishaid/allwebsitemedia/30whatwedo/climatechange/Vietnam-Country-Climate-Action-Reports-2016.pdf

Alexander DC (2017) Natural disasters: Routledge, Abingdon

Birkmann J, Garschagen M, Van Tuan V, Binh NT (2012) Vulnerability, coping and adaptation to water related hazards in the Vietnamese Mekong Delta. In: The Mekong Delta System (pp 245–289). Springer, Berlin

Bộ Y tế (2018) Quyết định phê duyệt Kế hoạch hành động ứng phó với biến đổi khí hậu của ngành y tế giai đoạn 2019–2030 và tầm nhìn đến năm 2050. (7562/QĐ-BYT). Hà Nội

Brunet G, Shapiro M, Hoskins B, Moncrieff M, Dole R, Kiladis GN, Morss R (2010) Collaboration of the weather and climate communities to advance subseasonal-to-seasonal prediction. Bull Am Meteor Soc 91(10):1397–1406

Chau VN, Holland J, Cassells S (2014) Institutional structures underpinning flood management in Vietnam. Int J Disaster Risk Reduct 10:341–348

Field CB, Barros VR, Mach K, Mastrandrea M (2014) Climate change 2014: impacts, adaptation, and vulnerability. Part B: regional aspects (vol 1). Cambridge University Press, Cambridge and New York

Hansen JW, Mason SJ, Sun L, Tall A (2011) Review of seasonal climate forecasting for agriculture in sub-Saharan Africa. Exp Agric 47(2):205–240

International Federation of Red Cross and Red Crescent Societies (2006) Vietnam: Typhoon Xangsane. Retrieved from https://reliefweb.int/sites/reliefweb.int/files/resources/61ACD1CCF8165A1B4925721F001FCE8D-ifrc-vnm-06nov.pdf

JANI—The Joint Advocacy Network Initiative (2010) Four on-the-spot motto in disaster management. JANI, Vietnam

Japan International Cooperation Agency (2015) Country report. Vietnam

JICA (2009) Assessment report on response to Typhoon Ketsana in the central region. Retrieved from https://reliefweb.int/sites/reliefweb.int/files/resources/22C22BBA4C8122CD49257696001A6A4D-Full_Report.pdf

Kreft S, Eckstein D, Melchior I (2016) Global climate risk index 2017: Who suffers most from extreme weather events? Weather-related loss events in 2015 and 1996 to 2015. Germanwatch Nord-Süd Initiative eV

Law on Natural Disaster Prevention and Control (2013) No. 33/2013/QH13, dated 19 June 2013

Luu C, Von Meding J, Kanjanabootra S (2017) Analysing flood fatalities in Vietnam using national disaster database and tree-based methods

Luu C, Von Meding J, Kanjanabootra S (2018) Flood risk management activities in Vietnam: a study of local practice in Quang Nam province. Int J Disaster Risk Reduct

Ministry of Natural Resources and Environment (2016) Climate change and sea level rise scenarios for Viet Nam—summaries for policy makers. Retrieved from http://vihema.gov.vn/wp-content/uploads/2015/12/02.-Tom-tat-Kich-ban-BDKH-va-NBD-cho-VN_2016-Tieng-Anh.pdf

National Centre for Hydro-Meteorological Forecasting (2015) Statistics of annual major storms struck Vietnam between 1961 and 2014

National Center for Hydro-Meteorological Forecasting (2017) National Disaster Risk in Vietnam in 2006–2016 and forecasting and warning system. https://www.apec-epwg.org/public/uploadfile/act/f15e3a390421e8a5719bb2c859049604.pdf

Navrud S, Huu Tuan T, Duc Tinh B (2012) Estimating the welfare loss to households from natural disasters in developing countries: a contingent valuation study of flooding in Vietnam. Glob Health Act 5(1):17609

Nguyen Van Thang (2010) Nghiên cứu ảnh hưởng của biến ổi khí hậu đến các điều kiện tự nhiên, tài nguyên thiên nhiên và đề xuất các giải pháp chiến lược phòng tránh, giảm nhẹ và thích nghi, phục vụ phát triển bền vững kinh tế xã hội ở Việt Nam. Mã đề tài KC.08.13/06-10

Nhu HV, Trang DTH (2011) Safe hospitals in emergencies and hospitals' resilience to climate change in Vietnam. http://www.wpro.who.int/vietnam/publications/safe_hospitals_hospitals_resilience_to_CC_assessment_report_en.pdf

OXFAM, Actionaid (2011) Participatory poverty monitoring in rural communities in Vietnam. http://oxfamblogs.org/vietnam/wp-content/uploads/2012/05/rural-poverty-monitoring-report_round-4_en_low-res.pdf

Reliefweb (2012) Storm Son Tinh leaves four dead. Retrieved from https://reliefweb.int/report/viet-nam/storm-son-tinh-leaves-four-dead

Reliefweb (2016) Vietnam braces for storm Mirinae. Retrieved from https://reliefweb.int/report/viet-nam/vi-t-nam-braces-storm-mirinae

Socialist Republic of Viet Nam (2011) The National strategy on climate change (was issued by Prime Minister Nguyen Tan Dung in Decision 2139/QĐ-TTg on December 05, 2011). Retrieved from http://chinhphu.vn/portal/page/portal/English/strategies/strategiesdetails?categoryId=30&articleId=10051283

The Ministry of Agriculture and Rural Development (2012) Báo cáo Đánh giá tác động của Dự thảo Luật Phòng tránh giảm nhẹ thiên tai.Bộ Nông nghiệp và Phát triển Nông thôn, Hà Nội

The World Bank (2018) Vietnam overview. Retrieved from http://www.worldbank.org/en/country/vietnam/overview

The World Bank, GFDRR (2011) Vietnam: vulnerability, risk reduction, and adaptation to climate change. http://sdwebx.worldbank.org/climateportalb/doc/GFDRRCountryProfiles/wb_gfdrr_climate_change_country_profile_for_VNM.pdf

Tran T, Neefjes K, Ta T, Le N (2015) Vietnam special report on managing the risks of extreme events and disasters to advance climate change adaptation. In: Vietnam Publishing House of Natural Resources, Environment and Cartography: Institute for Hydrology, Environment and Meteorology (IMHEM) and UN Development Programme (UNDP): Hanoi, Vietnam

Tran Thi Tuyet Hanh, Ha Van Nhu, Hoang Van Minh, Le Thi Thanh Huong, Nguyen Thi Trang Nhung, Tran Mai Kien, …, Tran Thi Van Anh (2018) Health vulnerability and adaptation assessment in Vietnam. Hanoi

UN Country Team in Viet Nam (2013) Viet Nam: Typhoon HaiYan Situation Report No. 1 (as of 11 November 2013. Retrieved from https://reliefweb.int/report/viet-nam/viet-nam-typhoon-haiyan-situation-report-no1–11-November-2013

UN Department of Human Affairs (1997) Vietnam Typhoon Linda Situation Report No. 4. Retrieved from https://reliefweb.int/report/viet-nam/vietnam-typhoon-linda-situation-report-no4

UN Vietnam, Oxfarm (2009) Responding to climate change in Vietnam: Opportunities for improving gender equality. https://policy-practice.oxfam.org.uk/publications/responding-to-climate-change-in-vietnam-opportunities-for-improving-gender-equa-112555

USAID (2017) Vietnam: Climate change risk profile. https://www.climatelinks.org/sites/default/files/asset/document/2017_USAID_Vietnam%20climate%20risk%20profile.pdf

Van Minh H, Anh TT, Rocklöv J, Giang KB, Trang LQ, Sahlen K-G, Weinehall L (2014) Primary healthcare system capacities for responding to storm and flood-related health problems: a case study from a rural district in central Vietnam. Glob Health Act 7(1):23007

World Health Organization (2010) Safe hospitals in emergencies and disasters: structural, non-structural and functional indicators. http://www.wpro.who.int/emergencies_disasters/documents/SafeHospitalsinEmergenciesandDisastersweboptimized.pdf

Chapter 18
Health Emergency and Public Involvement in the Philippines: Syndromic Surveillance Efforts and System Integration

Maria Regina Justina E. Estuar, Kennedy E. Espina, Christian E. Pulmano, Charles Ventura and Roy Miguel G. Romarate

Abstract In a country situated within the typhoon belt, the Philippines experience an average of 28 typhoons in a year with a minimum of two major typhoons causing great damage and loss of lives. Acknowledging that disaster preparedness, response, and rescue require a multidimensional approach, emergency clusters recognize the need for an ICT platform that will consolidate disaster information coming from various sources including the public. The Philippine government, through the Department of Health (DOH) and the Philippine Health Insurance Company (PHIC), has mandated all rural health units to select and use an electronic medical record (EMR) system for digitizing health records. The digitization of health records in primary care is the first step in achieving universal health care for the country at the same time providing a more efficient way of accessing health records during emergencies. This chapter discusses efforts and experiences in the development and deployment of eHealth systems at three levels: (1) digitizing health records (SHINE OS+), (2) syndromic surveillance (FASSSTER), and (3) seamless integration with a disaster management system (HDDX), all of which provide a preventive approach in health emergencies and extreme disasters. SHINE OS+ is a web- and mobile-based electronic medical and referral system that is used by rural health units to digitize and submit health data for Primary Care Benefit (PCB) program and eClaims. FASSSTER is an online syndromic surveillance tool that collects information from electronic medical records, other eHealth systems, and social media to develop a geospatial disease forecast model. However, the lack

M. R. J. E. Estuar (✉) · K. E. Espina
Ateneo de Manila University, Loyola Heights, Quezon City, Philippines
e-mail: restuar@ateneo.edu

C. E. Pulmano
Ateneo Block Chain Laboratory, Ateneo de Manila University, Quezon City, Philippines

C. Ventura · R. M. G. Romarate
Factset Philippines, Inc, Mckinley Hill, Taguig City, Philippines

E. Ying Yang Chan and R. Shaw (eds.), *Public Health and Disasters*, Disaster Risk Reduction, https://doi.org/10.1007/978-981-15-0924-7_18

of a unified disaster management system that provides health-related information motivated the development of HDX (Health Disaster Exchange) comprising of API endpoints to integrate health information in disaster management systems. The chapter ends with perspectives on use of ICT for health emergency and disaster risk management practices and how the public provides both personal and social contributions to zero casualty.

Keywords Syndromic surveillance · Health security systems · ICT in disaster · Interoperability in systems · Semantic API

18.1 Understanding Health Emergency Within the Context of a Developing Country

The 1991 Philippine Local Government Code has provided for a devolved approach in governance of municipalities and cities of the Philippines, with national government providing additional grants to local government units to create independent systems and processes in delivery of services which includes health and disaster management and mitigation (World Bank 1994). This approach has proven to be beneficial at the local level where communities make use of normative and cultural approaches in rescue and response. However, at a larger scale, when the predicted effect and damage of an incoming disaster are considered high risk, information exchange from communities to national agencies and vice versa becomes more relevant. The multidimensional value of human security data, however, poses a challenge specifically in identifying to which the information should be sent. In most cases, the immediate information needed is about rescue and response, and in cases, where there is a need for medical treatment, access to health-related information is of high priority. During disasters and extreme emergencies, in a developing country with limited access to high-valued communications, the use of standard information and communication technologies including mobile- and web-based systems is best managed with the use of information communication technologies (ICT). These systems can facilitate the real-time flow of information between and among response clusters including camp management and coordination, education, emergency telecommunications, food and non-food, international humanitarian relations, law and order, logistics, management of dead and missing, psycho-spiritual integration, protection, search, rescue (SSR) and retrieval, and water, sanitation and health (WASH). Information exchange is important. At present, the interoperability between and among systems remains weak (Raza 2018).

18.1.1 Contextualizing Disaster Management Governance in the Philippines

The Philippines is geographically situated within the typhoon belt allowing the country to be a victim to at least one destructive typhoon every year. In 2009, Republic Act 101211[1] was enacted into law just a few months before Typhoon Yolanda caused massive destruction in the country. Known as the Presidential Risk Reduction and Management Act, the RA 101211 provides a comprehensive approach to disaster risk reduction and management, requiring the formation of DRRM groups comprising of government agencies and civil society from the national level to the city or municipality level. In 2015, the Philippine government, through the National Disaster Risk Management Council (NDRRMC) of the Office of Civil Defense (OCD), formed the Rapid Emergency Telecommunications Team (RETT) or more commonly called ICT Bayanihan with the aim of using ICT to facilitate in the communication and information requirements of the different response clusters (Memorandum Circular 43s 2016). Local government units as well as national agencies are trained on incident command system (ICS) as a model to manage and mitigate effects of disasters. NDRRMC Memo Circular 4 (2012) provides guidelines in the use of ICS under the Philippine Disaster Response, Rescue and Mitigation (PDRRM) systems for both natural and human-induced hazards. This required the formation of Incident Management Teams (IMT) at the local, provincial, regional, and national levels. Moreover, in 2016, NDRRMC Memo Circular 43s. 2016 required that coordination of within members of the IMT be interoperable to facilitate in making policy decisions, establishing priorities, resolving critical issues, mobilizing and tracking resources, and in collection, analysis and disseminating situational reports and information (Memorandum Circular 43s 2016). In July 2018, the president of the Philippines approved the creation of a department that will oversee the management as well as improve disaster response efforts in the country.[2]

18.2 Context of PH in Digitization of Health Records and the Story of SHINE OS+

Access to health record during rescue and response especially for those who need treatment and special care in evacuation camps is vital to ensure proper treatment. With emphasis on the coordination of response and rescue during disasters and emergencies, the need for health information has not been made central to most

[1]Philippine Republic Act of 10121. http://www.ndrrmc.gov.ph/attachments/article/45/Republic_Act_10121.pdf.

[2]Peralta J. http://cnnphilippines.com/news/2018/07/11/Duterte-Cabinet-disaster-department-NDRRMC-SONA.html.

disaster management systems. Traditional disaster management systems provide situational reports, risk assessment, estimate of damages and deaths, and assessment of human security. Non-traditional systems include empowering citizens to contribute real-time disaster information. For these systems, there is a need to have a middle process of report validation. Social media services have provided a free alternative for the public to contribute disaster information as well as contribute to information verification. However, there is still a need to organize these efforts such that communities will only access one system for all information requirements.

Though there is an awareness of the need for use of ICT to facilitate the flow of information, including health information, adoption of these systems at the local level is far and few. The digitization of health records, however, is a different story. First presented in 2011 and updated in 2014, the Philippine eHealth strategic framework has been the guide efforts related to universal health care. Prominent are eHealth systems and solutions which aim to provide a more efficient alternative to the manual and paper-based systems. It is envisioned that by 2020, healthcare services as well as health information are more accessible and shared securely to improve the quality of care that every Filipino deserves to have.

An electronic medical record (EMR) is characterized as electronic facility-based recording of legal medical records. Through the use of ICT, paper-based records are transformed into digital records, which can help health practitioners in easier management of patient data, report generation, and patient health and clinical information tracking. They are good sources of health data as they can be very comprehensive in terms of collecting patient and consultation history. EMRs are usually handled by health practitioners (physicians, nurses, midwives, etc.) and as such give the information it generates the capability to be the gold standard for knowing the health status of different areas in a country. The shift to digital-based recording gives the country the potential to produce comprehensive electronic health records for each patient. With standards and interoperability mechanisms in place, various EMR systems can participate in health information exchanges and provide a single point of access for patient clinical and health data.

The most basic of all eHealth system solution, therefore, is providing rural health centers with access to affordable electronic medical record systems. Notwithstanding that an electronic medical record (EMR) is the basic unit of information needed during health emergencies and disasters, data from EMR is also necessary for community profiling as well as benefit claims. In the Philippines, the Department of Health has required all rural health units to use an electronic medical record with full implementation beginning 2017. Among several accredited providers of EMRS include iClinicsys (Integrated Clinic Information System by the Department of Health), Community Health Information Tracking System by the National Telehealth Center (CHITS), Wireless Access for Health (WAH), Segworks, eHATID, and SHINE OS+. As of the fourth quarter of 2017, these EMRs cover almost 50% of the underserved population.

Secured Health Information Network and Exchange (SHINE) OS+ has its roots in its first version in 2011, when smart communications deployed SHINE, a free electronic medical record and referral system, primarily to contribute to the eHealth roadmap of the country. Aside from the web-based system, a mobile version running on Nokia platform was also tested. In 2013, SHINE was transformed into SHINEOS+ through the partnership between smart communications and Ateneo Java Wireless Competency Center (AJWCC). SHINEOS+ was designed as an open-source electronic medical record, with its source code available for use by the developer community. Designed as a referral system, the SHINEOS+ architecture (Fig. 18.1) addresses the need for on-demand access to health record as well as provision of health services.

SHINEOS+ runs in several platforms including cloud-based platform (software as a service) for the Internet ready communities and offline platform for communities still challenged with the Internet connectivity. SHINEOS+ mobile is designed for doctors on the go, to be able to access health information as well as provide remote consultation when necessary. MySHINE is a patient-centric mobile application allowing the patient to have access to health data as well as approve sharing

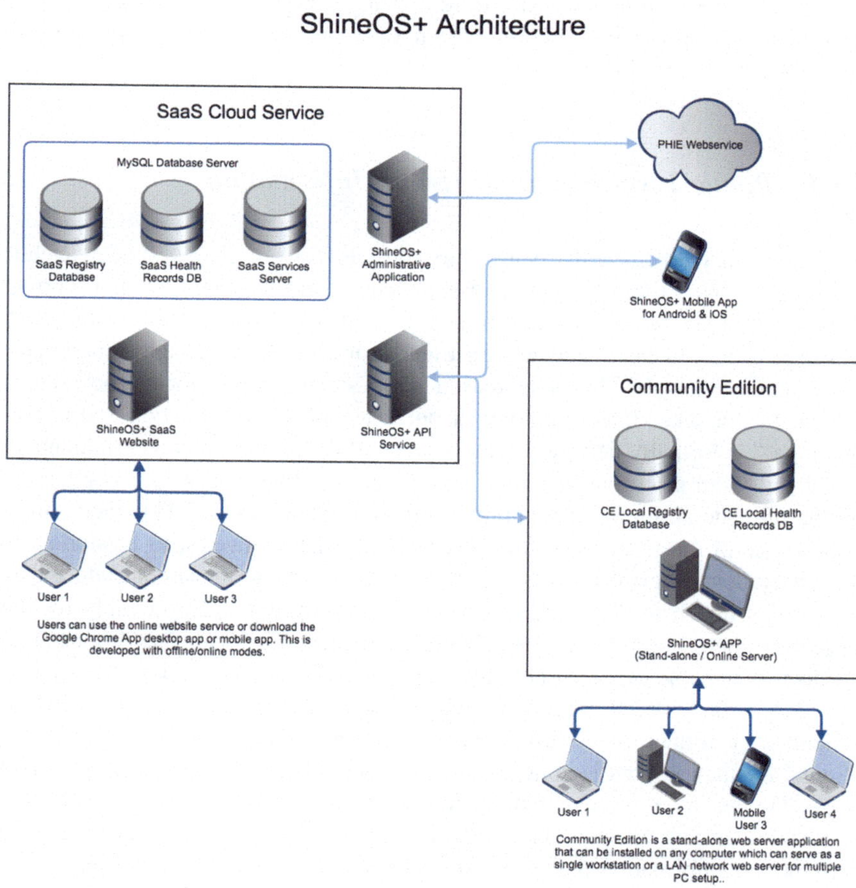

Fig. 18.1 SHINEOS+ enterprise architecture

of health records to other systems when necessary. The SHINEOS+ architecture provides a seamless ecosystem for flow of information necessary during health emergencies. At the individual level, this is sufficient. However, when health information is required at a national level, especially during major calamities and disasters, there is a need to incorporate health information in surveillance systems.

18.3 Context of Syndromic Surveillance and the Story of FASSSTER

From a public health perspective, aggregated health information from EMRs provides a landscape view of the health status in a particular area. With having more than one EMR provider, sharing of EMR data in aggregate form, therefore, provides a larger perspective on public health. Beyond the electronic medical record, other sources of health data coming from existing surveillance systems as well as social media data provide a multidimensional perspective in understanding health behaviors and spread of disease.

18.3.1 Public Involvement on Health Information

Social media is a tool currently being enjoyed by billions of people. It is an avenue where people share their thoughts and experiences, mostly based on their current situation. Through the use of micro-blogging (short and limited media) sites, such as Twitter, users can easily update their friends (or followers) through a single post about their current state. This is the reason why Twitter is a potentially good source of latent health data. Infodemiology, a portmanteau of "information" and "epidemiology," is formally defined as "the science of distribution and determinants of information in an electronic medium, specifically the Internet, or in a population, with the ultimate aim to inform public health and public policy." This then allows for the usage of electronic data from the internet, such as tweets, to be studied to arrive at epidemiological profiles for a population. The term was introduced by Gunther Eysenbach in 2009 and is still used specially today. Specifically for the Philippines, there are several ways Twitter is being utilized for health care with physicians being able to reach out to their clinical audience through Facebook pages (See: *Endocrine Witch* Facebook page), and researchers using collected Infodemiological data for syndromic surveillance as an example.

Social media text data is considered as unstructured, which makes it harder to process. Unlike structured data with distinct data types and prescribed formats, text data such as tweets are context-based, meaning further processing is needed to determine the information that it could generate. In terms of the collected health tweets, initial "filtering" of the context is done by searching for possible health-related keywords within the tweets. For example, looking for the word "sick" in a tweet helps in filtering down to a collection of tweets that *may* relate to having a fever. Although this is the case, the word "sick" may be used in multiple contexts as well, such as its usage in a colloquial manner (e.g., "this new song is sick!"). Another characteristic observed is the availability of two prominent languages, namely the English and the Filipino languages. The Philippines is known to have more than a hundred active languages being used by people in different parts of the country. Being considered as the official languages in the Philippines, Filipinos tweet more often in both English and Filipino—and in some cases, mixing both (informally known as *Taglish*). These characteristics (tweets being unstructured and being bilingual) make it hard to create algorithmic classification models that focus on specific contexts. In this regard, separate models for the two languages were created for the classification of health-related tweets.

The Ateneo Java Wireless Competency Center (AJWCC) is a research laboratory that collects health data from Twitter and News Sites in the Philippines. Through the use of the publicly available Twitter Streaming API, a tweet collector is set up to *listen* to mentions of infodemiological terms. In terms of the collection, many of the collected tweets isolated in the Philippines are in the Filipino language (e.g., "*Ang sakit ng ulo ko*" ["My head hurts"], "*Magkakaubo ata ako*" ["I think I'm going to have cough"], etc.). These are some tweets that people post nonchalantly in the social media platform. For a user, these are "harmless" tweets just to update their followers, but collectively—with the proper tools for data analysis and visualization—a concentration of these tweets may signify a change in the health status of a population in an area. Questions such as "*Is it possible that there is an outbreak in Quezon City, since there was a sudden surge of people tweeting they're having fever there?*" may arise. Twitter then becomes a significant source of information for decision makers in the local government and even for national agencies such as the Department of Health.

Although the tweets being collected can help generate useful information, it is important to note that the demographics of Twitter does not capture the entire population of the Philippines. It is estimated that people who are active in Twitter come from the middle to upper class and do not come from the lower income class. Although the Philippines is one of the countries known for spending the most number of hours in social media, Twitter is not the Web site where many Filipinos spend their time on. Due to promotions by telecommunications companies in the Philippines, many smartphone users are given free access to Facebook, which makes Facebook the more obvious option as a social media platform for many.

Added to this, there is also a limitation on the current search keywords being used for the collection of tweets. The list though is constantly updated to make sure that the list is as comprehensive as possible. When the tweets are visualized and observed on a map, it could be seen that there is a concentration of tweets in the Metro Manila area, the country's capital where the Filipino language is widely used. This shows that there is a bias in the collected tweets leaning toward the Metro Manila users. Being a country with more than a hundred known languages, it is a challenge to compile a list of relevant keywords to listen to, which could help in removing the previously mentioned bias. This limitation can be addressed by providing different translations of relevant keywords to cater to the different regions and provinces in the Philippines. Moreover, because of the diversity of languages in the Philippines, there are only two (2) classification algorithms that are currently developed, namely for the Filipino and English languages. The classification algorithm is developed to determine which tweets are infodemiological (health-related) and which are not.

Once the tweets are collected and processed, the FASSSTER web application *(Feasibility Analysis of Syndromic Surveillance Using a Spatio-Temporal Epidemiological ModeleR)* then accesses the tweets from AJWCC's tweet server through APIs. FASSSTER is an epidemiological web application that mainly forecasts and visualizes spreads of diseases in the Philippines through syndromic surveillance. Forecasts are made through the creation of disease models (dengue fever, measles, typhoid fever) using IBM's *Spatio-Temporal Epidemiological Modeler (STEM)*. STEM uses the compartmental modeling approach in creating the disease models, and syndromic surveillance then comes in by integrating data coming from different systems into the disease models.

FASSSTER is a three-year project (2016–2019), with the first two years dedicated into building disease models and the web application, as seen in Fig. 18.2. During the first year, base models for dengue fever, measles, and typhoid fever were created. The three diseases were chosen to be modeled for the reason that they differ in modes of transmission—vector-based (dengue fever), contact-based (Measles), and food-borne (typhoid fever). And because of this difference, the dynamics of the disease models were also different. For example, for dengue fever, the population of vectors (*Aedes aegypti mosquitoes*) were taken into consideration, and for typhoid fever, the chronic carriers were also included. Once the base models were done, for year two, the team dedicated their time in collecting new syndromic surveillance data from different systems, particularly from *electronic medical record* systems and social media (Twitter and News sites). For the final year, the deployment of the system will be prioritized while developing new disease models are also being created.

FASSSTER's web application was designed with the help of the epidemiologists from a regional office of the Department of Health (DOH). One of the main features is the dashboard (see Fig. 18.3), which the users will see first when they log into the system. The dashboard presents a summary of the data that the system currently has the number of the disease models uploaded and syndromic data statistics. Added to this, multiple charts were placed in the dashboard for quick visualization (STEM disease forecast, historical EMR data, historical syndromic surveillance data). As consulted from the epidemiologists, they prefer to also have data represented in a line chart added to the map view (see Fig. 18.4).

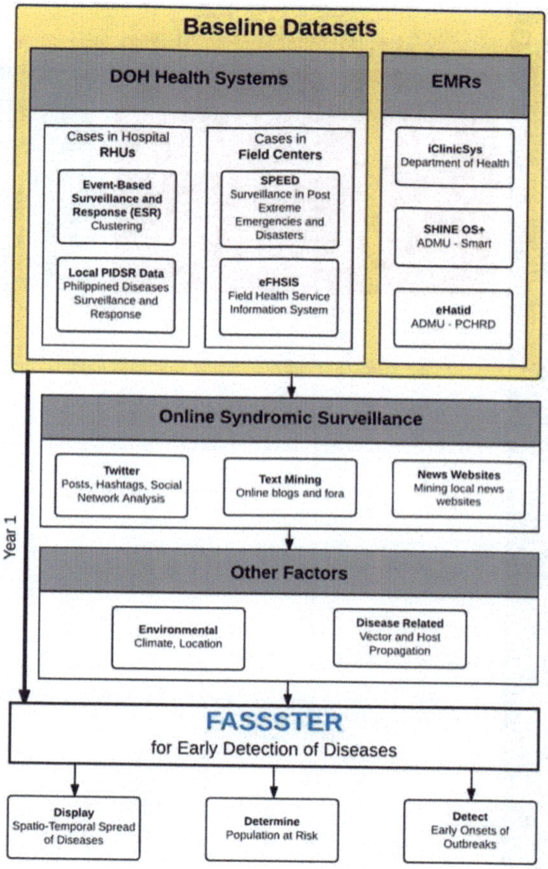

Fig. 18.2 FASSSTER framework

18.3.2 Stories of System Integration Efforts: Application of Interoperability

One of the more challenging aspects of developing ICT systems is ensuring that it is designed to be interoperable with other systems. Its relevance is seen when data from one system needs to be shared to another system to ensure more reliable and accurate information. There are several approaches to interoperability solutions, most standard of which is to establish an information exchange or data exchange by developing an interoperability layer through application programming interface (APIs) that allows other systems to extract information from another system. This section discusses two examples of interoperability which is used for syndromic surveillance.

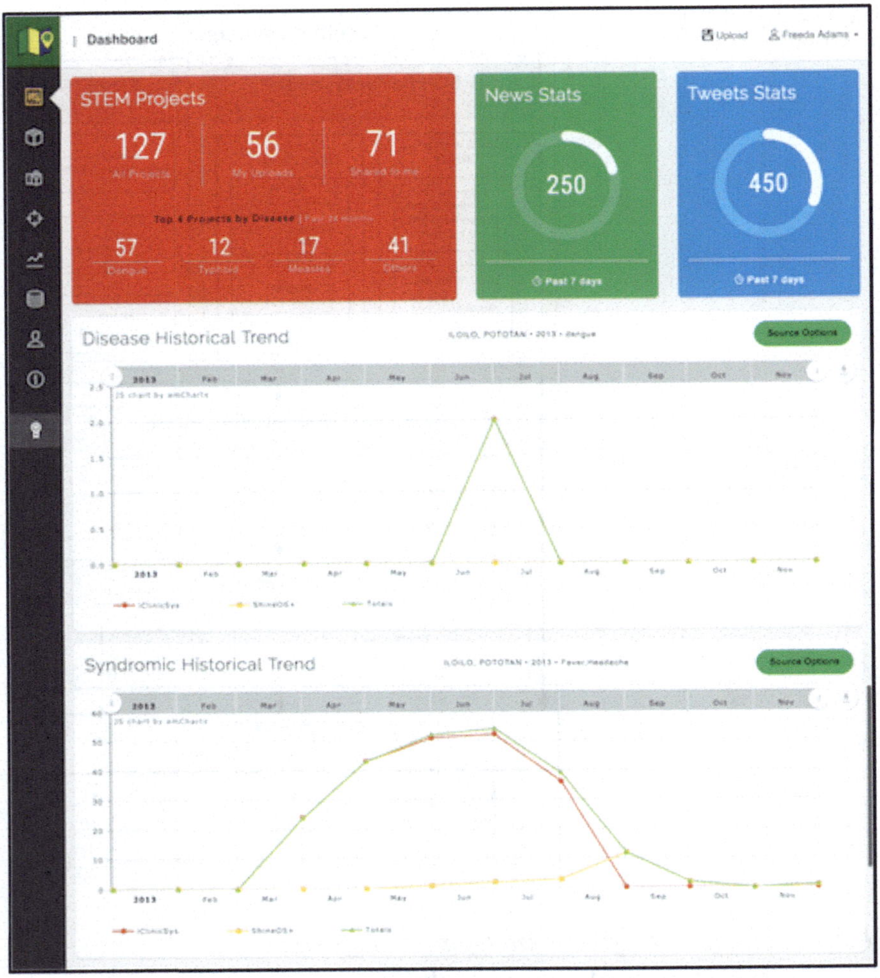

Fig. 18.3 FASSSTER dashboard

18.3.2.1 Integrating Tweets in FASSSTER

As mentioned in the previous section, the tweets collected mainly are in the Filipino and English languages, the two official languages in the Philippines. These tweets then show the *pulses* of the health status in the Philippines. Being in a social media platform, the tweets are posted by users in a nonchalant manner. This means that the users tweet without the direct intention to actually report their current health status to any governing bodies, such as the Department of Health (DOH). This then makes Twitter a *complementary* source of data for existing health systems, such as the DOH's *Philippine Integrated Disease Surveillance and Response (PIDSR)* system. Rather than a replacement for the systems that are already in place, the tweets

Fig. 18.4 FASSSTER map view

posted and collected give the DOH an additional data source that would potentially cover the population who decided to not get medical treatment in facilities, where their records should be recorded in the PIDSR system or in an EMR system.

The tweets are collected through a dedicated tweet server at the Ateneo Java Wireless Competency Center (AJWCC). A tweet server is constantly collecting tweets using keywords that relate to three specific diseases, namely dengue fever, typhoid fever, and measles. Some keywords that were used include the symptoms of the diseases (“*rashes,*” “*fever,*” etc.) in both Filipino and English languages. Added to this, indirect terms such as medicines relating to the diseases such as prescription medicines (“*Biogesic*” [a popular *Ibuprofen* brand]) and drugstores (“*Mercury Drug*” [the leading drugstore in the Philippines]) were also included. Using an R script, the Twitter streaming API is accessed to listen to tweets in real time and save the collected data to local files in the server. A script that runs every midnight then collates all the tweets and processes them to identify which ones are “infodemiological” (health-related) or not (Espina and Estuar 2017). This is done using a trained support vector machine classification algorithm. Once classified, two json files are generated from the tweets, namely “*Municipal Counts*” and “*Location Points.*” The “*Municipal Counts*” json file includes the number of tweets found within municipalities boundaries, and the “*Location Points*” json file includes the actual tweets and the exact latitude-longitude data of the tweet.

Tweets have been integrated with FASSSTER through a developed web service. The web service handles the transfer of tweet data from the tweet server to the FASSSTER system using *GET* requests. Several endpoints were developed (e.g., getting the json data, changing collection keywords, etc.) to allow for flexibility in the data access. On the side of the FASSSTER system, the json files received are processed using jQuery and then visualized on a spatio-temporal map. The “*Location Points*” json file is visualized daily, meaning all the tweets on a given day are laid out on a map. The users can then click on a point and see the actual tweet.

Fig. 18.5 Tweet location points visualization

With regard to privacy, only the tweets are visualized (see Fig. 18.5). No user data is revealed in the FASSSTER visualization. On the other hand, the "Municipal Counts" json file is visualized as a spatio-temporal choropleth map.

18.3.2.2 Application of EMR in FASSSTER

Since EMR systems capture digital patient clinical records, they are good sources of information that can be used for comprehensive statistics and report generation. For disease modeling, data in EMR systems can provide syndromic and diagnostic data that are captured in various times and geographic locations. They offer a wide array of data that can be valuable to syndromic surveillance and to development of disease models.

FASSSTER requests from EMR systems health and clinical data such as patient vitals, symptoms, and diagnoses. It is a standard, however, that the data being requested be anonymized to maintain the privacy and security of the patients involved. FASSSTER communicates with the EMR systems through the use of APIs. Usually, the EMR systems open endpoints in their system from which FASSSTER can retrieve data. After collecting data, further processing is applied to be able to produce spatio-temporal reports that are relevant to disease modeling and syndromic surveillance.

As an example of reports generated using EMR data, in Fig. 18.6, a visualization of monthly dengue disease trends was created based on the data that FASSSTER was able to pull from EMR systems. The following report counts dengue diagnosed cases that were recorded in the EMR systems. In Fig. 18.7, it follows the same process but instead of focusing on diseases, it focuses on analyzing symptoms. In these visualizations, users of the system, specifically the health administrators, are able to select different scenarios in the report that they want to produce.

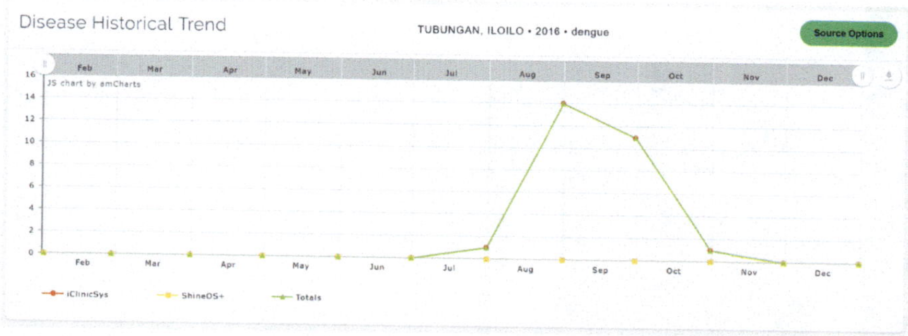

Fig. 18.6 Visualization disease historical trend based on EMR data

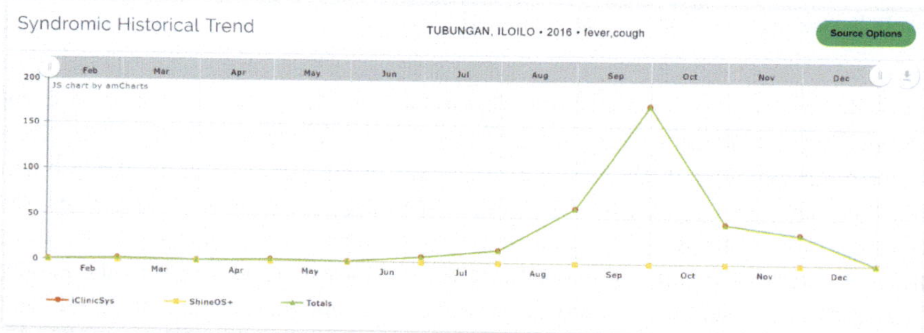

Fig. 18.7 Visualization syndromic historical trend based on EMR data

Aside from graphs and charts, spatio-temporal maps are also produced using EMR data. In Fig. 18.8, counts of cases of symptoms and diagnoses are displayed on a map. Different filters such as year, age range, gender, blood type, blood pressure, and body mass index can be applied to further narrow down the data that users want to display. These maps allow health administrators who have location-based visualizations of disease and symptoms behavior that can help them make more informed epidemiology-related decisions for disaster situations.

18.3.3 Toward Health Disaster Data Exchange

In 2009, three tropical storms caused massive flooding in the Philippines, specifically in National Capital Region (NCR) and Region IV-A. This event eventually led to the two world's biggest leptospirosis outbreak, with 3000 cases in hospitals around the country, and 200 deaths. Factors associated with population displacement, a common post-disaster scenario, are usually what makes the population

Fig. 18.8 Spatio-temporal map visualization of EMR data

vulnerable to disease outbreaks. Such factors include the availability of clean water for drinking and facilities for sanitation, proper hygiene and healthcare services, degree of crowding, and the current health status of the victims. Not to mention that temporary shelters can be used by the displaced population indefinitely. Thus, the consistent surveillance of the area and the efficient communication of data are crucial for preventing an outbreak of disease and ensuring the health of the people in the area. This can be achieved through the development and use of an efficient health disaster data exchange that can allow easy access of data in post-disaster situations.

Information technology has provided tools like electronic medical record systems to efficiently record and gather data. The efficiency of such tools provides an opportunity to use significant amounts of data for different purposes. The major problem is being able to gather the data from different systems and being able to do so accurately. Web services and RESTful APIs allow various systems to exchange information through different methods, each with their own advantages and disadvantages. Semantic interoperability can be applied as long as syntactic interoperability is first achieved, and use cases are properly identified with their corresponding data entities and terms.

Assuming interoperability exists between and among systems, EMRs have important health information that can be utilized by any disaster surveillance systems. Manual ways of extracting data from these EMRs may exist, but these methods are slow, inefficient, and may cause redundancy and errors in the data. Also, several EMRs are already being used in the country. The challenge then is being able to pool and collect data from the different existing EMRs in the country and making it available to various systems that can make use of such data. This may be done through the development of a web service.

Furthermore, the existing EMRs may vary in their implementations. However, they usually follow the same base model. This provides an opportunity to maximize the use of semantics in creating web service. Semantic interoperability is the highest level of health information technology interoperability. It allows different systems to exchange information and actually use the information exchanged by taking advantage of both structuring the data exchange and the codification of the data including vocabulary.

Semantic Interoperability

eBayanihan is an existing participatory platform that collects and monitors disaster-related information. It is participatory in the sense that it allows citizens to post and receive information regarding disaster situations. It also connects to relevant social media sites and requests help from skilled individuals and professionals who would be willing to volunteer. The system has a feature, called SHEREPO, that collects data from shelters and evacuation centers to create profiles that help determine post-disaster needs. What is lacking in this system is data coming from electronic medical records primarily because there are no standards, frameworks, and implementing rules and regulations in relation to sharing of health records to other non-health systems. However, there is opportunity to test its feasibility by developing a health disaster data exchange (HDDX).

Figure 18.9 depicts the implementation framework of the web service. A sample system is used in place of an EMR system to test the posting of data through one of the APIs developed in this study. The system also uses ShineOS+ and its APIs in testing the application's capability to gather data through the use of an EMR system's own API. A generic system simulating a disaster surveillance system is then used to test data requests from the developed service through the service API.

The web service is able to interact with the two systems via application programming interface (APIs). There are two ways of gathering data. First, EMR providers can provide APIs for their systems to allow the developed service to

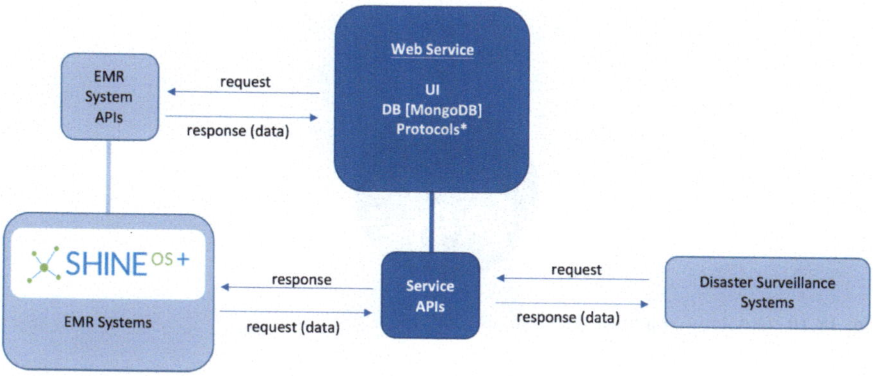

Fig. 18.9 Design for health disaster data exchange (HDDX)

gather data. The service makes a request to the EMR system's API and receives data as a response. Second, the developed service also has an API that allows systems to send, or in more technical terms, post data. The EMR system posts a request with the data and receives a response to confirm that the data has been posted. For sending data to external systems, the service has another API. External systems like disaster surveillance systems can make requests via this API and receive data based on the request as a response.

Semantics for data attributes and fields are developed after the basic functionality of the web service has been manually tested and debugged. Once the basic functions of the developed RESTful APIs are set up online, corresponding data entities for a disaster surveillance use case are identified. The corresponding data entities identified for the use case are the following: patient, location, and diagnosis.

The design for a semantic health disaster data exchange begins by creating an ontology of words per data entity. Figure 18.10 shows a graph of how each entity relates to one another. Modeling the ontology in this manner is necessary so that there is a seamless search for health information from an external health system within a disaster system.

Figure 18.11 shows an example of a semantic search for "women" with "fever." Using the ontology map, the system accesses a health information system, in this case, an EMR and searches for all records that use words related to "women" and "fever."

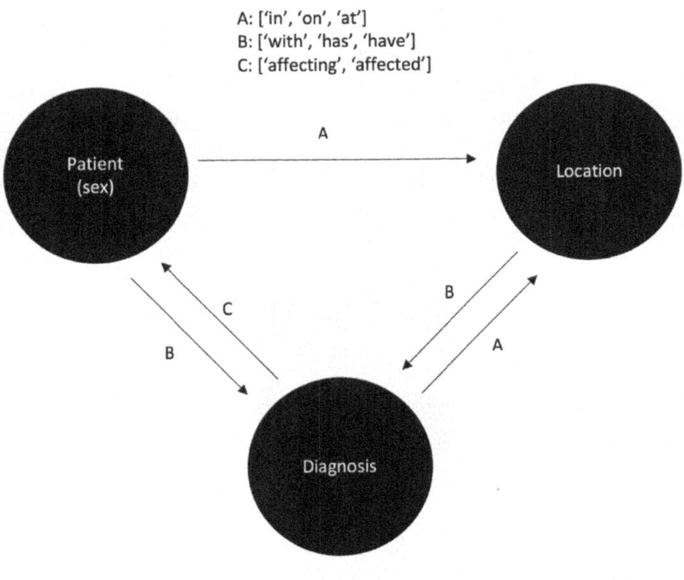

Fig. 18.10 Ontology graph for data entities

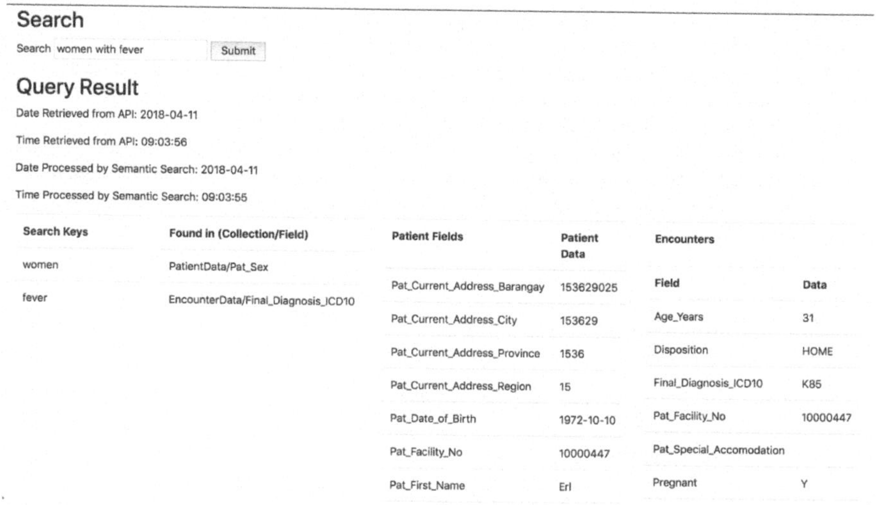

Fig. 18.11 Sample results after a semantic search

18.4 Conclusion and Insights

This chapter presents three systems that provide a solution to the need for available health information during disasters and extreme emergencies. In a developing country like the Philippines, challenged by intermittent internet infrastructure, ICT solution should be multi-platform allowing for continuous flow of information online and offline. The use of electronic medical records becomes an important source of information during disasters. However, on its own, it only provides efficient health services toward targeted individuals and communities. Another approach is to consider, therefore, the development of systems that allow for sharing of data. Within the eHealth ecosystem, as demonstrated by FASSSTER, information from EMRs, disease surveillance as well as infodemiological tweets can be used to derive an epidemiological model that can be used to forecast spread of disease. Moreover, the feasibility on the use of health data in other ecosystems such as disaster management systems shows that semantic approach makes the request for information seamless to its end users.

References

Espina K, Estuar MRJE (2017) Infodemiology for syndromic surveillance of dengue and typhoid fever in the Philippines. Procedia Comput Sci 121:554–561

Memorandum Circular 43s (2016) http://digitaleducation.net/ndrp/wp-content/uploads/2016/09/ICS-AND-RESPONSE-CLUSTERS.pdf

NDRRMC Memorandum Circular 4 (2012) http://ndrrmc.gov.ph/attachments/article/401/MEMO_No_4_s_of_2012_re_Implementing_Guidelines_on_the_use_of_ICS.pdf
Raza T (2018) Localizing disaster risk reduction and climate change adaptation in planners' and decision makers' agenda: technical comprehensive model, Quezon City, Philippines. Procedia Eng 212:1311–1318. https://doi.org/10.1016/j.proeng.2018.01.169
World Bank (1994) Philippines—Devolution and health services: managing risks and opportunities (English). World Bank. Washington, DC. http://documents.worldbank.org/curated/en/145991468776118864/Philippines-Devolution-and-health-services-managing-risks-and-opportunities

Chapter 19
Breakthrough on Health Data Collection in Disasters—Knowledge Arises in Asia Spread to the World

Tatsuhiko Kubo, Flavio Salio and Yuichi Koido

Abstract Aim of this chapter is to introduce recent development in methodology over health data collection during disaster in Asia. During acute phase of disaster, information gathering and analysis are key elements for better coordination and timely response. Previous case studies revealed that emergency medical teams (EMTs) sometimes are the only capacity able to report medical or more broadly health needs to a coordination body. Once standardization of the data reporting process from EMTs to EMT coordination cell (EMTCC), will allow for better coordination and for strengthening of the disease early warning system since EMTs will act as additional sentinel reporting sites. Following this concept, in 2015, based on the model of the Surveillance in Post Extreme Emergencies and Disasters (SPEED) system developed in the Philippines, a Japanese version of the SPEED, so-called J-SPEED has been developed and published in Japan. The J-SPEED was firstly activated during the Kumamoto earthquake which occurred on April 2016. During 48 days of the response, the 353 EMTs successfully sent 1828 daily reports to the EMTCC, including type, quantity, and geographical distribution of medical demand of 8089 patients. The case of Kumamoto earthquake revealed significant potential of the J-SPEED as a breakthrough methodology on health data collection during disaster. Based on these lessons, the WHO has developed SPEED-like system called the Minimum Data Set (MDS) as international standard. The MDS enables EMTs to serve as sentinels of the national system during emergencies and enhances data-based coordination within the EMTCC—contributing to strengthening capacity by improving practices and sharing knowledge toward the future disaster response.

T. Kubo (✉)
Hiroshima University, Hiroshima, Japan
e-mail: tkubo@hiroshima-u.ac.jp

F. Salio
World Health Organization, Geneva, Switzerland

Y. Koido
National Disaster Medical Center, Tokyo, Japan

E. Ying Yang Chan and R. Shaw (eds.), *Public Health and Disasters*,
Disaster Risk Reduction, https://doi.org/10.1007/978-981-15-0924-7_19

Keywords Emergency medical team · Epidemiology · Minimum data set

19.1 Preface

In this chapter, recent developments in methodology over health data collection during disasters in Asia will be introduced. This methodology was developed in the Philippines and shared with Japan contributing to significant improvement in its disaster risk management system. Subsequently, the methodology was shared internationally. The active progress and the future perspective of this movement from Asia will be introduced.

19.2 Fundamental Challenges and Potentials

A disaster is defined as a situation or event, which overwhelms local capacity, necessitating a request for national or international level of assistance (World Health Organization). This definition suggests that, health emergency and disaster risk management plans should always consider the relationship between preexisting local capacity and the different levels of surge capacity as fundamental element. Taking this into consideration, three main challenges have been identified, Firstly, considering various background of the relief teams and organizations come into affected site, maximum leverage of external capacity is quite a challenge. To provide effective support to affected population, it is necessary to set up instant mechanism to command and control those external capacities and maximize the potential to relief activity. To reduce burden of affected side by receiving surge of support itself, this mechanism is also essential. Secondly, that external capacity should be utilized with best harmonization with existing local capacity. Surge activities during disaster should focus on complementing and strengthening pre-existing services provided by local capacity in way not to hinder the restoration of services provided by local capacity. "Do no harm" principle, developed by Mary B. Anderson is underlying this issue (Anderson 1999). Thirdly, to coordinate those local and external capacities, a coordination body should be established in order to deliver essential services to the affected population, while the community continues to recover. This coordination body should be led by the local authority in order for the effective management of both local and external capacities. In order for this mechanism to function, information on the type, scale, and geographical distribution of needs is essential (Kubo 2018).

To tackle the challenges discussed above, sufficient information is needed; however, every disaster response encounters serious structural difficulties in information gathering. Experience has shown that various needs increase simultaneously when the corresponding local capacities are damaged and overwhelmed making access to the preexisting information difficult. That results in latent need for information, while at the same time, various relief teams rush into the affected area as external capacity. The teams generally are self-sufficient with ability to report on regular basis. However, in many countries, the absence of standard templates and a clear reporting system reduces the access to information; therefore, affecting the evidence-based process. In order to fulfill this information gap, three main essential elements have been identified and introduced hereafter. Firstly, in way to collect information from disaster sites, some teams which are acting as additional sentinels sites reporting to the coordination mechanism on regular basis are required. These sentinels need to be robust in terms of logistics and well trained, enough to be activated even during disaster. Secondly, relevant technique to make report should be provided to those sentinels. This technique should be concise and practical enough to be used during disaster. The core information for that reporting includes type, quantity of needs, and their geographical distribution. Thirdly, mechanism to connect information to coordination is essential. During disaster, reported information from sentinels should be directly used and contribute in enhancing current response. Otherwise, the reporting becomes unnecessary burden on sentinels, and sentinels would naturally stop reporting. To solve this issue, relevant mechanism to connect information to coordination is necessary (Emergency Medical Team Minimum Data Set Working Group 2017; Kubo 2018).

To fill this systemic information gap, especially in terms of health information, emergency medical team (EMT) has been identified to have significant potential. EMT is a "groups of health professionals and supporting staff providing health care specifically to disaster- and health emergency-affected populations" (World Health Organization). During disasters, many EMTs rush into the affected sites to supplement local medical capacity which is damaged due to the event. EMTs should be able to timely deploy and self-sufficient in terms of logistics. As part of the EMT minimum standards, EMT should report on regular basis on their activities and status of the assigned location. These natures of the EMT meet the best potential to be a sentinel for health needs information (Emergency Medical Team Minimum Data Set Working Group 2017).

Additionally, to form a part of the local health systems, EMT coordination cell (EMTCC) has been proposed by the World Health Organization (WHO) as a universal technical platform for coordinating EMT. The EMTCC is set within national health authority, such as Ministry of Health and will coordinate both external and local capacity to best fulfill the unmet identified health care needs (World Health Organization 2017).

Considering the three essential elements needed to fulfill the information gap during disaster, the EMT can be considered as ideal sentinels for health system and the EMTCC as universal mechanism to ensure information support operational decisions. This suggests that the additional element required is a practical and effective technique for reporting as part of the health needs information system for disaster management (Kubo et al. 2018).

19.3 The Breakthrough—The SPEED

The breakthrough regarding the discussed challenges and potentials has arisen from the Philippines. On November 8, 2013, the Philippines was hit by the typhoon Haiyan, locally known as super typhoon Yolanda. The scale of the typhoon was the category 5 by Saffir–Simpson scale with winds speed 195 mph, and storm surges of height of 10–20 ft. smashed into coastal communities. On November 11, "State of National Calamity" was declared and by next month a total of 2,146,341 families or 9,923,378 persons were affected in 11,880 barangays, 44 provinces, and 9 region (Phillipines Department of Health).

Following this situation, a number of EMTs rushed into the affected areas: a total of 75 Department of Health (DOH) national teams of the Philippines, 60 international EMT, and 23 local EMT were registered. (Phillipines Department of Health). During the initial acute phase, many EMTs tried to reach Tacloban city in Leyte Island being the most damaged by the storm surge. At the Eastern Visayas Regional Medical Center (EVRMC) in Tacloban city, coordination meeting was chaired by the DOH with the support of the WHO. As part of the meeting, every international EMT was suggested to provide a brief update over their activities though, most of the teams reported without using an appropriate format and reported only verbally, resulting in significant challenges for the DOH to obtain an overall picture of medical service provided by the EMTs. On November 8th, seeing that situation, a member of the Japan Disaster Relief (JDR) recommended to utilize national template of the Philippines, the Surveillance in Post Extreme Emergencies and Disasters (SPEED) form (Table 19.1). The SPEED form complies of 21 sets of syndromes which represent medical situation/resources in need. This form supports the health information system for emergencies and disasters through timely information essential for evidence-based coordination and a more focused and targeted response (Philippines Department of Health). The DOH chair immediately endorsed the recommendation provided by JDR, and the paper forms were distributed to the attending EMTs at that meeting. Implementing the SPEED standard reporting form made significant difference. Thanks to the great simplicity of the SPEED form, EMTs could start their daily reporting of their medical activity soon after its introduction. Namely, the international EMTs suddenly became additional sentinels of the national health information system of the Philippines, helping the DOH to understand the situation, and support evidence-based coordination. Practically, the implementation of the standard daily reporting form shortened meeting time. Further, the form provided framework for discussion and participants of the meeting could make discussion with common list of major health issues among affected patients (Kubo et al. 2014).

Table 19.1 Reporting form used for the SPEED

S.P.E.E.D. SYNDROMIC REPORTING FORM – 1

(HEALTH CENTERS, BHS AND EVACUATION CENTERS)

Province:________________**City/Municipality:**__________________**Barangay:**________________

Name of Reporting Health Facility: ____

Population size of Evacuation Centre: < 5 yrs____________ **≥ 5 yrs**____________

Name of Reporting Officer: ________________________ **Mobile Number:** __________________

Date the patients reported below were seen by a health professional: ________________

Date this report is submitted: ________________

HF Code (Health Facility Code):________________

SMS Format (daily): HF X MM/DD/YY POP 0 0 DISEASECODE 0 0 0 0

SMS Format (alert): HF X ALERT DISEASECODE 0 0 0 0

*** X is the HF code**

#	Disease Syndrome / Health event	<5 years old		≥5 years old	
		Cases	Deaths	Cases	Deaths
1	Fever **(FEV)**				
2	Cough, colds or sore throat with or without fever **(ARI)**				
3	Fever with rash **(MEA)**				
4	Fever with spontaneous bleeding (i.e. nose bleeding, gum bleeding) **(AHF)**				
5	12 months and over: sudden onset of fever (>38° C) with severe headache and stiff neck; <12 months: fever (>38°C) with bulging fontanel, or refusal to suckle **(MEN)**				
6	Fever with headache, muscle pains and any of the following: eye irritation, jaundice, skin rash, scanty urination **(LEP)**				
7	Yellow eyes or skin with or without fever **(AJS)**				
8	Fever with other symptoms not listed above **(FOS)**				
9	Loose stools, 3 or more in the past 24hrs with or without dehydration **(AWD)**				
10	Loose stools with visible blood **(ABD)**				
11	Open wounds and bruises/burns **(WBS)**				
12	Fractures **(FRS)**				
13	Skin disease **(SDS)**				
14	Animal bites **(ANB)**				
15	Eye itchiness, redness with or without discharge **(CON)**				
16	Spasms of neck and jaw (lock jaw) **(TET)**				
17	High blood pressure ≥140/90 **(HBP)**				
18	Known diabetes **(KDM)**				
19	Difficulty in breathing and wheezing **(AAA)**				
20	Floppy paralysis of the limbs which occurred recently in a child < 15 years who is previously normal **(AFP)**				
21	Visible wasting with or without bipedal pitting edema **(AMN)**				
22	Others (please specify): ________________________				

1. Keep proper records of daily consultations in register/ logbook
2. Use the daily consultation records to complete Form-1 and submit daily to the Main RHU ____________________ by 5pm until further instructions are given
3. This form is to be filled by BHW/community health workers, midwives, nurses and doctors in temporary clinics, evacuation centre clinics, BHS & Health Centres
4. HOTLINE (S) for immediate notification of alerts: ____________________________

Reprinted from the Department of Health, Philippines (2011)

This case during the super typhoon Yolanda response revealed great potential of the SPEED to be the ideal element in combination with the EMT and the EMTCC to establish health needs information system for disaster management.

19.4 Knowledge Transfer—J-SPEED

Based on the lessons from the Yolanda response, a Japanese version of SPEED—called J-SPEED—has been developed (Kubo et al. 2014) in two versions: J-SPEED (Table 19.2) for domestic response use by national EMTs such as Disaster Medical Assistance Team (DMAT); and JDR-SPEED for international response use by the JDR. Since the original SPEED system implemented in the Philippines was designed to work within the context of the Philippines, modifications were made to reflect the Japanese context to the J-SPEED. These modifications will be introduced hereafter.

Table 19.2 Reporting form of the J-SPEED

Disaster Medical Activities, Reporting System

J-SPEED Reporting Form (Ver1.0)

Reporter's Information		【Team/Position/Name】:	【Reachable mobile number】:
		【Date of the Activity】:	【e-mail address】:
		【Name of the Field site/Location】:	【Team origin】: □internal, effected area □internal, non-effected area □internal, different region □external
		【Activity Plan of the next day】: □same area □different area □end of mission □not decided	【Organization】: □DMAT □National Hospital Organization □Japanese Red Cross □JMAT □( )
Memo		Things to report to the Emergency Medical Team Coordiantion Cell (EMTCC)	

※How to: 1. Determine the vetical column according to the age group. 2. Count the number of the cases apply to each symptoms/health issue. 3. Check only sex and main cause for death.
※Every team should report numbers below DAILY to the Coordination Cell.

	No	Symptoms / Health issues		0 year		1-8 years		9-74 years (exclude pregnancy)		75 years-		pregnant		Total	
				alive	death	alive	death	alive	death	alive	death	alive	death	alive	death
Sex / N of patient	1	Male	Male												
	2	Female	Female												
Severity	3	More than moderate (Triage Category≥ Ⅱ)	Unable to walk (caused by disaster related injury)												
	4	Needs of transfer	Needs to be referred to a hospital												
Trauma/ Enviromental disorder	5	Wound	Wound, organ injury												
	6	Fracture	Fracture or suspected fracture												
	7	Burn	Skin or inhalation burn												
	8	Drowning/Near-drowning	Near-drowning, hypothermia, any event of drowning												
	9	Crush syndrome	Prolonged compression + obnubilation/incontinence/oliguria												
Advanced care	10	Dialysis	AKI/CKD in need of dialysis												
Cardio-vascular	11	DVT/ suspected PE, CI, or MI	Dyspnea, chest pain, syncope, swelling of lower leg after												
Infectious Disease	12	Fever	High Temperature												
	13	Acute Resperatory Infection	Cough, chill, sore throat, fever, etc.												
	14	Gatrointestinal Infection	Diarrhea, vomitting												
	15	Suspected Measles	Fever, rash												
	16	Suspected Tetanus	Lockjaw												
Skin	17	Dermatosis (exclude wounds & burns)	Dermatosis (exclude wounds & burns)												
Chronic Disease	18	Hypertension	SBP>160 or DBP>100												
	19	Bronchial Asthma	Dyspnea and wheeze												
Psychological	20	Disaster stress-related syptoms	Insomnia, headache, dizziness, anorexia, gastric pain, constipation, etc.												
	21	Urgent need of psychological support	Suicide attempt, disturbance of behaviour or mental status												
Public Health	22	Ungent need of nursing care	Long-term care, Disability, Mental Disorder or Retardation												
	23	Urgent need of water & food	Lack of minimum drinking water (3L/day) or food												
	24	Urgent need of nutritional support	Lack of food ready for allergy/halal food, severe malnutrition												
	25	Interrupted essential medications	Interruption of essential medications due to the disaster												
	26	Non-disater related illness	Needs of medical intervention but not related to the disaster												
Additional Syndrome	27														
	28														
	29														
	30														

Modified from Joint Committee for Disaster Medical Record of Japan (2015)

19.4.1 Replacing the Syndromes

To reflect the Japanese context on the J-SPEED, some syndromes were replaced. Firstly, syndromes which were not observed in Japan, such as tropical diseases, were excluded. Secondly, the original SPEED does not include the number of patients in the reporting form, thus sex (male/female) was added to the syndromic counting list. By counting the sex, total number of patients can be obtained at the J-SPEED. Thirdly, based on lessons learnt from the Great East Japan Earthquake of 2011, syndromic counting for public health issues—including the urgent need for mental health care, nursing care, water and food, nutritional support, and essential medications—was added. Lastly, J-SPEED counts "Relation to the disaster" so to capture change in characteristics of patients by counting number of patients whose health event has no relation to disaster (Kubo et al. 2014; Joint Committee for Disaster Medical Record of Japan 2015).

19.4.2 Connection to the Medical Record

The SPEED of the Philippines was introduced to Japan through the field experience of the Japan Disaster Relief, namely the EMT. This background makes the adaption of the technics specifically to the field of EMT. This includes connecting the SPEED reporting and medical record recording. A tick box column with the items of syndromic counting of the J-SPEED was added to the standard disaster medical record in Japan with the aim of improving the quality of the documentation with limited additional workload (Fig. 19.1). This connection between recording (standard medical record of EMT) and reporting (J-SPEED), enables medical staff of the EMT to count the syndromes reducing the change of omissions (Joint Committee for Disaster Medical Record of Japan 2015).

19.4.3 Expanding the Syndromic Counting

Following the national EMT movement of the J-SPEED, the international EMT, JDR has further developed the JDR-SPEED (i.e. J-SPEED for international disaster relief operations). The major progress was that a count for procedures was added. The original idea for adding this part was the quantification of supplied services as achievement of medical assistance provided, and at the same time estimating the required medical resources (Kubo 2016).

This concept was adopted from the Case-Mix System. The Case-Mix System is a patient classification system focused on clinical similarities and homogeneity in medical resources. Examples of the Case-Mix System include diagnosis-related groups (DRG) in the USA, AR-DRG in Australia, and diagnosis procedure

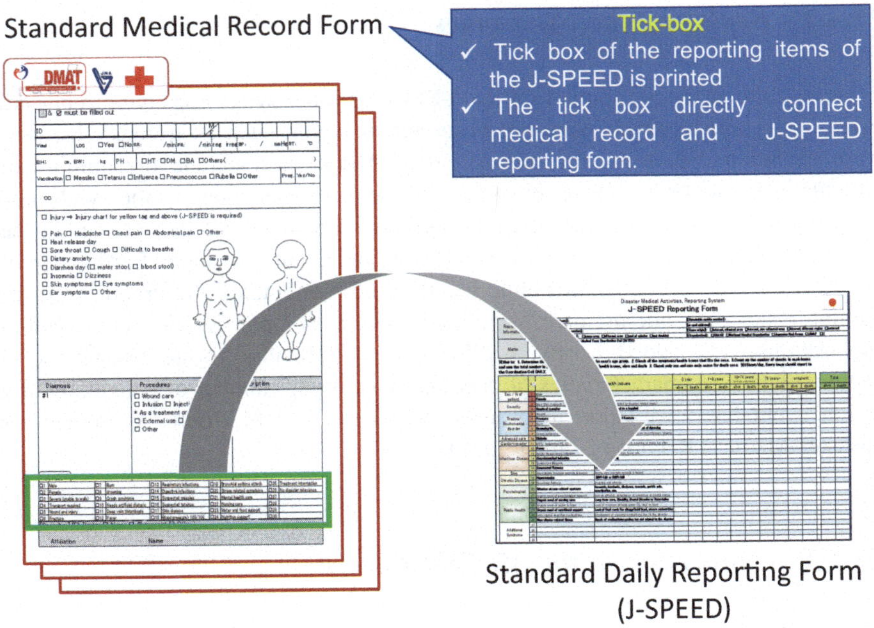

Fig. 19.1 Standard recording and reporting form of EMT in Japan. Modified from Joint Committee for Disaster Medical Record of Japan (2015)

combination (DPC) in Japan. Most of the debate over the Case-Mix System is focused on hospital efficiency and medical expenses; however, the original idea behind this system is a patient classification based on clinical similarities and homogeneity in medical resources. In the same line, SPEED/J-SPEED attempt to capture the required medical resources and suggest the types of medical expertise and equipment needed (Kubo 2016).

The Case-Mix System utilizes diagnosis and medical procedures as a set for classification, while SPEED/J-SPEED only reports diagnoses (syndromes). Information on procedures allows for a better picture of the medical demand. For example, considering trauma cases, the basic diagnostic information (e.g. the type of injury) alone is of limited use in helping decision makers ascertain the condition of patients and the seriousness of the injuries, and adjusting the response based on the medical demand. Providing information on procedures such as hospital care, general anesthesia, and hemodialysis together with the diagnosis support the above-mentioned process (Kubo 2016).

19.5 Mobilization of the J-SPEED

The J-SPEED was firstly activated at the Kumamoto earthquake in 2016. That revealed the great usefulness of J-SPEED. The item ticking/tally sheet system enable standardized EMT reporting with very limited workload. Indeed, 1828 daily reports from 353 EMTs successfully sent to EMTCC during 47 days of response. This massive number of standardized daily reporting was observed for the first time among many disaster response experiences of Japan. Further, the reported information was success to visualize daily number of patients and their geographical distribution (Fig. 19.2). This information included detecting minor outbreak of gastrointestinal infectious disease in a temporary shelter area, and some urgent mental health need like suicide commitment. Importantly that urgent information induced proper support within a day. Also, quantitative data of the J-SPEED make communications among stakeholders much efficient and support data-based decision making. For example, Fig. 19.3 revealed clear daily trend in number of patients and proportion of patients whose visit was diagnosed as non-relation to disaster by medical doctors of EMTs. Local officers played a key role in coordinating EMTs despite are not always specialists in EMT mobilization. J-SPEED supported those local officers by providing data by EMT activity, showing situational shifts and helping EMT coordination to become more evidence and consensus based (Kubo et al. 2017).

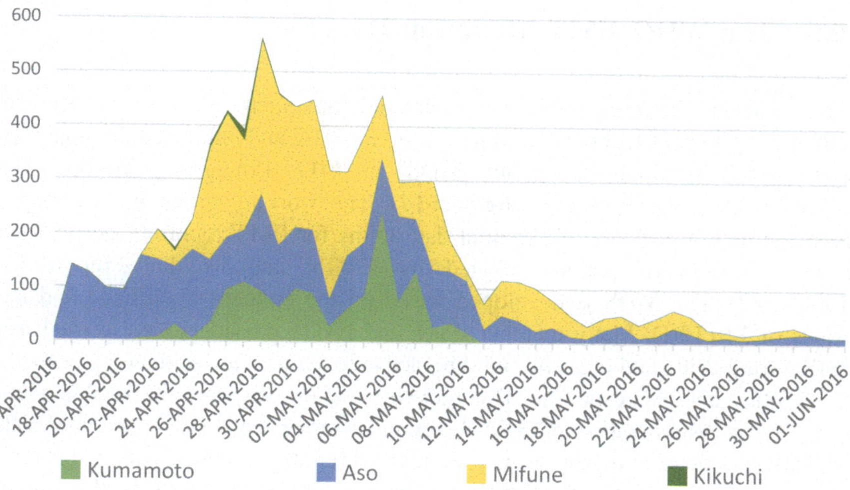

Fig. 19.2 Number of patients by area during the Kumamoto earthquake 2016. Modified from Kubo (2018) Copyright 2018

Fig. 19.3 Trend in number of patients and proportion of non-relation to disaster in Kumamoto earthquake 2016. Modified from Kubo (2018) Copyright 2018

Date (Sunday/Holiday)	Total patients	Non-relation to disaster	
	N	N	%
26-APR-2016	427	32	7.5%
27-APR-2016	392	52	13.3%
28-APR-2016	563	89	15.8%
29-APR-2016	460	77	16.7%
30-APR-2016	435	52	12.0%
01-MAY-2016	448	74	16.5%
02-MAY-2016	317	53	16.7%
03-MAY-2016	315	37	11.7%
04-MAY-2016	384	64	16.7%
05-MAY-2016	457	111	24.3%
06-MAY-2016	297	91	30.6%
07-MAY-2016	299	67	22.4%
08-MAY-2016	300	100	33.3%
09-MAY-2016	190	55	28.9%
10-MAY-2016	132	36	27.3%
11-MAY-2016	73	23	31.5%
12-MAY-2016	116	49	42.2%
13-MAY-2016	114	34	29.8%
14-MAY-2016	102	37	36.3%
15-MAY-2016	79	38	48.1%
16-MAY-2016	52	24	46.2%
17-MAY-2016	32	17	53.1%

19.6 The WHO EMT Minimum Data Set

These practices in Asia, induced international movement. To meet the universal demand, WHO set a technical working group in 2016, and published the final report on the EMT Minimum Data Set (MDS) in 2017 (Emergency Medical Team Minimum Data Set Working Group 2017). The working group has defined the MDS, namely a package of essential data items for EMT reporting derived from medical records of patients treated by EMTs, and its daily report form (Table 19.3). The MDS is developed based on the concept of SPEED and techniques of J-SPEED. Many member states and EMTs providers have started using MDS during drills and capacity building activities in way to facilitate its utilization in the future disaster responses. This movement is mostly active in the Association of Southeast Asian Nations (ASEAN) region. Among the ASEAN member states, medical assistance is a key priority despite, different capacities within the region and the absence of a standardized regional cooperation structure. In this context, Japan International Cooperation Agency (JICA) has been cooperating with Thailand's National Institute for Emergency Medicine and the Ministry of Public Health, since 2016 to implement the Project for strengthening the ASEAN Regional Capacity on Disaster Health Management (ARCH Project) (Nakajima and Silapunt 2017). Aims of the ARCH project are strengthening coordination among EMTs of

Table 19.3 The WHO EMT MDS Daily report form

Insert MOH Logo

EMT-MDS Daily Reporting Form (Ver1.0)

Team information

a	Organization name:
b	Team name:
c	Type 1 mobile / Type 1 fixed / Type 2 / Type 3 / Specialized cell
d	Contact Person(s) name(s):
e	Phone No.:
f	Email:
g	Estimated date of departure (dd/mm/yyyy):

h	Date of activity (dd/mm/yyyy):
i	Time of reporting (dd/mm/yyyy/hh:mm (24h)):
	Location †
j	State etc. (admin1)
k	City etc. (admin2)
l	Village etc. (admin3)
m	Facility name:
n	Geo-tag (Lat) (Long)

† Submit one form per one activity day and location. For Admin 1 = e.g. State, Province, Governorate: Admin 2 = e.g. County, District, City, Municipality; Admin 3 = e.g. Sub-district, Village, Payam.

Daily Summary — Number of patient / Bed Count

o	Patients	Total Number of new consultation ‡	
p	Patients	New admission (=MDS40)	
q	Patients	Live Birth	
r	Bed	Total bed capacity	
s	Bed	Empty inpatient bed (Non-ICU)	
t	Bed	Empty Intensive Care Unit Bed (ICU)	
u			

MDS statistics ‡

No		Item	
36	Outcome	Discharge without medical follow-up	
37	Outcome	Discharge with medical follow-up	
38	Outcome	Discharge against medical advice	
39	Outcome	Referral	
41	Outcome	Dead on arrival	
42	Outcome	Death within facility *	
43	Outcome	Requiring long term rehabilitation *	
44	Relation	Directly related to event	
45	Relation	Indirectly related to event	
46	Relation	Not related to event	
47	Protection	Vulnerable child *	
48	Protection	Vulnerable adult *	
49	Protection	Sexual Gender Based Violence (SGBV) *	
50	Protection	Violence (non-SGBV) *	

‡ Consider 24 hours period from midnight or other agreed cut off time for reporting. MDS statistics report outpatient consultations, inpatient admissions, as well as preformed procedures (MDS No. 30-35), outcomes (MDS 36-43) and contextual issues (MDS No. 44-50) newly counted during the reported period. MDS No. 43 is a subset of MDS No. 37-38. MDS No. 47-50 are a subset of MDS No. 53.

Demographic — MDS statistics

No		Age Categories	<1	1-4	5-17	18-64	65-	Total
1	Sex & Age	Male						
2	Sex & Age	Female non-preg.						
3	Sex & Age	Female pregnant						

Health Events and Procedure — MDS statistics

No		Health Events	<5	>=5	Total
4	Trauma	Major head / spine injury			
5	Trauma	Major torso injury			
6	Trauma	Major extremity injury			
7	Trauma	Moderate injury			
8	Trauma	Minor injury			
9	Infectious disease	Acute respiratory infection			
10	Infectious disease	Acute watery diarrhea			
11	Infectious disease	Acute bloody diarrhea			
12	Infectious disease	Acute jaundice syndrome			
13	Infectious disease	Suspected measles			
14	Infectious disease	Suspected meningitis			
15	Infectious disease	Suspected tetanus			
16	Infectious disease	Acute flaccid paralysis			
17	Infectious disease	Acute haemorrhagic fever			
18	Infectious disease	Fever of unknown origin			
19	Additional §				
20	Additional §				
21	Additional §				
22	Additional §				
23	Emerg.	Surgical emergency (Non-trauma)			
24	Emerg.	Medical emergency (Non-infectious)			
25	Other key diseases	Skin disease			
26	Other key diseases	Acute mental health problem			
27	Other key diseases	Obstetric complications			
28	Other key diseases	Severe Acute Malnutrition (SAM) *			
29	Other key diseases	Other diagnosis, not specified above			

No		Procedure	<5	>=5	Total
30	Procedure	Major procedure (excluding MDS32)			
31	Procedure	Limb amputation excluding digits *			
32	Procedure	Minor surgical procedure			
33	Procedure	Normal Vaginal Delivery (NVD)			
34	Procedure	Caesarean section			
35	Procedure	Obstetrics others			

Needs and Risks

Free text reporting to EMTCC / MOH on the following issues.

51	Immediate report	Unexpected death *	☐
52	Immediate report	Notifiable disease *	☐
53	Immediate report	Protection issues #	☐
54	Immediate report	Critical incident to EMT and/or community	☐
55	Immediate report	Any other issue requiring immediate reporting	☐
56	Community Risks	WASH	☐
57	Community Risks	Community / suspected over infectious disease	☐
58	Community Risks	Environmental risk / exposure	☐
59	Community Risks	Shelter / Non food items	☐
60	Community Risks	Food insecurity	☐
61	Operational constraints	Logistics / operational support	☐
62	Operational constraints	Supply	☐
63	Operational constraints	Human resources	☐
64	Operational constraints	Finance	☐
65	Operational constraints	Others	☐

Detailed comment for (No.)

Detailed comment for (No.)

Detailed comment for (No.)

Detailed comment for (No.)

* Line list (including detailed information) should be submitted with this MDS form to relevant authorities. § Additionals are used for context specific reporting items indicated by the relevant authorities e.g. Malaria / Dengue / TB / Leptospirosis / Rabies / Hazmat etc. # Protection issues to be reported confidentially to appropriate authority or protection cluster in locally agreed manner.

Reprinted from WHO EMT MDS Working Gropu Report (2017)

ASEAN member states and training EMT personnel. As part of the project, the MDS has been repeatedly tested during drills, importantly as fast as before the official endorsement of WHO, which achieved in February 2017. The MDS is currently under process to be adapted as part of ASEAN regional standard tools.

19.7 Value of the MDS

In terms of medical information management, the EMT MDS is expected to be practical solutions for the universal challenge suggested in Sect. 19.2 of this chapter. Firstly, by using the MDS, the bases of EMTs can be utilized as additional "sentinel reporting sites" in affected area. The MDS, which is of universal use irrespective of regions or the dispatching organization, enables those sentinels to collect and submit standardized health data. This reporting system will contribute in managing those external capacities in a standardized manner. Secondly, the MDS is expected to be a tool to facilitate decision-making process and harmonize local and external capacity. The MDS can support the call for external support through the indication of excess medical needs in relation to existing local capacity. In addition, as revealed during the Kumamoto earthquake response, the proportion of patients who were diagnosed to had "no relation to disaster" provide useful information to determine timing for EMT demobilization. Thirdly, the MDS enable local authority to coordinate both local and external capacities by one methodology. The MDS is designed as a tool for the coordination of EMTs from both local and external capacities, and the MDS is expected to support and enhance data-based coordination and maximize support to affected people. Since the EMTCC would be placed within the local health authorities system, decisions such as mobilization and demobilization of EMTs would be supported by information related to the EMTs operation. As revealed by the J-SPEED in Japan, the MDS is expected to provide actual data and empower those authorities (Kubo 2018; Emergency Medical Team Minimum Data Set Working Group 2017).

In addition to providing solutions to the above three universal issues, MDS is expected to advance emergency medical assistance toward next generation. One is to break language barrier during international disaster relief. MDS data is free from language. Suspected Measles (English) and "Mashin utagai" (Japanese) are synonymous as data item No. 13 on the MDS. MDS will be translated into many languages and will promote the collaboration between National EMT and International EMT. The other aspect is that, the MDS enable research. Currently, number of quantities studies in the field of disaster medicine is limited. When searching the PubMED by search term ""Disaster Medicine/statistics and numerical data"[Mesh]," only 14 literatures can be found today (search on September 2018). At the Kumamoto earthquake 2016 in Japan, more than 8089 clinical data of patients treated by EMT were left by activation of the J-SPEED. Also in Japan, the West Japan heavy rain disaster in 2018 left 3620 clinical data. Based on these data, comparative studies such as difference in composition of diseases can be conducted, which will reveal difference in diseases related to earthquake or heavy rain. This case in Japan predicts the future of the MDS, that it will be possible to conduct research compare data by types of disaster and country-by-country. That will accelerate the progress of disaster medicine. Finally, the MDS will contribute in developing electronic disaster medical tool. At this moment, the gold standard of disaster information tool is paper based; however, there are obvious advantages, such as remote reporting and instant analysis in electronic tool. In order to set up electronic tools to manage disaster information, the most critical issue will be how to connect systems developed by relevant organizations. Generally, relief organizations side seems to be ambitious in system development. However, as long as those relief organizations are independent from local authorities, they can never become system administrators. The system which will contain crisis data must be administrated by local authorities. So the challenge would be how to connect electronic tool of external relief organizations to systems which is administrated by local authorities. This is difficult, unless the information itself is standardized. The MDS contributes to this issue by providing a basic information design for both local authorities and relief organizations. The introduction of electronic tools will lead to more swift and effective relief coordination and further to breaking language barriers more easily and allow more comprehensive research after disaster.

19.8 Conclusions

On this chapter, recent developments in methodology over health data collection during disaster essential for EMT coordination were discussed. The origin of this movement is the SPEED of the Philippines. Japan following the positive implementation of the SPEED during 2013 super typhoon Yolanda response, has developed the J-SPEED. The J-SPEED was firstly used during the Kumamoto earthquake response in 2016 and contributed in evidence-based coordination of EMTs. Based on the concept of the SPEED and techniques of the J-SPEED, the WHO working group has developed the MDS. The MDS has been endorsed by the WHO in 2017. It is an international standard tool to incorporate EMTs as sentinels of national health system during disaster, facilitate evidence-based coordination, and further build capacity based on data and share knowledge to improve the future disaster responses.

References

Anderson MB (1999) Do no harm: how aid can support peace or war. Lynne Rienner Publishers, Boulder, Colo.

Emergency Medical Team Minimum Data Set Working Group (2017) Minimum data set for reporting by Emergency Medical Teams. World Health Organization

Joint Committee for Disaster Medical Record of Japan (2015) Standard disaster medical record—report

Kubo T (2016) Literature review on the medical records of the emergency medical team and its progress toward standardization and digitization. WHO Kobe Centre

Kubo T (2018) Disaster Medical Record/J-SPEED—towards national implementation and international sharing of lessons learned from the Great East Japan Earthquake and the Kumamoto Earthquake. Prehospital Care 31:64–68

Kubo T, Takada Y, Seto H, Otomo H, Tomioka G, Nakase K, Hirao T (2014). Development of the health status reporting system during disaster: J-SPEED. Japanese J Disaster Med 19

Kubo T, Yamauchi S, Sato E, Kondo H, Koido Y (2017) Achievements and issues from the initial activation of the J-SPEED at the Kumamoto earthquake. Japanese J Disaster Med 21:557–657

Kubo T, Natsukawa T, Igarashi Y, Nakayama E, Tomioka J, Koido Y (2018) Essence of the WHO EMT Minimum Data Set. Japanese J Disaster Med 22:86–586

Nakajima Y, Silapunt P (2017) Overview of the project for strengthening the ASEAN regional capacity on disaster health management (ARCH Project). Prehospital Disaster Med 32:S63–S64

Phillipines Department of Health. Super Typhoon Yolanda

Philippines Department of Health. Surveillance in post extreme emergencies and disasters (SPEED)

Philippines Department of Health (2011). Surveillance in post extreme emergencies and disasters (SPEED) operations manual for managers

World Health Organization. Emergency medical teams. http://www.who.int/hac/techguidance/preparedness/emergency_medical_teams/en/

World Health Organization (2017) Emergency medical team coordination cell (EMTCC) coordination handbook. World Health Organization

World Health Organization (2018) Definitions: emergencies. World Health Organization. Accessed 17th April 2018. http://www.who.int/hac/about/definitions/en/

Chapter 20
The Development of Health Vulnerability Index with Open Access Data

Emily Ying Yang Chan, Holly Ching Yu Lam and Zhe Huang

Abstract Most existing disaster vulnerability indicators only emphasized economic and social vulnerability. Important underlying health risks such as non-communicable disease are not included in common included in vulnerability measures. There is a need of the development of an integrated health vulnerability index and how the proposed vulnerability index may be incorporated into an all-hazard-based disaster risk index. In China, the Belt and Road Initiative (BRI) has led to the interest of advancing the development and tracking of such scientific tools to enable disaster risk reduction initiatives. The author team has attempted to identify the relevant health risks indicators to support disaster risk map development.

Keywords Disaster in Asia · Disaster prevention hierarchy · Health-EDRM · Information and communications technology (ICT)

20.1 Introduction

Disasters have brought huge losses in human health and economy globally. According to *Economic Losses, Poverty and Disasters, 1998–2017* issued by the CRED and UNISDR in 2018, climate-related and geophysical disasters alone have taken lives from 1.3 million people and have affected 4.4 billion people in the world in 1998–2017 (CRED and UNISDR 2018). Asia, similar to previous years, suffered from the highest disaster occurrence (more than 40% of the total) (CRED 2018), while China, India, Indonesia and the Philippines were four of the top five countries most frequently hit by natural disasters over the last decade (Guha-Sapir et al. 2017). Due to climate change, both frequency and intensity of disasters have been

E. Y. Y. Chan (✉) · H. C. Y. Lam · Z. Huang
JC School of Public Health and Primary Care, The Chinese University of Hong Kong, Shatin, Hong Kong
e-mail: emily.chan@cuhk.edu.hk

E. Ying Yang Chan and R. Shaw (eds.), *Public Health and Disasters*, Disaster Risk Reduction, https://doi.org/10.1007/978-981-15-0924-7_20

predicated to increase in the twenty-first century (IPCC 2014). A robust disaster vulnerability assessment tools and disaster risk reduction plans are important in saving lives and reducing losses in future.

Understanding disaster risk in all its dimensions is the first priority for disaster risk reduction action in the Sendai Framework, the successor instrument to the Hyogo Framework for Action endorsed by the UN General Assembly on disaster risk reduction for policies and practices for disaster risk management (UN 2015). Disaster risk can be conceptualized as a function of hazard, exposure and vulnerability (Kron 2005). According to UNISDR (2009), risk is defined as the harmful consequences resulting from interactions between hazards, exposure and vulnerable conditions. Hazard refers to a potentially damaging event that may cause negative health impacts; exposure refers to people present in hazard zones that are subject to potential health losses; vulnerability refers to the characteristics and circumstances of a community that make it susceptible to the damaging effects of a hazard. Disaster risk assessment can be understood as quantifying these three components among the population.

There are major technical gaps in how to describe vulnerability, particularly to health risks when constructing disaster risk indexes (Chan and Murray 2017). Existing vulnerability indicators/indexes mostly focus on economic and social vulnerability (Brooks et al. 2005; Cutter et al. 2003; UNDP 2004). Most health vulnerability indexes were developed after 2010 and were related to human health vulnerability towards climate-related disasters such as heatwave (Reid et al. 2009; Wolf and McGregor 2013; Zhu et al. 2014), flooding (Phung et al. 2016), dengue fever (Pastrana et al. 2014) and climate change (Malik et al. 2012). In addition, as the data used for index construction were largely based on the country's own capacity in data collection, multi-country comparison is often difficult as countries may have different data collection methods and capacities. The Index For Risk Management (INFORM), a collaborative work between interested parties, and the World Risk Index, a joint work with the Integrated Research on Disaster Risk (IRDR), are two complex global disaster risk indexes that have accounted for health vulnerability (Inter-Agency Standing Committee [IASC] and the European Commission 2015; Welle and Birkmann 2015). However, the current indexes do not include important health affecting factors such as chronic diseases. Chronic disease is an important aspect to be considered in disaster risk management as discontinuous treatment and medicine, which is possible during a disaster event, can lead to adverse health consequences among chronic disease patients. For instance, the provision of insulin may sustain the well-being and survival of diabetes patients (Chan 2017).

Under the influence of globalization, the spread of health risks is borderless, and the prevention and control of health emergencies (e.g. disasters) need to be managed collaboratively. China's Belt and Road Initiative (BRI) was initiated in 2013 aiming to connect the Asian, European and African continents and their adjacent seas and to establish and strengthen partnerships among the countries along the Belt and Road (NDRC 2015). Among these BRI countries, various types of disasters occur frequently, and the widespread damage and destruction caused by disasters disrupted seriously the functioning of a society and pose a major socio-economic development challenge for the Belt and Road Initiative. The BRI also provides a health cooperation platform to handle regional health emergencies and offers medical assistance and disseminates experience in the field of health care (Hu et al. 2017). Understanding of disaster risk and vulnerability for the countries along the Belt and Road is crucial for resource allocation. Yet, currently available health vulnerability indexes may not apply to the countries within Belt and Road Initiative.

Health-emergency disaster risk management (Health-EDRM) is an academic paradigm representing the intersection of health and disaster risk reduction that covers systematic analysis and management of health risks surrounding emergencies and disasters (Lo et al. 2017). The scientific enquiry highlights within this chapter fall into the primary Health-EDRM intervention category (prevention/preparedness) in the system (country) level. Using open access data sources, this chapter describes the development of a health vulnerability index, which can help track health risks and vulnerability. The tool can also be incorporated into an all-hazard-based disaster risk index globally. Of note, the developed health vulnerability index described in this chapter used open access data and proposed indicators available in most countries, which make disaster risk comparison between countries possible.

20.2 Methods

A two phase methodology approach was used to develop the health vulnerability index. *Phase* 1 of the approach focuses on the development of the health vulnerability index which includes an extensive literature review to identify relevant published indicators to construct the health vulnerability. *Phase* 2 involves a two-stage dimension reduction statistical method to identify the weighting for indicators included for the health vulnerability index development.

20.2.1 Phase I: Data Scoping and Variable Selection

Variable selection criteria include (1) indicator that is conceptually relevant to health vulnerability or may capture Health-EDRM risks of the community, (2) have been identified/suggested in relevant organizations or literature (e.g. WHO, UN, INFORM model (Inter-Agency Standing Committee [IASC] and the European Commission 2015) and World Risk Index (Welle and Birkmann 2015)) and (3) available for open access from reliable sources (e.g. WHO and the World Bank) for all study regions. Since subsequent factor analysis cannot be performed with missing values, countries with missing values were excluded in the subsequent analysis.

20.2.2 Phase II: Statistical Model for the Health Vulnerability Index

As this study made no assumption on the weighting for indicators, to determine the weightings and to explore the importance of underlying dimensions to the overall vulnerability, a two-stage dimension reduction statistical method was used. This method also increases robustness (Cutter et al. 2003) and allows monitoring of the changes in the weighting of indicators over time. In stage 1, factor analysis (FA) was used as the primary statistical procedure for dimension reduction. The observed and correlated indicators were assumed to be adequately explained by a lower number of unobserved and uncorrelated factors. Stage 2 modelling was based on the result of FA, and the selected health indicators were used to produce a more compact representation of the indicators (factors).

Stage 1: Selected indicators are standardized and then included in the FA. The matrix of factor loadings was estimated via maximum likelihood method, and the number of factors extracted should contribute cumulatively to the explanation of the overall variance by more than 60% (Nardo et al. 2008). Chi-square test was used to examine whether the number of factors, k, is sufficient to account for the observed covariance (Everitt and Hothorn 2011). A non-significant Chi-square test result ($P \geq 0.05$) indicates that k is sufficient to explain the observed covariance. The explanation power increases when k increases. To obtain the most efficient model, the model with the smallest k that yielded a non-significant Chi-square test was chosen. The initial k tried was 1, and then, k was increased by one at a time. The process is repeated until p-value of Chi-square test ≥ 0.05 (Everitt and Hothorn 2011). Factors identified from FA are sometimes expressed as a compound with a relatively large number of non-zero weighting indicators which may make factor interpretation hard. To make interpretations of factors easier, Varimax rotation (Kaiser 1958) was conducted to obtain as few large loadings and as many near-zero loadings as possible.

Stage 2: the development of health vulnerability index (HVI) based on the latent factors derived from the FA. Each latent factor has factor loading on every health indicator, measuring the correlation between the health factor and the health indicator. The calculation of the weights of the selected health indicators was from the rotated matrix of factor loadings (Nardo et al. 2008; Li et al. 2018): (1) the proportion on each latent factor of the total unit variance was extracted; (2) intermediate weights of all the health indicators were calculated from the factor loadings corresponding to the latent factor; (3) the proportion on each latent factor multiplies by intermediate weights of all the health indicators on each latent factor to generate the weights for all the selected health indicators. Finally, the weights were multiplied by the corresponding standardized health indicator and were added together for every country's HVI. A higher value indicates a more vulnerable country. R version 3.4.1 was used. The method of constructing Health Vulnerability Index has been improved and described more thoroughly in Chan et al. (2019).

20.3 Health Vulnerability Index

Based on the three evaluation criteria, nine health indicators were identified and included in the final index development (see Table 20.1).

The proposed health vulnerability assessment index covers five key components include disease burden (with two dimensions of infectious diseases and chronic diseases), health and well-being (with two dimensions of maternal health and health for children under 5), curative health service, preventive health service and health vulnerability associated with demographic pattern. Nine related indicators are proposed for formulating the final vulnerability index, namely (1) proportion of population below 15 years and above 65 years, (2) under-five mortality ratio, (3) maternal mortality ratio, (4) prevalence of tuberculosis, (5) the age-standardized raised blood pressure, (6) physician ratio, (7) hospital bed ratio, (8) coverage of MCV1 and 9) the DTP3 vaccines. Potential implication of how this health vulnerability index might be applied to create a more comprehensive risk index is discussed in Chan et al. (2019).

Meanwhile, Table 20.2 shows a comparison between the proposed health vulnerability index and the health vulnerable dimension as adopted in the Index For Risk Management (INFORM) and the World Risk Index which are the two commonly used global disaster risk indexes that have incorporated health-related components for vulnerability. The INFORM is a global risk assessment index collaborating with the United Nation (UN) and is adopted in the Global Risk Map (https://globalriskmap.terria.io/About.html) (including 190 countries). It has included tuberculosis prevalence, HIV prevalence, malaria death rate and under-five mortality as vulnerability indicators and have included physician density in capacity coping indicator. The World Risk Index (171 countries considered) presented by Birkmann and Welle and the Integrated Research on Disaster Risk (IRDR) team (http://www.irdrinternational.org/2016/03/01/word-risk-index/) (Welle and Birkmann 2015) have combined susceptibility, lack of coping and adaptive capacity to vulnerability. They considered dependency ratio for susceptibility, physicians and hospital beds ratio for coping capacity and private and public medical expenditure for adaptive capacity. Although the INFORM model and the Word Risk Index were built with sophisticated calculations and considered variables from different aspects, such as health-related component, economic status, political environment and infrastructure, in risk assessments, important health vulnerability burdens which might reflect preparedness and potential health risk and vulnerability in times of crisis such as non-communicable diseases were not included.

Table 20.1 Key indicators of health vulnerability and their relevance

Dimension of health vulnerability	Indicator	Conceptual relevance to health vulnerability
Vulnerable age[a]	1. Population ages 0–14 and population ages 65 and above (% of total)	Age extremes (children and elderly) are known to be more vulnerable to health risks and less resilient to disasters. The proportion of age extremes is an important component of "dependency ratio". Those people are more likely to accumulate post disaster health and service needs
Premature death[b]	2. Under-five mortality rate (probability of dying by age five per 1000 live births)	It is a leading indicator of health. The decline of under-five mortality rate is a target of Sustainable Development Goals (SDGs). It is closely linked to maternal health
Preventable death[b]	3. Maternal mortality ratio (per 100 000 live births)	It is a leading indicator of health. The decline of maternal mortality ratio is a target of Sustainable Development Goals (SDGs). In addition to preventable deaths, this indicator reflects the capacity of health systems to effectively prevent and address the complications occurring during pregnancy and childbirth
Vaccination gap[b]	4. Measles-containing vaccine first-dose (MCV1) immunization coverage gap among 1-year-olds (%)	Standard Expanded Program on Immunization (EPI) for common preventable childhood communicable diseases for < 1 year old. Coverages of vaccination may be used to monitor immunization services and to guide disease eradication and elimination efforts, and are good indicators of health system performance MCV1: Measles is one of the most infectious and mortality diseases in displaced camps DTP3: Tetanus is a common preventable infection associated with injury or wounds
	5. Diphtheria tetanus toxoid and pertussis (DTP3) immunization coverage gap among 1-year-olds (%)	
Chronic diseases status[b]	6. Raised blood pressure (SBP $\geq$ 140 OR DBP $\geq$ 90), age-standardized (%)	It is a proxy indicator for chronic non-communicable disease. Hypertension and heart disease are some of the leading causes of mortality and morbidity globally. Disease status and potential activity limitations among adults can impair one's ability to prepare for, respond to or recover from a disaster
Infectious disease[b]	7. Incidence of tuberculosis (per 100,000 population per year)	Tuberculosis (TB) is the second leading infectious cause of death and one of the most burden-inflicting diseases in the world. SDGs include ending the TB epidemic by 2030. The incidence of tuberculosis gives an indication of the burden of TB in a population

(continued)

Table 20.1 (continued)

Dimension of health vulnerability	Indicator	Conceptual relevance to health vulnerability
Coping capacity[b]	8. Hospital beds (per 10,000 population)	Resources in health systems indicate the level of access to care and the provision of quality medical care, which are highly correlated with live-saving and health status
	9. Physicians density (per 1000 population)	

Sources [a]Data collected from the World Bank; [b]Data collected from the World Health Organization; Adapted from Chan et al. (2019)

Table 20.2 Health-related components considered in the INFORM model, the World Risk Index and the index developed in this study

Key Components	Dimension(s)	INFORM	World Risk Index	Indicators in the proposed index
Disease Burden	Infectious diseases	Tuberculosis prevalence	/	Tuberculosis prevalence
		Estimate % of adults (>15) living with HIV		/
		Malaria death rate		
	Chronic diseases	/	/	Age standardized raised blood pressure
Health and Well-being	Maternal health	Maternal mortality	/	Maternal mortality
	Health for children under 5	Under-5 mortality	/	Under 5 mortality
		Malnutrition in children under 5		/
Curative Health Service	Healthcare services	Physician ratio	Physician ratio	Physician ratio
		/	Hospital bed ratio	Hospital bed ratio
		Per capita expenditure on private and public health care	Public medical expenditure; Private medical expenditure	/
Preventive Health Service	Immunization	Measles immunization coverage	/	Coverage of MCV1 and DTP3 vaccines
Health vulnerability associated with demographic pattern	Dependency ratio	/	Proportion of population <15 years old and > 65 years old	Proportion of population <15 years old and > 65 years old

Source Chan et al. (2019)

20.4 Conclusion

This chapter presents a health vulnerability index that aims to capture health-emergency and disaster risk at a country level to enhance disaster risk assessment and monitoring. The suggested health vulnerability index covers five components and seven health vulnerability dimensions including infectious disease, chronic disease, maternal health, health for children under 5, healthcare services (physician ratio and hospital bed ratio, immunization (coverage of MCV1 and DTP3 vaccines) and emphasis had been placed in the health vulnerability, well-being and potential needs which might be associated with the increasing population living with chronic diseases. This new index has attempted to incorporate key health dimensions, with open access data source, into an existing hazard-based disaster risk map. Due to its open access data use, the index may facilitate understanding of health vulnerability at country level as well as enable cross-country comparison. With more comprehensive health-related disaster risk assessment, decision makers may engage in better resource arrangement and capacity planning to meet the Health-EDRM needs for the disaster-affected regions.

Acknowledgements The authors thank Prof. Peng Cui for his support and advice in this collaborating project and study outcome discussions. The content of this chapter was based on re-adaptation of the published paper by the authors (Chan et al. 2019).

Funding This research was funded by Global Health and Disaster Studies and Research Development Fund, CUHK, CLP/OUT//2017, OAL China, CUHK and International partnership program of Chinese Academy of Sciences (grant no. 131551KYSB20160002).

References

Brooks N, Adger WN, Kelly PM (2005) The determinants of vulnerability and adaptive capacity at the national level and the implications for adaptation. Glob Environ Change 15(2):151–163

Chan EYY (2017) Public health humanitarian responses to natural disasters. Routledge, London

Chan EYY, Murray V (2017) What are the health research needs for the Sendai Framework? Lancet 390(10106):e35–e36

Chan EYY, Huang Z, Lam HCY, Wong CKP, Zou Q (2019) Health vulnerability index for disaster risk reduction: application in belt and road initiative (BRI) region. Int J Environ Res Public Health 16(3):380

CRED (2018) Natural disasters in 2017: lower mortality, higher cost. Retrieved from: https://cred.be/sites/default/files/CredCrunch50.pdf

CRED, UNISDR (2018) Economic losses, poverty & disasters, 1998–2017. Retrieved from: https://www.unisdr.org/files/61119_credeconomiclosses.pdf

Cutter SL, Boruff BJ, Shirley WL (2003) Social vulnerability to environmental hazards. Social Sci Quart 84(2):242–261

Everitt B, Hothorn T (2011) An introduction to applied multivariate analysis with R. Springer Science & Business Media, New York

Guha-Sapir D, Hoyois PH, Wallemacq P, Below R (2017) Annual disaster statistical review 2016: the numbers and trends. Centre for Research on the Epidemiology of Disasters (CRED), Brussels, Belgium

Hu R, Liu R, Hu N (2017) China's belt and road initiative from a global health perspective. Lancet Glob Health 5(8):e752–e753

Inter-Agency Standing Committee (IASC) and the European Commission (2015) Interpreting and applying the INFORM global model. Retrieved from http://www.inform-index.org/LinkClick.aspx?fileticket=M-RXb0tKsjs%3d&tabid=101&portalid=0&mid=447

IPCC (2014) Climate change 2014: impacts, adaptation, and vulnerability. Retrieved from: http://www.ipcc.ch/report/ar5/wg2/

Kaiser HF (1958) The varimax criterion for analytic rotation in factor analysis. Psychometrika 23 (3):187–200

Kron W (2005) Flood Risk = hazard • values • vulnerability. Water Int 30(1):58–68

Li VO, Han Y, Lam JC, Zhu Y, Bacon-Shone J (2018) Air pollution and environmental injustice: are the socially deprived exposed to more PM2. 5 pollution in Hong Kong? Environ Sci Policy 80:53–61

Lo STT, Chan EYY, Chan GKW, Murray V, Abrahams J, Ardalan A et al (2017) Health emergency and disaster risk management (health-EDRM): Developing the research field within the Sendai framework paradigm. Int J Disaster Risk Sci 8(2):145–149

Malik SM, Awan H, Khan N (2012) Mapping vulnerability to climate change and its repercussions on human health in Pakistan. Globalization and health 8(1):31

Nardo M, Saisana M, Saltelli A, Tarantola S, Hoffman A, Giovannini E (2008) Handbook on constructing composite indicators. OECD, Paris, France

NDRC (National Development and Reform Commission) (2015) Vision and actions on Jointly Building Silk Road Economic Belt and 21st-Century Maritime Silk Road. Retrieved from: http://en.ndrc.gov.cn/newsrelease/201503/t20150330_669367.html

Pastrana MEO, Brito RL, Nicolino RR, de Oliveira CSF, Haddad JPA (2014) Spatial and statistical methodologies to determine the distribution of dengue in Brazilian municipalities and relate incidence with the health vulnerability index. Spat Spatio-Temporal Epidemiol 11:143–151

Phung D, Rutherford S, Dwirahmadi F, Chu C, Do CM, Nguyen T, Duong NC (2016) The spatial distribution of vulnerability to the health impacts of flooding in the Mekong Delta, Vietnam. Int J Biometeorol 60(6):857–865

Reid CE, O'neill MS, Gronlund CJ, Brines SJ, Brown DG, Diez-Roux AV, Schwartz J (2009) Mapping community determinants of heat vulnerability. Environ Health Perspect 117 (11):1730–1736

Shi P, Kasperson R (2015) World atlas of natural disaster risk. Springer, Heidelberg

UN (2015) Sendai Framework for Disaster Risk Reduction 2015–2030. Retrieved from: https://www.unisdr.org/files/43291_sendaiframeworkfordrren.pdf

UNDP (2004) Reducing disaster risk: a challenge for development. United Nations Development Programme, New York

UNEP, UNISDR (2013) Global risk data platform. Retrieved from: https://preview.grid.unep.ch

UNISDR (2009) Terminology on disaster risk reduction. UNISDR, Geneva, Switzerland

Welle T, Birkmann J (2015) The world risk index–an approach to assess risk and vulnerability on a global scale. J Extrem Events 2(01):1550003

Wolf T, McGregor G (2013) The development of a heat wave vulnerability index for London, United Kingdom. Weather Clim Extrem 1:59–68

Zhu Q, Liu T, Lin H, Xiao J, Luo Y, Zeng W et al (2014) The spatial distribution of health vulnerability to heat waves in Guangdong Province, China. Global Health Action 7(1):25051

Chapter 21
Technical Review of Health Emergency and Disaster Risk Management Literature in Rural Asia

Emily Ying Yang Chan, Asta Yi Tao Man and Holly Ching Yu Lam

Abstract Despite Asia being the continent with the highest frequency of natural hazards, there is limited disaster literature that explores the existing interventions and practices regarding disaster risk reduction, especially with health as a focus. Rural demography was chosen as the focus due to its known vulnerability. Furthermore, this chapter breaks down Health-EDRM into primary, secondary, and tertiary interventions to assist in identifying the gaps for action. The lack of monitoring and evaluation is highlighted along with the need to provide disaggregated data in meteorological and hydrological hazards.

Keywords Asia · Health-EDRM · Natural hazards · Disaster preparedness · Rural communities

21.1 Introduction

To have an overview of the existing Health-EDRM interventions that are being practiced in rural Asia, a technical review was conducted for the three most common natural hazards in Asia: geophysical, meteorological, and hydrological hazards. Rural demography was chosen to highlight the populations who tend to be the most vulnerable. Furthermore, most existing literature was written about urban contexts, which have different healthcare needs and disease patterns from their rural counterparts. This chapter was developed from '*Scientific evidence on natural disasters and health emergency and disaster risk management in Asian rural-based area*' in the *British Medical Bulletin*. It has been updated to include peer-review articles published from January 2000 to December 2018.

E. Y. Y. Chan (✉) · A. Y. T. Man · H. C. Y. Lam
JC School of Public Health and Primary Care, The Chinese University of Hong Kong, Shatin, Hong Kong
e-mail: emily.chan@cuhk.edu.hk

E. Ying Yang Chan and R. Shaw (eds.), *Public Health and Disasters*, Disaster Risk Reduction, https://doi.org/10.1007/978-981-15-0924-7_21

21.2 Geophysical Hazards

Earthquakes are one of the most researched natural hazards, it is reflected by the sheer volume of published articles. In particular, majority of disaster epidemiology is focused on the 2008 Sichuan (Wenchuan) Earthquake and the 2004 Indian Ocean Tsunami. The cause of the hazard is due to the ground shaking from seismic waves, usually found along the Pacific Rim. Mortality is commonly caused by buildings collapsing along with soft-tissue injuries and fractures of the extremities (Doocy et al. 2016). Existing chronic health conditions such as hypertension and diabetes have been noted in elderly patients in the Sichuan Earthquake. Furthermore, infants who were provided with the same nutritional provision as adults showed signs of acute malnutrition, stunted growth, and wasting two years later (Dong et al. 2014). While infectious disease outbreaks are uncommon, certain environments such as cramped temporary shelters have been linked to communicable disease outbreaks. When they do occur, they are usually after the initial response phase and respiratory infections and diarrheal diseases were identified in China and Iran (Babaie et al. 2014; Cartwright et al. 2017). Other reports of infectious diseases include measles and meningitis in the wake of the 2005 Pakistan Earthquake and 80% rise in water-borne infections within the first six months of the 2015 Nepal Earthquake (Kouadio et al. 2014; Uprety et al. 2017). Mental health problems were found to be associated with loss of farmland (property damage), loss of kin, and exposure to traumatic events. Due to local-specific civil conflict, men in Indonesia had a higher prevalence of depression, anxiety, and somatic symptoms compared to women (Marthoenis et al. 2016; Musa et al. 2014). On the topic of resilience, there has yet to be a consensus on whether the cultural worldview of minority communities play an important role in how they interpret and react to stress compared to the Han Chinese (Xu et al. 2011).

For Health-EDRM interventions regarding geophysical hazards, previous disaster experience in itself is not associated with individual disaster risk reduction (DRR) knowledge or practices unless the individual or household was severely impacted. Health education interventions in China and Thailand found individual/household-level possession of disaster kits were well received as a disaster preparedness concept but many lacked the skills or knowledge of what items were needed (Chan et al. 2014; Witvorapong et al. 2014). Community-level disaster training of frontline and rescue workers is also gravely lacking as many were used to working in urban settings. While short-term clinical training was taught, the clinical outcome impact of such training has not been evaluated. On the policy-level, a communicable disease surveillance system was set up in Iran used in the 2012 East Azerbaijan Earthquakes and China's Health and Family Planning commission of Yunnan Province also took measures to prevent outbreaks of malaria in high-risk areas after the 2014 earthquakes (Babaie et al. 2014; Feng et al. 2016).

21.3 Meteorological Hazards

Cyclones, or commonly known as typhoons in Asia, are non-frontal storms with spirals of strong winds and rain. Previous reviews found mortality was directly caused by drowning and trauma, especially if winds exceeded 155 miles/h, storm surges were above 10 m, and torrential rainfall was present (Doocy et al. 2013). Common forms of injury range from lacerations, contusions, and blunt force trauma. Despite the frequency of the hazard in Asia, rural populations often have low awareness of the health risk associated with typhoons. There is also a lack of information regarding the demography breakdown of the morbidity and mortality figures. Research does, however, highlight the risk of water security as many typhoon-affected areas are coastal regions, putting an estimated 25 million people at risk of clean water contamination by seawater (Hoque et al. 2016). Risks of vector-borne diseases (e.g., malaria and dengue fever) due to pools of stagnant water have also been highlighted in China and the Philippines (Dolhun 2013; Zheng et al. 2017). Limited research has been done on mental health but studies after Typhoon Haiyan expressed the need to address existing psychiatric issues as they were left untreated as medication supplies were interrupted. Alcohol use is also a concern as it is used as a coping mechanism, which may have exacerbated the already high-alcohol consumption rates in the Philippines (Hoque et al. 2016). Mental health is still a highly stigmatized issue in Asia; therefore, identification and treatment of PTSD, anxiety, depression, etc. are often difficult.

Parallel to earthquake-related education, literature suggests that people thought DRR or typhoon-related education was needed. This was reflected as only around 50% study participants from rural China knew to keep emergency food and drugs, and to check electrical circuits (Zhang et al. 2017). Indeed, the low prevalence of practicing DRR measures was associated with a lack of technical know-how. On a community-level, early warning systems have been developed and identified by many people, however, there are high proportion of individuals who do not adhere to the warnings or carry out protective measures (Jiang et al. 2011). Strengthening disaster management through communications with social media and increasing rehabilitative/chronic disease service delivery have been suggested in the Philippines (Ganchoon et al. 2018). That being said, global partnerships (ESCAP/WMO Typhoon Committee) have been established to facilitate time-sensitive data exchange to minimize blind spots during typhoon tracking (Quevillon et al. 2016). Of note, meteorological hazards also affected the availability of drugs. Typhoon Haiyan caused massive disruption for tuberculosis (TB) drug treatment in the Philippines despite its endemic status and potential implication for multiple-drug resistant TB development (Rosalind et al. 2015). Without systematic backup of health records and treatment plans, it made disease management post-impact challenging.

21.4 Hydrological Hazards

Floods are a dynamic hazard as they have many causes, from storm surges to torrential rainfall/monsoons to melting snow (Doocy et al. 2013a, b). In fact, it is often a cascading hazard associated with typhoons. Drowning is the main cause of death, but other disaster epidemiology characteristics are unclear. High mortality has been associated with flash floods and rapid urbanization as it has eroded the natural and historical methods of flood prevention (Tran and Shaw 2007). Morbidity issues reported in Thailand include skin infections, cold symptoms, and snake bites (Srikuta et al. 2015). Extreme ages were also highlighted in China and India where wasting under five-years-old was higher in flooded communities compared to non-flooded areas. Severe wasting was even more prevalent for repeated floods (Rodriguez-Llanes et al. 2016). In particular, flash floods have been occurring more frequently in the Himalayan region of Pakistan where there is already a high prevalence of TB, malaria, and polio (Shabir 2013). Immediate communicable diseases were acute diarrhea and acute respiratory infections. Of interest, a study from China noted that previously eliminated infectious diseases had also re-emerged (Wu et al. 2008). Similar to typhoons, very few studies have looked into mental health outcomes. In China, PTSD was found across all age groups, especially for people who low social support. Mental health problems in Thailand were associated with agricultural losses and food insecurity and India found associations between floods and symptoms of depression and anxiety (Srikuta et al. 2015; Wind et al. 2018).

Floods often reoccur in the same area, therefore system-level post-disaster recovery and rehabilitation efforts should take into account future hazards. Individual and household-level preparedness was low in rural China and major knowledge gaps in geographic locations and risk factors for displacement were found (Lee Poyi 2012). Health education for homemade oral rehydration solution was retained 12 months after the intervention, which shows Health-EDRM interventions were well received. More tailored disaster management has to be applied to different communities. For example, Vietnam reported the role of primary healthcare workers in emergency response was not defined, certain communities in Pakistan did not have access to any toilet facilities six-months post-flood, and while emergency shelters were utilized in India, the social hierarchy meant certain social class (caste) of people were turned away from shelters (Doocy et al. 2012; Ray-Bennett 2009; Van Minh et al. 2014).

21.5 Intervention for Health, Emergency and Disaster Risk Reduction

Health-EDRM is a catchment term for many intervention or practices surrounding health and disaster. It can be conceptualized and examined according to primary, secondary, and tertiary hierarchy levels. Respectively, they minimize and mitigate health risk before the onset of disasters, immediately after the occurrence of disasters, and those already affected by the disaster (Table 21.1). Review indicated although efficacies of primary interventions, such as vaccinations, have been examined in the context of armed conflicts and displaced populations the effectiveness for natural hazards was rarely reported. Low tetanus vaccination rates have been reported in China but the general lack of baseline information for vaccination coverage is likely to create an opportunity for an otherwise preventable disease outbreak (Chan 2008). Publication of secondary interventions tends to focus on evacuation, owning disaster kits ('grab bags'), and awareness of disaster risk. Most of those DRR practices have not been quantified or evaluated beyond knowledge assessments of selected rural villages. Nevertheless, a seven-year follow up was conducted and found personal/household hygiene and waste management knowledge was maintained (Chan et al. 2018). A review of 'grab bags' has also identified apart from knowledge gaps, the availability and accessibility of the items to pack are also a barrier (Pickering et al. 2018). Furthermore, whether these interventions were accepted or used by affected people during a disaster has not been systematically evaluated. Tertiary interventions tend to revolve around medical treatment such as patient management and disease control. Community livelihood and resilience should be emphasized during times of non-disasters. In China, one paper reported despite the push for equal development, the number of doctors in rural areas has declined instead of improving. Monitoring hazards which affect agricultural output such as floods and droughts was suggested as system-level recommendations (Fang et al. 2018). To have evidence for directing and supporting best practices, data on morbidity and mortality should be disaggregated by demography and diagnosis, especially in floods and typhoons. PTSD overwhelmingly dominates the mental health research field but no treatment options were reported let alone assessed. Better clinical management and deployment of other medical disciplines are common suggestions but are not standard practice due to the lack of evidence-based support in Asia. Furthermore, marginalized groups such as gender and indigenous populations should not be left behind. Health risk identified for girls and women include the lack of sanitary products and private areas to change, male community health providers who were unwilling to talk about reproductive health issues, and human trafficking post-disaster due to mass displacements through kidnapping or tricked by promises of migrant work (Gyawali et al. 2017; Krishna et al. 2018; Sohrabizadeh et al. 2018).

Table 21.1 Current evidence on various levels of Health-EDRM Interventions in rural Asia from reviewed literature

		Mentioned Health-EDRM Interventions[b]		
		Primary	**Secondary**	**Tertiary**
		Measures aim to prevent the onset of disaster or minimize the risks that might contribute to the occurrence of emergency and disasters risks	*Aims to prevent potential harm and health risks associated immediately after the occurrence of a disaster*	*Aims to minimize the impact and damage after disaster. Targets specifically at people who might already suffer from the disaster impact*
Levels of preparedness[a]	Individual/ household	**Health and disaster risk literacy** • Knowledge of one's medical symptoms and treatment plan • Training in first aid • Aware of personal risk • Understand warning and evacuation **Vaccination** • Access to clinics and awareness of vaccine-preventable diseases	**Health risk management** • Own disaster kits[c] • Have personal stock of medication and drugs for a short-time period[c] **Warning and evacuation** • Activate household-based emergency plans for response and evacuation	**Health response** • Use items from the disaster kit • Make and drink ORS • Apply first aid to self or others • Follow through with emergency/evacuation plan
	Community	**Health and disaster risk literacy** • Disseminate disaster-related information through education and community outreach[c] **Vaccination** • Have stock of essential drugs and vaccines in clinics and hospitals	**Health risk management** • Manage water, sanitation, and hygiene (WASH), and food security • Perform vector-borne disease control measures • Ensure continuous waste management **Warning and evacuation** • Communicate risks in various channels • Activate community disaster response plans	**Health response** • Triage injuries and ensure hospital and clinics collect relevant information • Provide psychological first aid • Maintain routine medical/healthcare services
	System/ policy	**Health and disaster risk literacy** • Include health in disaster contingency plans • Training and exercise in disaster management[c] • Activate timely multi-hazard early warning system **Vaccination** • Establish routine childhood and emergency vaccination programs[c]	**Health risk management** • Coordinate between various emergency response services, government agencies, and NGOs **Warning and evacuation** • Streamline multidisciplinary programs • Detect the risks of disease outbreak accurately	**Health response** • Training health facilities to treat disaster-specific injuries/diseases • Ensure medical infrastructure can withstand impact with backup energy sources

Modified from Chan et al. (2019), copyright CC-BY-NC-ND

[a],[b]Levels of disaster preparedness and health interventions categorized from Chan (2017). [c]Studies that evaluated disaster preparedness interventions in Asia

21.6 Conclusion

Overall, this review indicated most evidence-based research was related to earthquakes and was also from China. In fact, it is difficult to review papers on rural populations as the urban/rural demography is not specified in most of the disaster literature. Despite the amount of health education interventions and trainings conducted, whether the same interventions under different hazards would have the same outcome has not been explored. For example, is household ORS more widely used in floods, typhoons, or earthquakes? With urbanization becoming a growing trend, the migrant population is also expanding. The disaster risks of migrants should account for dynamic migratory patterns in multiple locations, especially as rural and urban settings carry different risks.

The previous paradigm on health and DRR relied upon government-level responses to save lives, which may be insufficient. An all-of-society engagement approach is needed to bridge the gaps between the delivery of disaster interventions and the targeted recipients. Contingency plans must factor in vulnerabilities of older people, ethnic minorities, illiterate populations, people with disabilities and NCDs, those who face gender-based discrimination, and other marginalized communities. An all-hazards approach also needs to be considered as hazards have cascading risks and are rarely single events, for example, many typhoons also cause flooding (Aitsi-Selmi et al. 2016). Limitations of this review are that it excluded non-English, non-electronic, and grey literature such as governmental and non-governmental reports. Also, only three types of hazards were explored. However, by identifying individual/household, community, and system/policy levels of Health-EDRM, this review presents clear topics for future clinical and public health research.

Acknowledgements The content of this chapter was based on re-adaptation of the published paper by the authors (Chan et al. 2019).

References

Aitsi-Selmi A, Murray V, Heymann D, Mccloskey B, Azhar EI, Petersen E et al (2016) Reducing risks to health and wellbeing at mass gatherings: the role of the Sendai Framework for Disaster Risk Reduction. Int J Infect Dis. https://doi.org/10.1016/j.ijid.2016.04.006

Babaie J, Fatemi F, Ardalan A, Mohammadi H, Soroush M (2014) Communicable diseases surveillance system in East Azerbaijan Earthquake: strengths and weaknesses. PLoS Curr 6. https://doi.org/10.1371/currents.dis.9085e38035f25b34f093f357ac2c3973

Cartwright C, Hall M, Lee CK (2017) The changing health priorities of earthquake response and implications for preparedness: a scoping review. Public Health 150:60–70. https://doi.org/10.1016/j.puhe.2017.04.024

Chan EYY (2008) The untold stories of the Sichuan earthquake. Lancet, vol 372

Chan EYY (2017) Public health humanitarian responses to natural disasters. London: Routledge

Chan EYY, Kim JH, Lin C, Cheung EYL, Lee PPY (2014) Is previous disaster experience a good predictor for disaster preparedness in extreme poverty households in remote muslim minority

based community in China? J Immigr Minor Health 16(3):466–472. https://doi.org/10.1007/s10903-012-9761-9
Chan EYY, Ho JYE, Huang Z, Kim JH, Lam HCY, Chung PPW et al (2018) Long-Term and immediate impacts of health emergency and disaster risk management (Health-EDRM) education interventions in a rural chinese earthquake-prone transitional village. Int J Disaster Risk Sci 9(3):319–330. https://doi.org/10.1007/s13753-018-0186-5
Chan EY, Man AY, Lam HC (2019) Scientific evidence on natural disasters and health emergency and disaster risk management in Asian rural-based area. Br Med Bull. https://doi.org/10.1093/bmb/ldz002
Dolhun E (2013) Aftermath of Typhoon Haiyan: the imminent epidemic of waterborne illnesses in Leyte, Philippines. Disaster Med Public Health Prep 7(6):547–548. https://doi.org/10.1017/dmp.2013.114
Dong C, Ge P, Ren X, Zhao X, Wang J, Fan H, Yin SA (2014) Growth and anaemia among infants and young children for two years after the Wenchuan Earthquake. Asia Pac J Clin Nutr 23 (3):445–451. https://doi.org/10.6133/apjcn.2014.23.3.03
Doocy S, Catlett C, Wadhwani C, Sauer L, Kirsch TD (2012) Impact of the 2010 Pakistan floods on rural and urban populations at six months. PLoS Curr. https://doi.org/10.1371/4fdfb212d2432
Doocy S, Daniels A, Murray S, Kirsch TD (2013a) The human impact of floods: a historical review of events and systematic literature review. PLOS Curr Disasters 1:1–32. https://doi.org/10.1371/currents.dis.f4deb457904936b07c09daa98ee8171a.Authors
Doocy S, Dick A, Daniels A, Kirsch TD, Hopkins J (2013b) The human impact of tropical cyclones: a historical review of events 1980-2009 and systematic literature review citation abstract authors. PLOS Curr Disasters 1:1–25. https://doi.org/10.1371/currents.dis.2664354a5571512063ed29d25ffbce74
Doocy S, Daniels A, Packer C, Dick A, Kirsch TD (2016) The human impact of earthquakes : a historical review of events 1980 2009 and systematic literature review. PLOS Curr Disasters, 1–46 (October 2013). https://doi.org/10.1371/currents.dis.67bd14fe457f1db0b5433a8ee20fb833.Authors
Fang YP, Zhu FB, Qiu XP, Zhao S (2018) Effects of natural disasters on livelihood resilience of rural residents in Sichuan. Habitat Int 76(April):19–28. https://doi.org/10.1016/j.habitatint.2018.05.004
Feng J, Xia Z, Zhang L, Cheng S, Wang R (2016) Risk assessment of malaria prevalence in Ludian, Yongshan, and Jinggu Counties, Yunnan Province, After 2014 Earthquake Disaster. Am J Trop Med Hyg 94(3):674–678. https://doi.org/10.4269/ajtmh.15-0624
Ganchoon F, Bugho R, Dy R, Calina L, Gosney J (2018) Philippine academy of rehabilitation medicine emergency basic relief and medical aid mission project (November 2013–February 2014): the role of physiatrists in Super Typhoon Haiyan. Disabil Rehabil 40(18):2217–2225. https://doi.org/10.1080/09638288.2017.1326534
Gyawali B, Keeling J, Kallestrup P (2017) Human trafficking in nepal: post-earthquake risk and response. Disaster Med Public Health Prep 11(2):153–154. https://doi.org/10.1017/dmp.2016.121
Hoque MA, Scheelbeek PFD, Vineis P, Khan E, Ahmed KM, Butler AP (2016) Drinking water vulnerability to climate change and alternatives for adaptation in coastal South and South East Asia. Clim Change, 136(2):247–263. https://doi.org/10.1007/s10584-016-1617-1
Jiang L-P, Yao L, Bond EF, Wang Y-L, Huang L-Q (2011) Risk perceptions and preparedness of typhoon disaster on coastal inhabitants in China. Am J Disaster Med 6(2):119–126
Kouadio IK, Aljunid S, Kamigaki T, Hammad K, Oshitani H (2014) Anti-infective Therapy Infectious diseases following natural disasters: prevention and control measures. Expert Rev Anti-Infective Therapy. https://doi.org/10.1586/ERI.11.155
Krishna RN, Ronan KR, Alisic E (2018) Children in the 2015 South Indian floods: community members' views. https://doi.org/10.1080/20008198.2018.1486122
Lee Poyi CU of HK (2012). Impact of floods in China : an update literature review. In: The 13th World Congress on Public Health, p 10105

Marthoenis M, Yessi S, Aichberger MC, Schouler-Ocak M (2016) Mental health in Aceh-Indonesia: a decade after the devastating tsunami 2004. Asian J Psychiatr 19 (2016):59–65. https://doi.org/10.1016/j.ajp.2016.01.002

Musa R, Draman S, Jeffrey S, Jeffrey I, Abdullah N, Halim NAM et al (2014) Post tsunami psychological impact among survivors in Aceh and West Sumatra, Indonesia. Compr Psychiatr 55(Suppl 1):S13–S16. https://doi.org/10.1016/j.comppsych.2012.12.002

Pickering CJ, O'Sullivan TL, Morris A, Mark C, Mcquirk D, Chan EYY et al (2018) The promotion of "Grab Bags" as a disaster risk reduction strategy, 1–24. https://doi.org/10.1371/currents.dis.223ac4322834aa0bb0d6824ee424e7f8.Authors

Quevillon RP, Gray BL, Erickson SE, Gonzalez ED, Jacobs G (2016) Helping the helpers: assisting staff and volunteer workers before, during, and after disaster relief operations. J Clin Psychol 72(12):1348–1363. https://doi.org/10.1002/jclp.22336

Ray-Bennett NS (2009) The influence of caste, class and gender in surviving multiple disasters: a case study from Orissa, India. Environ Hazards 8(1):5–22. https://doi.org/10.3763/ehaz.2009.0001

Rodriguez-Llanes J, Ranjan-Dash S, Mukhopadhyay A, Guha-Sapir D (2016) Flood-exposure is associated with higher prevalence of child undernutrition in rural eastern India. Int J Environ Mental Res Public Health 13(2):210. https://doi.org/10.3390/ijerph13020210

Rosalind C, Garfin AM, Woojin, Hall JL (2015) Restarting the tuberculosis programme Post-Haiyan. Western Pacific Surveill Response 6(1):96–101. https://doi.org/10.5365/wpsar.2015.6.2.HYN

Shabir O (2013) A summary Case report on health impacts and responses to Pakistan floods of 2010. PLOS Curr Disasters 2010:1–37. https://doi.org/10.1371/currents.dis.cc7bd532ce252c1b740c39a2a827993f

Sohrabizadeh S, Jahangiri K, Khani Jazani R (2018) Reproductive health in the recent disasters of Iran: a management perspective. BMC Public Health 18(1):1–8. https://doi.org/10.1186/s12889-018-5311-2

Srikuta P, Inmuong U, Inmuong Y, Bradshaw P (2015) Health vulnerability of households in flooded communities and their adaptation measures: case study in Northeastern Thailand. Asia Pac J Public Health 27(7):743–755. https://doi.org/10.1177/1010539514568709

Tran P, Shaw R (2007) Towards an integrated approach of disaster and environment management: a case study of Thua Thien Hue province, central Viet Nam. Environ Hazards 7(4):271–282. https://doi.org/10.1016/j.envhaz.2007.03.001

Uprety S, Hong PY, Sadik N, Dangol B, Adhikari R, Jutla A et al (2017) The effect of the 2015 earthquake on the bacterial community compositions in water in Nepal. Front Microbiol 8 (DEC), 1–11. https://doi.org/10.3389/fmicb.2017.02380

Van Minh H, Tuan Anh T, Rocklöv J, Bao Giang K, Trang LQ, Sahlen K-G et al (2014) Primary healthcare system capacities for responding to storm and flood-related health problems: a case study from a rural district in Central Vietnam. Global Health Action 7:23007. https://doi.org/10.3402/GHA.V7.23007

Wind TR, Joshi PC, Kleber RJ, Komproe IH (2018) The impact of recurrent disasters on mental health: a study on seasonal floods in Northern India. https://doi.org/10.1017/S1049023X13000290

Witvorapong N, Muttarak R, Pothisiri W (2014) Social participation and disaster risk reduction behaviors in Tsunami Prone Areas. PLoS ONE 10(7):1–20. https://doi.org/10.1371/journal.pone.0130862

Wu X-H, Zhang S-Q, Xu X-J, Huang Y-X, Steinmann P, Utzinger J et al (2008) Effect of floods on the transmission of schistosomiasis in the Yangtze River valley, People's Republic of China. Parasitol Int 57(3):271–276. https://doi.org/10.1016/j.parint.2008.04.004

Xu J, Zhang Y, Chan J, Li N, Yang Y, Li J (2011) A comparison of the acute stress reactions between the Han and Tibetan ethnic groups in responding to devastating earthquakes. Int J Psychiatry Med 42(2):167–180

Zhang W, Wang W, Lin J, Zhang Y, Shang X, Wang X et al (2017) Perception, knowledge and behaviors related to typhoon: a cross sectional study among rural residents in Zhejiang, China. Int J Environ Res Public Health 14(5):1–12. https://doi.org/10.3390/ijerph14050492

Zheng J, Han W, Jiang B, Ma W, Zhang Y (2017) Infectious diseases and tropical cyclones in southeast China. Int J Environ Res Public Health, 14(5). https://doi.org/10.3390/ijerph14050494

Chapter 22
Health-EDRM: Lessons Learnt in Asia

Emily Ying Yang Chan, Chi Shing Wong and Rajib Shaw

Abstract The chapter provides an overview and summary of how the 14 Asian case studies in the book fit into and be employed to illustrate the prevention-centred Health-EDRM framework, as well as the common Health-EDRM topics in the region. Traditionally, health emergency disaster risk management approaches may assume risks are stable and risks might be management in a linear manner. Yet, the rapid changes in infrastructure built environmental and human behavioural patterns with modernization have rendered health risk management in emergency and disaster risk with a constant struggle. However, there remain some challenges, which include redundant research, lack of a strategic research agenda, limited development of multi-sectoral and interdisciplinary approaches, deficiency in the science–policy–practice nexus, the absence of standardized terminology and meagre coordination among stakeholders.

Keywords Disaster in Asia · Disaster prevention hierarchy · Health-EDRM · Information and communications technology (ICT)

22.1 Overview of Cases from Asia

Chapters 8–21 consist of case studies that cover a wide range of countries in South, East and Southeast Asia (Bangladesh, India, Nepal, Pakistan, China, Japan, Vietnam and the Philippines) with specific context which illustrates how health risks might be conceptualized, examined and addressed. Each of these cases illustrates various components of the Health-EDRM framework. In addition to addressing different types of health risk prevention along with the prevention

E. Y. Y. Chan (✉) · C. S. Wong
JC School of Public Health and Primary Care, The Chinese University of Hong Kong, Shatin, Hong Kong
e-mail: emily.chan@cuhk.edu.hk

R. Shaw
Graduate School of Media and Governance, Keio University, Fujisawa, Japan

E. Ying Yang Chan and R. Shaw (eds.), *Public Health and Disasters*, Disaster Risk Reduction, https://doi.org/10.1007/978-981-15-0924-7_22

hierarchy for emergencies and disasters, the importance of communication and data sharing and exchange within and across various sectors in the Health-EDRM framework across the disaster prevention hierarchy is also emphasized in various studies. As highlighted in the Health-EDRM framework, more need to be done to integrate the health sector into the national disaster management systems in the region.

Primary disaster prevention, with its cost-effectiveness, emerges as a major strategy of Health-EDRM in the region facing demographic, epidemiological and health risks transitions, particularly in less developed countries. Chapter 8 describes the disaster context of Bangladesh's (flood- and cyclone-prone, urban poverty) as well as the range of Health-EDRM issues of water-borne and vector-borne diseases, acute respiratory infection, dengue fever, malnutrition, mental health, domestic violence, chronic diseases and drinking water. Even in non-emergency period, communicable diseases remain a major health risk and concern in the country. Particularly after floods and cyclones, the extreme events would further disrupt the health and lifeline infrastructures. Moreover, the poverty-related unplanned rapid urbanization and the growth of slum areas add to the health risk, particularly in response and post-disaster settings. With urban flooding which tends to exacerbate by climate change, the authors described the complexity in addressing health risks. The main preventive Health-EDRM measures described in this chapter focuses on the primary level, including strengthening primary healthcare system development and poverty reduction. *Matsuyama* et al. highlight that community clinics were found to be pivotal in providing bottom-up health resilience. This chapter also suggests that special attention should be paid to disadvantaged groups most vulnerable to disaster health risk. Taking into account of the epidemiological and health risk transitions in Bangladesh as a developing country, the issue of non-communicable diseases (NCDs) should also be addressed as a primary preventive measure of disaster health risk.

Related to the same context, Bangladesh, Chap. 9 examines water resource management, which is another key topic in Health-EDRM as water is a non-negotiable life-sustaining component that maintains survival and well-being. In addition to describing the socio-demographic pressure of population growth, urbanization and industrialization, *Abedin* et al. describe the range of water security issues, drinking water shortage, sea-level rise and intrusion of saltwater in Bangladesh which is a slow onset disaster brought about by climate change. Authors argued that the provision of information about safe drinking source via information and communications technology may be the most suitable primary disaster prevention strategy for tackling this kind of slow onset disaster.

Chapter 10 describes the Health-EDRM at the latest policy and institutional levels in the People's Republic of China. As a new unified government disaster management system formed in 2018, the Ministry of Disaster Management aims to centralize the disaster management functions previously scattered across various government departments and agencies, with its new focus on primary disaster prevention. Chan et al. argue that although the new structure should enable better Health-EDRM, the review showed that similar to Bangladesh, health as the main component in disaster preparedness, response and rehabilitation, is yet to be fully incorporated into the country's unified disaster management system. The current health sector in the context is often involved only at a later stage of disaster response.

Chapter 11 provides an overview of the disaster profile, health and public health risks in India, another major country in the continent with a large population size and being disaster-prone. Similar to other developing countries in the region like Bangladesh mentioned in earlier chapters, India has been undergoing demographic, epidemiological and developmental transition during the past decades. While disaster management and health have yet to be better integrated, the achievement of better health outcomes in disaster hinges on the primary disaster prevention strategy of improving its health infrastructure, such as capacity training, the establishment of mobile clinical units, upgrading of trauma care and improvement of hospital safety. Specifically, *Krishnan* et al. show how the country's National Disaster Management Authority laid down guidelines for hospital safety in 2016, which reflects the global WHO disaster risk reduction policy towards safe hospital in emergencies, to mainstream disaster prevention, mitigation, preparedness and response activities in the health sector. This is an important effort to promote multidisciplinary approach central to the Health-EDRM framework.

Chapter 12 examines the specific topic of water contamination in West Bengal of India. As a Health-EDRM risk, arsenic contamination, saltwater intrusion, depleting groundwater reserves, falling water table and water-related hazards like flooding and drought are the major causes of the water problem in the study area. *Ray* et al. show how major education and communication initiatives were mounted up to address the slow onset water-related disaster, which are primary disaster prevention strategies recommended as affordable and sustainable solutions to the health risk associated with water contamination. The authors also described bottom-up initiatives of building mini infrastructure, e.g. compulsory installation of rainwater harvesting systems in government offices and institutions, treatment for microbiological contaminants of surface water, and installation of alternate tube wells marked as safe.

Chapters 13 and 14 discuss how Japan, a developed country in the region, engaged in tertiary prevention activities in Health-EDRM. In Chap. 13, *Miyagawa* et al. describe the management and prevention of health risks arising from long-term sheltering arrangements for the disaster-affected community. The authors review how early detection of and response to the health risks associated with long-term evacuation and living in temporary housing were done during post-Great East Japan Earthquake evacuation. They show the use of information and communications technology in Health-EDRM and the importance of collaborative effort among healthcare and non-healthcare workers in emergency response and health risk reduction. *Komino* delineates the range of health issues and risks associated with the in Fukushima following the 311 Great East Japan Earthquake and subsequent tsunami in Chap. 14. Public health secondary prevention strategies of disease monitoring and screening were established to minimize health risks arising from the primary incidents of geophysical and nuclear disasters. The author suggests the continuation of health surveys for early diagnosis, and treatment is essential (particularly for nuclear disasters with long-term health implications) for any Build Back Better efforts in line with the Sendai Framework as well as the Health-EDRM framework, so it is the pre-planning of DRR strategies and contingency planning with the involvement of a wide range of stakeholders in society, both from health and non-health sectors.

Cases of how tertiary Health-EDRM disaster prevention may be implemented are discussed in Chap. 15, which focuses on another major earthquake, the 2015 Nepal Earthquake, in the region. *Kanbara* et al. show the usefulness of the information and communications technology in tertiary disaster prevention of epidemiological surveillance in a developing country context. In particular, the authors highlight the important role that local nurses and allied health workers had played in this post-disaster communicable disease prevention strategy. This case also illustrated how the author teams attempted to develop the health security index and the application of action research methodology for better documentation.

Chapter 16 examines how disaster preparedness for Health-EDRM is done in a rural context in another developing country in South Asia, Pakistan. Although rural health centres were identified as having great potential to serve as the key primary healthcare infrastructure for health and disaster risk reduction in rural Pakistan, *Shah* et al. show the current systematic limitations, in terms of infrastructure, training of human resources and availability of financial resources, for the assumption of adopting rural health centres as the key infrastructure to support primary Health-EDRM disaster prevention efforts, particularly for rural floods.

Chapter 17 provides an overview of the scene of Health-EDRM in the relatively less developed country in Southeast Asia, Vietnam. *Linh* et al. describe the country's disaster profile of flooding and typhoon-proneness and issues around health and disaster risk management in the country. Approaches of primary disaster prevention in health risks of the country in terms of government DRR policies and the infrastructure of safe hospitals are discussed.

Technology use in information and communications are an important frontier in disaster responses and Health-EDRM in the coming decades. Chapter 18 describes how information technology developed during non-emergency times might be useful in facilitating response and decreasing health risks amidst disasters in the Philippines. As a disaster-prone developing country in Southeast Asia, *Estuar* et al. examine how the country's digitization of health records in the primary care system might facilitate rapid medical information and surveillance access during and after disasters resulting from its strong wind events, i.e. typhoons. The theme of integrating health into the disaster management system in terms of information exchange is also highlighted in this study.

Chapter 19 explains the data collection methodology, as a measure of disaster prevention, in Japan. *Kubo* describes the data collection mechanism is a kind of primary disaster prevention in terms of capacity building to prepare for disaster health risk. He also shows how the standardization of the data reporting process allows better coordination and the strengthening of the disease early warning system. Referring to the Surveillance in Post Extreme Emergencies and Disasters

(SPEED) of the Philippines, he also discusses the possibility of building an international standard for data collection mechanism with the Japanese version of the SPEED (J-SPEED) and how these systems might facilitate the building of the Minimum Data Set (MDS) concept as advocated by the World Health Organization. This is a fine example of cross-country knowledge transfer and international cooperation in line with the collaborative essence of Health-EDRM. It also reminds global and national policymakers in Health-EDRM of the relevance of the developing country experience for more developed countries.

Chapter 20 shows how academic research might facilitate the conceptualization of health emergency and disaster risks across national boundaries. Chan et al. explain the constraints, methodology gaps and current challenges in quantifying health and disaster risks. Specifically, with the challenges of different reporting systems, there is a lack of tools to facilitate cross-country comparison of health vulnerabilities and risks to facilitate the detection of trends and monitor of Health-EDRMs in the region and across the globe. The authors share their latest research findings in developing the health vulnerability risk maps, which aims to examine the health risks and vulnerability of the countries along China's Belt-and-Road Initiative region.

Chapter 21 reviews existing Health-EDRM interventions that are being practised in rural Asia for the three most common natural hazards in Asia: geophysical, meteorological and hydrological hazards. This chapter looks into the primary, secondary and tertiary interventions in the Health-EDRM framework to assist in identifying the gaps for action. The lack of monitoring and evaluation and the need to provide disaggregated data in meteorological and hydrological hazards are identified.

22.2 Summary of Key Findings

As summarized in Table 22.1, Health-EDRM spans across a wide range of topics from diseases to water, infrastructure, information and communications technology, and government policies and institutions. The case studies in this book suggest that the more cost-effective primary disaster prevention strategies seem to better cater for the context of developing countries in the region and for slow onset disasters like the water-related ones, while secondary and tertiary prevention strategies are more affordable in developed countries and more crucial for sudden onset high impact disasters like those arising from earthquakes and nuclear accidents. Improvement of infrastructure—from modest and mini ones of rural health centres and community clinics to larger-scale safe hospitals, integration of health and non-health sectors, as well as collection, reporting, provision and technology of information and communications emerge as the key lessons to be further studied in the nascent field of Health-EDRM.

Table 22.1 An overview of Health-EDRM lessons in Asia

Chapter	Country	Hazards	Health-EDRM topics	Prevention hierarchy	Insights
8	Bangladesh	Flooding Cyclone	Communicable diseases NCDs Health and lifeline infrastructure Unplanned rapid urbanization Poverty	Primary prevention: Healthcare system development Poverty reduction NCDs	Importance of utilization of existing infrastructure (community clinics) to maximize community resources in disaster risk reduction (DRR) and health
9	Bangladesh	Drinking water shortage Sea-level rise Intrusion of saltwater	Water resource management	Primary prevention: Provision of information	How important health risk reduction information (e.g. location of provision of safe drinking water) and communications technology might be used in Health DRR
10	People's Republic of China	All-hazard	Health-EDRM at policy and institutional levels	Primary prevention	How "Health" might be missed in disaster management system reform
11	India	All-hazard	Safe hospital	Primary prevention: Health infrastructure improvement, safe hospital	Disaster management gaps in country which is undergoing demographic, epidemiological and developmental transition Health yet to be better integrated Even with the multidisciplinary approach, health might be missed
12	India	Water shortage Flooding Drought	Water contamination	Primary prevention: Education Communication Infrastructure	Building of mini DRR infrastructure Affordable education and communication initiatives
13	Japan	Earthquake	Long-term sheltering post-disaster	Tertiary prevention	How information and communications technology might be used in supporting Health DRR Health and non-health collaboration
14	Japan	Earthquake Tsunami Nuclear accident	Health survey post-nuclear accidents	Public health secondary prevention as tertiary disaster prevention: Disease monitoring and screening	Secondary prevention with repeated health-related surveys will be useful to assess health risks and development related response to reduce risks Health and non-health collaboration

(continued)

Table 22.1 (continued)

Chapter	Country	Hazards	Health-EDRM topics	Prevention hierarchy	Insights
15	Nepal	Earthquake	Information and communications technology Role of nurse Communicable disease prevention Health security index Action research methodology	Tertiary prevention: Epidemiological surveillance	Usefulness of information and communications technology in tertiary disaster prevention
16	Pakistan	Flooding	Infrastructure	Primary prevention	Importance of utilization of existing infrastructure (rural health centres) to maximize community resources in DRR and health
17	Vietnam	Flooding Typhoon	Government DRR policies Safe hospital	Primary prevention	Safe hospital Initiatives might be implemented. (WHO advocates as DRR strategies for health)
18	The Philippines	Typhoon	Information technology	Primary prevention	Digitized health records Integrating health into the disaster management system for facilitate risk communication and risk health reduction effort
19	The Philippines, Japan	All-hazard	Data collection mechanism	Primary prevention	Standardization of data reporting International cooperation in reducing risk through science and technology
20	Belt-and-Road Initiative region	All-hazard	Health vulnerability risk map	Primary prevention	Quantification of health vulnerabilities and risks Disaster risk reduction effort through developing scientific tool to allow cross-country comparison of health emergency and disaster risks
21	Asia	All-hazard	Interventions and practices	Primary, secondary and tertiary prevention	The lack of monitoring and evaluation The need to provide disaggregated data in meteorological and hydrological hazards

Reference

Chan EYY, Wright K, Parker M (2019) Health-emergency disaster risk management and research ethics. Lancet 393:112–113

Chapter 23
Future Perspectives of Health-EDRM and Risk Reduction in Asia

Emily Ying Yang Chan and Rajib Shaw

Abstract This chapter wraps up the discussions of the emerging framework of Health-EDRM by highlighting the future challenges of risk reduction under this framework, including constantly changing disaster and health risks, resilience building, disaster information dissemination and its potential disparity, and transient populations.

Keywords Disaster risk reduction · Health Emergency and Disaster Risk Management (Health-EDRM) · Information · Resilience · Transient population

Asia, the most disaster-prone region in the world also hosting a number of large population-based countries, is a global leading region in economic progress, technology advancement, adaptation and urbanization. With the rapid socio-economic developments, community reality changes, the dichotomy of categorizing communities into rural and urban is no longer true as communities tend to be in prolonged transitional stages in its socio-economic realities. Traditionally, health emergency disaster risk management (Health-EDRM) approaches may assume risks are stable and risks might be managed in a linear manner. Yet, the rapid changes in infrastructure, built environment and human behavioural patterns with modernization have rendered health risks management in emergency and disaster risk a constant struggle. Population who are living in these countries, varying in their stages of development, and demographic, epidemiological and environmental risk transitions, face a range of new social order and living arrangement that alter drastically the disaster and health risks they have to experience. Emergency and risk mapping becomes a dynamic process and would require regular re-evaluation and re-calibration. With more reliance on lifeline infrastruc-

E. Y. Y. Chan (✉)
JC School of Public Health and Primary Care, The Chinese University of Hong Kong, Shatin, Hong Kong
e-mail: emily.chan@cuhk.edu.hk

R. Shaw
Graduate School of Media and Governance, Keio University, Fujisawa, Japan

E. Ying Yang Chan and R. Shaw (eds.), *Public Health and Disasters*, Disaster Risk Reduction, https://doi.org/10.1007/978-981-15-0924-7_23

ture and mobile population, traditional risk evolved into modern risks that tend to be more complex and challenging to deal with at the individual level (Chan 2017, 2018). There are major gaps in understanding emergency impact patterns, the threshold for response planning, and community and individual behaviours in preparedness.

Emergencies, regardless of the causation of the events, will cause a large number *of injury and non-fatal casualties*. Implementation of disaster risk reduction effects depends on community resilience. Many communities have some capacity to respond to regular emergencies and crises, but with the increasing frequencies of climate-related disasters and increased concentration of people clustering around the cities, the community resilience will continue to face pressure in coping. Although traditional vulnerable groups such as extremes of age and people with disability and non-communicable diseases remain, a new format of information dissemination means that ensuring connectivity of the most vulnerable subgroups will be important to prevent technology discrimination. Outreaching to communities and preventing population enclaves arising from their inaccessibility to information will be crucial. In addition, addressing and managing health risks with transient residents displaced population or tourists remain uncharted. Scientific evidence is urgently needed, but there are also major infrastructural gaps to facilitate a reliable ethical evaluation, implementation and evaluation of emergency-related research (Chan et al. 2019).

The intersection of health and disaster risk reduction (DRR) has emerged in recent years as an interdisciplinary field of paramount human consequences. Throughout several landmark UN agreements adopted in 2015–2016, including the Sendai Framework for Disaster Risk Reduction 2015–2030, the 2030 Sustainable Development Goals (SDGs), the Paris climate agreement, and the New Urban Agenda (Habitat III), health is recognized as an inevitable outcome and a natural goal of disaster risk reduction, and the crossover of the two fields is no doubt essential for the successful implementation of the Sendai Framework. Health emergency and disaster risk management, as a joint venture of this crossover, have emerged as an umbrella field that encompasses emergency and disaster medicine, DRR, humanitarian response, community health resilience and health system resilience (Chan and Murray 2017; Lo et al. 2017) and the WHO has launched its strategic framework in 2019 (WHO 2019). Book presents the current knowledge landscape of health emergency and disaster risk management (Health-EDRM) to achieve disaster risk reduction. Nevertheless, the fragmented nascent field of Health-EDRM remains to be developed into a coherent enterprise. Key challenges include redundant research, lack of a strategic research agenda, limited development of multi-sectoral and interdisciplinary approaches, deficiency in the science–policy–practice nexus, the absence of standardized terminology and meagre coordination among stakeholders (Kayano et al. 2019; Lo et al. 2017). This book hopes to provide a timely and invaluable resource for undergraduate and postgraduate students, researchers, scholars and frontline practitioners as well as policymakers from across the component domains of Health-EDRM and facilitate the advancement of Health-EDRM discipline in the coming years.

References

Chan EYY (2017) Public health humanitarian responses to natural disasters. Routledge, London

Chan EYY (2018) Building bottom-up health and disaster risk reduction programmes. Oxford University Press, Oxford

Chan EYY, Murray V (2017) What are the health research needs for the Sendai Framework? Lancet 390(10106):e35–e36. https://doi.org/10.1016/S0140-6736(17)31670-7

Chan EYY, Wright K, Parker M (2019) Health-emergency disaster risk management and research ethics. Lancet 393:112–113

Kayano R, Chan EY, Murray V, Abrahams J, Barber SL (2019) WHO thematic platform for health emergency and disaster risk management research network (TPRN): report of the kobe expert meeting. Int J Environ Res Public Health 16(7):1232

Lo STT, Chan EYY, Chan GKW, Murray V, Abrahams J, Ardalan A, Kayano R, Yau JCW (2017) Health emergency and disaster risk management (Health-EDRM): developing the research field within the Sendai Framework Paradigm. Int J Disaster Risk Sci. 8:145–149. https://doi.org/10.1007/s13753-017-0122-0

World Health Organization (2019) Health emergency and disaster risk management framework. World Health Organization, Geneva

The manufacturer's authorised representative in the EU is Springer Nature Customer Service Centre GmbH, Europaplatz 3, 69115 Heidelberg, Germany. If you have any concerns regarding our products, please contact ProductSafety@springernature.com

Printed and bound by CPI Group (UK) Ltd, Croydon, CR0 4YY
25/09/2026
02222894-0012